Lecture Notes of the Unione Matematica Italiana

Volume 31

U M I Unione Matematica Italiana

Fabrizio Catanese
Editor

Varieties of Nodal Surfaces, Coding Theory and Discriminants of Cubic Hypersurfaces

Springer

Unione Matematica Italiana

Editor
Fabrizio Catanese
Mathematisches Institut
Universität Bayreuth
Bayreuth, Bayern, Germany

ISSN 1862-9113 ISSN 1862-9121 (electronic)
Lecture Notes of the Unione Matematica Italiana
ISBN 978-3-032-05898-0 ISBN 978-3-032-05899-7 (eBook)
https://doi.org/10.1007/978-3-032-05899-7

Mathematics Subject Classification: 14C30, 14J28, 14J25, 14J70, 14N25, 16G99, 32G20, 32Q15

This Springer imprint is published by the registered company Springer Nature Switzerland AG
The registered company address is: Gewerbestrasse 11, 6330 Cham, Switzerland

Dedicated to Ciro Ciliberto on the occasion of his 70th birthday.

Preface

This research monograph, while appearing as a collection of chapters with different authorship, is indeed meant to be, as shown in the Introduction (written by the main author), a coherent exposition of the different aspects and methodologies necessary for the study of Severi Varieties of nodal surfaces: projective geometry, homological algebra, coding theory, lattice theory, theory of algebraic surfaces and computer algebra. Some chapters contain some appendices, giving further information which is of interest by itself, especially from the point of view of coding theory.

We pass now to a brief description of the main topics treated in this research monograph: we exploit two binary codes associated to a projective nodal surface (the strict code $\mathcal{K}$ and, for even degree d, the extended code $\mathcal{K}'$) to investigate the 'Nodal Severi varieties' $\mathcal{F}(d, \nu)$ of nodal surfaces in $\mathbb{P}^3$ of degree d and with ν nodes, and their incidence hierarchy, relating partial smoothings to code shortenings.

Our first main result (the case of surfaces of degree $d \leq 3$ being elementary) solves a question which dates back over 100 years: it says that the irreducible components of $\mathcal{F}(4, \nu)$ are in bijection with the isomorphism classes of their extended codes $\mathcal{K}'$, that these are exactly all the 34 possible shortenings of the extended Kummer code $\mathcal{K}'_{Kum}$ ($\nu \leq 16$), and that a component is in the closure of another if and only if the code of the latter is a shortening of the code of the former.

We then extend this result to the classification of all the other varieties of nodal K3 surfaces, of all possible degrees $d = 2h$: again the connected components of the Nodal Severi variety $\mathcal{F}_{K3}(d, \nu)$ of nodal K3's of degree d and with ν nodes are in bijection with the isomorphism classes of their extended codes $\mathcal{K}'$; we classify the codes $\mathcal{K}'$, showing that their occurrence depends on the congruence class of d modulo 16; we also see that some unexpected families occur.

In 1979 Beauville used coding theory to prove that $\mu(5)$, the maximum number of nodes that a nodal surface in $\mathbb{P}^3$ of degree 5 can have, equals 31; hence the 'Togliatti quintics', the discriminants of Togliatti cubics (cubic hypersurfaces in $\mathbb{P}^5$ with 15 nodes) are the quintics with $\mu(5) = 31$ nodes.

For surfaces of degree $d = 5$ in $\mathbb{P}^3$ we determine (with one possible exception which remains an open question) all the possible codes $\mathcal{K}$ of nodal quintics, and show that we do not get all the codes with weights 16 and 20 satisfying the inequality pointed out by Beauville. Moreover, for several cases of $\mathcal{K}$, we show the irreducibility of the corresponding open set of $\mathcal{F}(5, \nu)$: in particular, using the previous results on nodal K3 surfaces, we show the irreducibility of the respective families of Togliatti quintic surfaces and of Togliatti cubic fourfolds in $\mathbb{P}^5$.

The fourth main result, in the final part of this paper, shows that a 'Togliatti-like' description holds for surfaces of degree 6 with the maximum number of nodes $\mu(6) = 65$: they are discriminants of cubic hypersurfaces in $\mathbb{P}^6$ with 31 nodes (also of cubics with 32 nodes), and we have an irreducible 18-dimensional family of them.

For degree $d = 6$, our main result is based on these novel auxiliary results:

(1) The study of the half-even sets of nodes on sextic surfaces
(2) The investigation of discriminants of cubic hypersurfaces X, and the description of the relation between the nodal singularities of X and the nodal singularities of the discriminant surface Y
(3) The computer-assisted proof that, for $\nu = 65$, both codes $\mathcal{K}, \mathcal{K}'$ are uniquely determined
(4) The explicit description of these interesting codes, starting from the Barth sextic with 65 nodes, ending up with its relation to the Doro-Hall graph (in particular, we find explicit discriminant and determinant realizations of the Barth sextic)

The codes of sextic surfaces are investigated, with a partial classification in the case of 'many' nodes.

Bayreuth, Germany — Fabrizio Catanese
Jinju-si, South Korea — Yonghwa Cho
Limerick, Ireland — Stephen Coughlan
Milano, Italy — Davide Frapporti
Bayreuth, Germany — Michael Kiermaier
Roma, Italy — Alessandro Verra
Bayreuth, Germany — Sascha Kurz

Acknowledgements Fabrizio would like to thank Michael Lönne for useful advice concerning the proof of Theorem 6, and Ciro Ciliberto for answering some queries. We would all like to heartily thank Alfred Wassermann for his initial contribution which lead to the classification of the codes of 65-nodal sextics, achieved in [65], see Chap. 5.

Thanks also to the referee for some advice concerning the exposition.

The authors acknowledge support of the ERC 2013 Advanced Research Grant - 340258 - TADMICAMT.

Introduction

Fabrizio Catanese

In this book we shall first of all consider irreducible hypersurfaces X of degree d in complex projective space $\mathbb{P}^N := \mathbb{P}^N_{\mathbb{C}}$: these are the zero sets of an irreducible homogeneous polynomial $F(x_0, \ldots, x_N)$ of degree d,

$$X := \{x := (x_0, \ldots, x_N) \in \mathbb{P}^N | F(x_0, \ldots, x_N) = 0\}.$$

The singular locus is the subset $\mathrm{Sing}(X) := \{x | \frac{\partial F}{\partial x_i} = 0 \ \forall i = 0, \ldots, N\}$, and a singular point $x^* \in X$ is said to be a **node** if the quadratic part of the Taylor development of F at x^* is of maximal rank ($= N$, in view of the Euler relation $\sum_i x_i \frac{\partial F}{\partial x_i} = dF$), that is, the $(N+1)$ gradients $\nabla \frac{\partial F}{\partial x_i}$ have rank $= N$ at x^*.

A hypersurface X is said to be **nodal** if its singular points are just nodes.

Definition 1 We propose in this paper the name of **Nodal Severi variety** $\mathcal{F}_n(d, \nu)$ for the locally closed set, inside the projective space parametrizing all hypersurfaces in $\mathbb{P}^{n+1}$, consisting of the nodal hypersurfaces X of degree d and dimension n in $\mathbb{P}^{n+1}$ having exactly ν nodes as singularities.

Severi in fact [86] studied the case $n = 1$, where one has to add the hypothesis that X is irreducible, and it turns out that $\mathcal{F}_1(d, \nu)$ (which is nonempty if and only if $\nu \leq \frac{(d-1)(d-2)}{2}$), is irreducible (the assertion, made in an appendix, Anhang F, of [86], had an incomplete proof, see [51]).

Several authors (see for instance [32] and [35]) extended these investigations to the case of nodal curves lying on smooth surfaces; here, instead, we mainly want to focus on the case of hypersurfaces and varieties of dimension at least two.

We have then the following (mainly unanswered) questions:

Question 2

(1) Basic question: when is $\mathcal{F}_n(d, \nu)$ non-empty ?

The above question relies on the fact that, for fixed dimension n, the number ν will be bounded by a function of the degree d, and this leads to another question:

(2) defining

$$\mu_n(d) := \max\{\nu | \mathcal{F}_n(d, \nu) \neq \emptyset\},$$

determine $\mu_n(d)$ explicitly.

In turn, knowing the maximum possible number of nodes $\mu_n(d)$ that a hypersurface of degree d can have, does not answer the first question, since one may ask a weaker question:

(3) in particular, for which values of (n, d) do we have the **no-gaps phenomenon** that $\mathcal{F}_n(d, \nu)$ is non-empty for all $\nu \leq \mu_n(d)$?

In turn, once we know that such a Severi variety $\mathcal{F}_n(d, \nu)$ is non empty, one may ask

(4) when is $\mathcal{F}_n(d, \nu)$ irreducible? How to determine its irreducible components?
(5) When is $\mathcal{F}_n(d, \nu)$ smooth?

Concerning the last question (5), one can give a sufficient condition for smoothness, namely the condition of unobstructedness (which will be used in the paper), meaning that the Severi variety is the fibre of a submersion. This criterion is used for instance by Dimca and Kloostermann [42, 61][1] to show that $\mathcal{F}_2(d, \nu)$ is smooth for $d = 5, 6, 7$, and has pure codimension ν in the space of degree d surfaces. In [61] examples are given, showing that $\mathcal{F}_2(d, 4(d-1))$ has a non-empty singular locus, for $d \geq 9$.

Already quite interesting, difficult and wide open is the case of surfaces: $n = 2$. In this case we shall often denote $\mathcal{F}_2(d, \nu)$ simply by $\mathcal{F}(d, \nu)$, and $\mu_2(d)$ by $\mu(d)$.

To give a vague idea of the state of the art, let us give the following definition:

Definition 3 Let X be a hypersurface of degree d in $\mathbb{P}^{n+1}$ with isolated singularities, and let

$$s(X) := |\operatorname{Sing}(X)|$$

be the number of singular points, and

$$\operatorname{Milnor}(X) := \sum_{P \in X} \operatorname{Milnor}(X_P)$$

the sum of the Milnor numbers at the singular points; recall here that the Milnor number is the dimension of $\mathcal{O}_{X,P}/(\frac{\partial F}{\partial x_0}, \dots, \frac{\partial F}{\partial x_{n+1}})$, and that the Milnor number equals 1 only for the nodes.

We can then consider the maximal number of singular points

$$s_n(d) := \max\{s(X) | X \subset \mathbb{P}^{n+1}, \deg(X) = d, \ |\operatorname{Sing}(X)| < \infty\},$$

and similarly $\operatorname{Milnor}_n(d)$.

For $n = 2$, we use the simpler notation $s(d) := s_n(d)$, $\operatorname{Milnor}(d) := \operatorname{Milnor}_2(d)$.

---

[1] We became aware of their results after everything was written up.

We clearly have a chain of inequalities:

$$\mu_n(d) \le s_n(d) \le \mathrm{Milnor}_n(d).$$

The inequality $\mu_n(d) \le \mathrm{Milnor}_n(d)$ is strict, since Miyaoka proved [71] $\mu(d) \le \frac{4}{9}d(d-1)^2$, while we can show by easy examples (Example 2.18) that Milnor$(d) \ge \frac{1}{2}d(d-1)^2$.

Also the inequality $s_n(d) \le \mathrm{Milnor}_n(d)$ is strict (but not by such an asymptotically large difference as before) as we can see comparing general inequalities by Bruce [17] for $s_n(d)$ with these examples (see again Example 2.18).

Concerning the first inequality, one may ask, assuming that life is simple, the following:

Question 4 Is it true that $\mu(d) = s(d)$, or even $\mu_n(d) = s_n(d)\ \forall n$?

It is in general very difficult to determine the numbers

$$\mu_n(d), s_n(d), \mathrm{Milnor}_n(d).$$

Chmutov [34] showed the existence, for each degree d, of nodal surfaces X of degree d with $\nu(X)$ equal to a polynomial of degree 3 in d (depending on the congruence class of d modulo 6), and with leading coefficient $\frac{5}{12}d^3$, so that that asymptotically $\mu_n(d)$ grows like at least $\frac{5}{12}d^3$.

By Miyaoka's inequality and Chmutov's examples the ratio $\frac{\mu_n(d)}{d^3}$ satisfies

$$\frac{15}{36} - o(d) \le \frac{\mu_n(d)}{d^3} \le \frac{16}{36},$$

and it is an intriguing question to ask for the determination of the two numbers in the interval $[\frac{15}{36}, \frac{16}{36}]$, $\liminf(\frac{\mu_n(d)}{d^3})$ and $\limsup(\frac{\mu_n(d)}{d^3})$.

We conclude this general discussion simply saying here that $s(d) = \mu(d)$ for $d \le 4$, and that the exact values of $\mu(d) = \mu_2(d)$ are known only for $d \le 6$,

$$\mu(d) = 0, 1, 4, 16, 31, 65 \text{ for } d = 1, 2, 3, 4, 5, 6.$$

- $d = 1, 2$: trivial
- $d = 3$: so-called Cayley cubic, first apparently found by Schläfli, see [31, 38, 82]
- $d = 4$: Kummer surfaces [62] (first examples found by Fresnel, 1822)
- $d = 5$: Togliatti quintics, see [91] for the inequality $\mu(5) \ge 31$, and Beauville [9] for the inequality $\mu(5) \le 31$
- $d = 6$: Barth [8] $\mu(6) \ge 65$, Jaffe and Ruberman [53] $\mu(6) \le 65$ (the proof was then simplified by Jonny Wahl, see [99] and [77]).

Also, it turns out that in this range $d \leq 6$ (as we shall explain in this article) the answer to questions 1 and 3 is that there are no gaps (in view of [28, 79, 80] the only new result here is for the case $d = 5$, see Corollary 1.27):

Corollary 5 *If $n = 2$, $d \leq 6$, then we have the 'no gaps' phenomenon for nodal surfaces: $\mathcal{F}_2(d, \nu)$ is non-empty for all $\nu \leq \mu(d)$. This means: for each ν with $1 \leq \nu \leq \mu(d)$ there exists a nodal surface X of degree d in $\mathbb{P}^3$ with precisely ν nodes.*

The answer to question 4 is classical for $n = 2, d \leq 3$: the Nodal Severi varieties are smooth and irreducible.

While the case $d = 4$, in spite of a lot of classical work [57, 79, 80], was until now open (see the remarks in the first three lines of page 349 of [45]). We are here able to classify the irreducible components of $\mathcal{F}_2(4, \nu)$, for $\nu \leq 16$, we have indeed our main theorem for nodal quartics:

Theorem 6

(I) The subset of the Nodal Severi variety $\mathcal{F}(4, \nu)$ corresponding to the nodal quartic surfaces with ν nodes and fixed binary code $\mathcal{K}''$ is a smooth connected component.

(II) The possible codes $\mathcal{K}''$ appearing are all the shortenings of the extended Kummer code $\mathcal{K}''_{Kum}$.

(III) A component of $\mathcal{F}(4, \nu)$ is in the closure of another component of $\mathcal{F}(4, \nu')$ if and only if $\nu \geq \nu'$ and the second code is a shortening of the first.

The main thrust of the above theorem is of being a conceptual statement, as opposed to hardly readable and very long computer generated lists, but indeed it leads to a very explicit classification, and the possible codes $\mathcal{K}''$ appearing are listed in Sect. 1.8 (11 nontrivial codes in their minimal embedding); the classification[2] shows for instance that the Nodal Severi variety $\mathcal{F}(4, \nu)$ has exactly one irreducible component for $\nu = 0, 1, 2, 3, 4, 5, 14, 15, 16$, exactly two irreducible components for $\nu = 6, 7, 13$, exactly three irreducible components for $\nu = 8, 9$, exactly four irreducible components for $\nu = 11, 12$, exactly five irreducible components for $\nu = 10$. Hence its stratification is made of 34 strata.

A direct consequence is:

Corollary 7 *Let Y be a nodal quartic surface in $\mathbb{P}^3$, with ν nodes, where $\nu \leq 15$.*

Then the fundamental group $\pi_1(Y^*)$ of its smooth locus is finite and isomorphic to its code $\mathcal{K} \cong (\mathbb{Z}/2)^k$.

For $\nu = 16$ $\pi_1(Y^*)$ is the semidirect product $\mathbb{Z}^4 \rtimes \mathbb{Z}/2$ corresponding to multiplication by -1.

The main difference from the case of curves is given here by topology: given a nodal hypersurface X, we consider the nonsingular part of X, $X^* := X \setminus \mathrm{Sing}(X)$.

[2] Again, everything is done by hand using linear algebra.

In the case of curves we just delete 2ν points from a Riemann surface of genus $g = \frac{(d-1)(d-2)}{2} - \nu$, so the topology is detected by the integers d, ν.

For the case of surfaces Y, instead, we have four important topological invariants of the connected components of $\mathcal{F}(d, \nu)$. The first two are the fundamental group $\pi_1(Y^*)$, and its abelianization $H_1(Y^*, \mathbb{Z})$, which is a vector space over $\mathbb{Z}/2$ and a quotient of $(\mathbb{Z}/2)^\nu$. The second two are similarly defined: letting Y' be the complement in Y^* of a smooth hyperplane section, we have the fundamental group $\pi_1(Y')$, and its abelianization $H_1(Y', \mathbb{Z})$, a quotient of $(\mathbb{Z}/2)^\nu \oplus (\mathbb{Z}/d)$ (see Corollaries 1.5 and 1.13).

While the fundamental groups are intriguing groups, and difficult to compute for $d \geq 5$, these homology groups are somewhat easier to calculate, and through them coding theory comes into play, as first pioneered in the seminal paper [9].

Definition 8 Let Y be a nodal surface in $\mathbb{P}^3$, and let $\mathcal{S}$ be its singular set

$$\mathcal{S} := \mathrm{Sing}(Y) = \{P_1, \ldots, P_\nu\} :$$

then the **strict code** $\mathcal{K} \subset (\mathbb{Z}/2)^{\mathcal{S}}$ associated to Y is defined as the image of the injective linear map dual to the surjective linear map $(\mathbb{Z}/2)^{\mathcal{S}} \to H_1(Y^*, \mathbb{Z})$.

If d is even, the **extended code** $\mathcal{K}' \subset (\mathbb{Z}/2)^{\mathcal{S}} \oplus (\mathbb{Z}/2)H$ associated to Y is defined via the injective linear map dual the surjective linear map obtained from the surjection $(\mathbb{Z}/2)^{\mathcal{S}} \oplus (\mathbb{Z}/d)H \to H_1(Y', \mathbb{Z})$ by setting $\frac{d}{2}H = 0$. It is a **bicoloured code**, as we are now going to explain.

In order to fully understand the statement of Theorem 6 above, one needs to recall some basic concepts and results from coding theory, for this we refer to the first section of Chap. 1.

The main use of coding theory is based on the determination of the cardinalities of strictly even sets of nodes (weights of vectors in $\mathcal{K}$), and of half-even sets of nodes (vectors in the projection $\mathcal{K}''$ of $\mathcal{K}' \setminus \mathcal{K}$ inside $\mathbb{F}_2^{\mathcal{S}}$), which was initiated in [21] and [9], and continued in several papers [20, 29, 44], and which has lead to the complete classification of the $\mathcal{K}$-weights for surfaces of degree up to 6 (we did not yet achieve the complete classification for the $\mathcal{K}'$-weights of sextic surfaces, but we obtain here, concerning the half-even sets, new results sufficient for our purposes).

In degree $d = 4$, the weights for $\mathcal{K}$ can only be 8, 16, and those for $\mathcal{K}''$ can only be 8, 16, 6, 10. While the classification of the codes $\mathcal{K}$, even for more general K3 surfaces, is well known since long time, the classification of the codes $\mathcal{K}'$ is a much harder and newer task, because of the occurrence of more than 4 possible weights (which depend on the congruence class of d modulo 8).

The main theorem for quartics, Theorem 6, is based on several ingredients: first of all elementary coding theory, then the relation between unobstructedness and shortenings (Theorem 1.23), then Nikulin's theorems on primitive embeddings of lattices in the K3 lattice [74] and finally the Torelli theorem for K3 surfaces (see [76] and [10]).

The relation with lattices occurs as follows: for nodal surfaces Y, with minimal resolution $\tilde{Y}$, the lattice $\Lambda := H^2(\tilde{Y}, \mathbb{Z})$ contains the sublattice

$$\mathcal{L} = \bigoplus_i \mathbb{Z}E_i \oplus \mathbb{Z}H.$$

Now, the extended code $\mathcal{K}'$ determines the saturation $\mathcal{L}^{sat} := \Lambda \cap \mathbb{Q}\mathcal{L}$ of $\mathcal{L}$, which is primitively embedded inside $\Lambda := H^2(\tilde{Y}, \mathbb{Z})$.

In the 1970s Nikulin developed a theory of primitive embeddings of lattices, generalizing results of Milnor et al. to the non unimodular case. Nikulin's results offer criteria for existence and unicity of such embeddings.

These results are very useful for the class of K3 surfaces, of which resolutions of nodal quartic surfaces are a special case.

If we fix a 'potential' K3 code $\mathcal{K}'$, it turns out, using the results of Nikulin on integral quadratic forms that there is at most one primitive embedding of $\mathcal{L}^{sat} := \Lambda \cap \mathbb{Q}\mathcal{L}$ inside Λ. Some delicate calculations show that the existence of such an embedding depends on the congruence class of d modulo 16, while the potential K3 codes depend only on the congruence class of d modulo 8.

In this way the theorem for quartics extends also to all nodal K3 surfaces as follows (see Theorem 1.73 for fuller details and a complete classification).

Theorem 9 *The connected components of the Nodal Severi variety $\mathcal{F}_{K3}(d, \nu)$ of nodal K3' s of degree d (d must be an even number) and with ν nodes are in bijection with the isomorphism classes of their extended codes $\mathcal{K}'$ (equivalently, of the pair of codes $\mathcal{K} \subset \mathcal{K}'' \subset \mathbb{F}_2^{\nu}$).*

The weights t of vectors in $\mathcal{K}'' \setminus \mathcal{K}$ satisfy the following congruence:

$$2t - d \equiv 0 \ (mod\ 8).$$

Apart from sporadic cases, which are however obtained via the projection from a node of a nodal K3 surfaces, we have several main stream cases:

(0) If $d \equiv 0 \ (mod\ 8)$ then $t \in \{4, 8, 12\}$ and we get all the shortenings $\mathcal{K}''$ of the K8 code (generated by the strict Kummer code $\mathcal{K}_{Kum}$ and by the characteristic function of an affine plane π).

(2) If $d \equiv 2 \ (mod\ 8)$ and $t \in \{5, 9\}$, we get all the shortenings $\mathcal{K}''$ of the code generated by the linear functions on $\mathbb{F}_2^4$ and by the quadratic function $x_1 y_1 + x_2 y_2 + 1$.

(4-6) If $d \equiv 4 \ (mod\ 8)$ and $t \in \{6, 10\}$, we get for $\mathcal{K}''$ all the shortenings of the extended Kummer code $\mathcal{K}''_{Kum}$.

(4-4) If $d \equiv 4 \ (mod\ 8)$ and there is some $t = 2$, we get all the shortenings of the special code $\mathcal{K}'$ occurring for $\nu = 14$, $k' = 4$.

(6) If $d \equiv 6 \ (mod\ 8)$, we get all the shortenings $\mathcal{K}'$ of the two codes occurring for $\nu = 15$: the first $\mathcal{K}'$ is the strict Kummer code, in the second $\mathcal{K}'$ there is a weight 4 *vector.*

In part I we also give a complete list of all the codes $\mathcal{K}'$ ($\mathcal{K}''$) which occur, making the above theorem suitable for applications.

An interesting consequence is the following (see Theorem 1.74).

Theorem 10 *Let Y be a nodal K3 surface with ν nodes. Then, for $\nu \leq 15$, the fundamental group $\pi_1(Y^*)$ of the smooth locus of Y is isomorphic to the strict code $\mathcal{K} \cong (\mathbb{Z}/2)^k$ associated to Y.*

For $\nu = 16$ Y is a Kummer surface and $\pi_1(Y^*)$ is the semidirect product $\mathbb{Z}^4 \rtimes \mathbb{Z}/2$.

In the third part of the paper we proceed to study the Nodal Severi varieties $\mathcal{F}(d, \nu)$ for surfaces of degree $d = 5$. In this degree the possible weights can only be 16, 20, and this fact was one basic tool which allowed Beauville in his seminal paper [9] to show that $\mu(5) = 31$, and that if $\nu = 31$, then $\mathcal{K}$ is the biduality code of $\mathcal{K} \cong \mathbb{F}_2^5$.

His first crucial observation consisted in what we call here the **B-inequalities= Basic inequalities = Beauville inequalities**, which are the lower bounds $k \geq \nu - \frac{1}{2}b_2(d)$, $k' \geq \nu + 1 - \frac{1}{2}b_2(d)$, see Proposition 1.19).

Using a result of [21], namely that every quintic surface with ν nodes and with an even set of nodes of cardinality 16 is the discriminant of a cubic hypersurface $X \subset \mathbb{P}^5$ with $\nu - 16$ nodes, for the projection onto $\mathbb{P}^3$ with centre a line $L \subset X^*$ (contained in the smooth locus of X, as we show here), Beauville was able to show that every quintic with 31 nodes is a Togliatti quintic, that is, it is the discriminant of a cubic hypersurface $X \subset \mathbb{P}^5$ with 15 nodes.

We prove here a new result which is not unexpected, and actually we have believed for long time to hold true: both Togliatti quintics (quintic surfaces with 31 nodes) and Togliatti hypercubics (cubic hypersurfaces $X \subset \mathbb{P}^5$ with 15 nodes) form respective irreducible families.

Again this result is based also on similar ingredients to the previous one on quartics, in particular it relies on the Torelli theorem for K3 surfaces, this time applied to complete intersections of type (2, 3) inside $\mathbb{P}^4$.

We actually prove more:

Theorem 11 *The Nodal Severi varieties $\mathcal{F}_2(5, 31)$ and $\mathcal{F}_4(3, 15)$ are irreducible.*

For $\nu \geq 29$ the weight 20 does not appear in the code $\mathcal{K}$ and the correspondence given by taking discriminants of cubic hypersurfaces yields a bijection between the respective irreducible components of $\mathcal{F}_2(5, \nu)$ and $\mathcal{F}_4(3, \nu - 16)$.

These are listed in Sect. 1.7.1.

In the Nodal Severi variety $\mathcal{F}_2(5, 28)$ there is exactly one irreducible component which corresponds to the case where $\mathcal{K}$ contains two codewords of weight 20.

All the codes $\mathcal{K}$ of nodal quintic surfaces are just:

(1) either the shortenings of the biduality code (simplex code), coming from smoothings of the Togliatti quintics, or
(2) the shortenings of the unique 2-dimensional code with two codewords of weight 20,
(3) possibly the unique 2-dimensional code with (exactly) two codewords of weight 16.

Moreover, $\mathcal{F}_2(5, 20)$ contains as irreducible component the variety of quintic symmetroids ($dim(\mathcal{K}) = 1$, with the weight 20), and at least two other components (corresponding to shortenings of the simplex code, one with $dim(\mathcal{K}) = 1$ and the weight 16, the other with $dim(\mathcal{K}) = 0$).

The Appendix of Chap. 3 completely classifies all the codes with weights 16, 20, and those among them which satisfy the B-inequality (a necessary condition for the codes of nodal surfaces), hence the codes which could potentially be codes of some nodal quintic surface.

Our main result is that all these do actually occur as codes of some quintic surface, with one exception (third in the list) and the possible exception of one case (sixth in the list) which we cannot resolve, leaving therefore an interesting open question.

The fourth part of the paper and most of the second part, devoted to cubic hypersurfaces and their discriminants, are concerned with our initial aim: the study of the codes of nodal sextics, especially of those with 65 nodes, the maximum allowed.

The main underlying idea was that the method of discriminants of cubic hypersurfaces should also work for $d = 6$, and we show that this is the case, even if in a rather subtle way.

Recall that Barth constructed [8] a sextic with 65 nodes, and admitting an icosahedral symmetry; we show here, by the way, that the Barth sextic admits as group of automorphisms exactly the Icosahedral group $(\mathbb{Z}/2) \times \mathfrak{A}_5$.

Our research began with describing first the codes $\mathcal{K}_B$, $\mathcal{K}'_B$ of the Barth sextic, in an explicit way. Later on Sascha Kurz was able to show, (see Chap. 5) via extensive computer calculations (using that the weights of $\mathcal{K}$ lie in the set $\{24, 32, 40, 56\}$, and that $\mathcal{K}$ can be extended to $\mathcal{K}'$) that every sextic surface with 65 nodes has the same extended code as the Barth sextic.

Using the partial classification of half-even sets of nodes on sextics that we achieve here in Sect. 4.1[3], the knowledge of the code for $\nu = 65$, and the study of discriminants performed in some sections of the paper (Chaps. 2 and 4), we are able to show the following theorem:

Theorem 12 *Every sextic Y with 65 nodes is the discriminant of a cubic hypersurface $X \subset \mathbb{P}^6$ with 31 nodes for the projection of X onto $\mathbb{P}^3$ with centre a plane L intersecting* $\mathrm{Sing}(X)$ *in* 3 *nodes.*

Y can be also realized as the discriminant of a cubic hypersurface $X \subset \mathbb{P}^6$ with 32 nodes for the projection of X onto $\mathbb{P}^3$ with centre a plane L intersecting $\mathrm{Sing}(X)$ *in* 2 *nodes.*

The subtle and quite interesting point is that, unlike the case of quintics, where we project using Togliatti cubics, those with the maximal number of nodes, we cannot use here the Segre cubic $X_S \subset \mathbb{P}^6$, which has the maximal number ($= 35$) of nodes: since all planes contained in X_S are subsets of a three dimensional subspace contained in X_S.

We also show that the Segre cubic X_S is locally rigid (up to projective equivalence) in the variety $\mathcal{F}_5(3, 35)$.

[3] A similar classification of cardinalities of even sets of nodes on surfaces of degree 7 could pave the way to determining $\mu(7)$, but the enterprise looks not easy to fulfill.

We conjecture that every cubic hypersurface $X \subset \mathbb{P}^6$ with 35 nodes is projectively equivalent to X_S, and we pose similar questions, for instance regarding the Segre cubic hypersurfaces in projective spaces of even dimension.

The crucial question about the irreducibility of $\mathcal{F}_2(6, 65)$ motivates the analogous question about the irreducibility of $\mathcal{F}_5(3, 32)$.

Unlike the case of nodal quartics and quintics, where the binary codes occurring are easy to describe in terms of elementary linear algebra, for the case of the Barth sextic (the case of 65 nodal sextics), the description of the codes $\mathcal{K}_B, \mathcal{K}'_B$ is rather intriguing. As discovered by Michael Kiermaier after Yonghwa Cho's explicit description of these codes, these are associated to the Doro-Hall graph, which we briefly describe now.

Theorem 13 *The code $\mathcal{K}_B$ of the Barth sextic Y_B is associated to the following graph Γ, whose vertices are the nodes of Y_B, and where an edge joins two vertices P, P' if and only if the line $\overline{PP'}$ intersects Y_B in 4 distinct points.*

For each vertex P, there are exactly 10 vertices at distance 1, 30 at distance 2, and 24 at distance 3. The code $\mathcal{K}_B$ is generated by the characteristic functions of the 3-spheres of Γ (sets of points at distance 3 from one vertex).

The graph Γ above turns out to be the so-called Doro-Hall graph, whose group of automorphisms is equal to the group of collineations

$$\mathbb{P}\Sigma L(2, 25) := \mathbb{P}SL(2, \mathbb{F}_{25}) \rtimes (\mathbb{Z}/2)$$

(the non-normal subgroup is the one generated by the Frobenius automorphism of $\mathbb{F}_{25}$, yielding a semi-projectivity).

Doro was the first to construct this graph Γ, whose set of vertices is the conjugacy class of the Frobenius automorphism, with an edge connecting two elements if and only if they commute. Later Hall [50] showed that this was the largest of the class of the so-called locally Petersen graphs.

Since the Petersen graph is the graph associated to the vertices of the projective dodecahedron (quotient of the dodecahedron by the antipodal map), and the dodecahedron is the dual polyhedron to the icosahedron, it is natural to ask whether it is possible to find a direct explanation, starting from the explicit equation of the Barth surface exhibiting its $\mathfrak{A}_5$-symmetry, of the fact that the code $\mathcal{K}_B$ is the Doro-Hall code associated to the graph Γ as we explained above [4].

The scripts supporting the computer-assisted proofs are available at the arXiv-webpage of the manuscript, see [33].

---

[4] A picture of the Doro-Hall graph can be found in https://www.wolframalpha.com/input?i=Hall+Graph, but it is not easy to see the local dodecahedral symmetry, since the number of vertices is 65, and the number of edges is 325, hence it is too intertwined.

Contents

Chapter 1
Binary Codes and the Components of the Varieties of Nodal K3 Surfaces

Fabrizio Catanese

Abstract We prove that the Nodal Severi variety parametrizing nodal quartic surfaces has a stratification comprising 34 irreducible components, whose incidence relation is described explicitly via a 'genealogy tree'. Each component corresponds to a so-called shortening of the extended Kummer code.

It is well-known that a quartic surface is a K3 surface: we then completely extend the above results to the Nodal Severi varieties parametrizing nodal K3 surfaces of all possible degrees.

Moreover, we prove that the fundamental group of the smooth locus of a nodal K3 surface is determined by its associated code.

We start by briefly reviewing the basics of coding theory, emphasizing the connections to nodal surfaces, and explaining how to associate a pair of binary codes $\mathcal{K}, \mathcal{K}'$ to a nodal surface; main tools of this chapter are the Torelli theorem and Nikulin's theory of primitive embeddings of lattices, while a key novel idea is to exploit the relationship between shortenings of codes and unobstructed deformations of nodal surfaces.

1.1 Coding Theory Notation and Basics

(i) Given a field $\mathbb{F}$, an $\mathbb{F}$-linear code is an $\mathbb{F}$-vector subspace $\mathcal{K}$ of the vector space $\mathbb{F}^{\mathcal{S}}$ of the $\mathbb{F}$-valued functions on a given set $\mathcal{S}$; **the dimension of** $\mathcal{K}$ **shall be denoted by** k.

Two such codes $\mathcal{K}_1, \mathcal{K}_2$ are equivalent if there is a bijection $\mathcal{S}_1 \cong \mathcal{S}_2$ between the two sets carrying $\mathcal{K}_1$ to $\mathcal{K}_2$.

We have a **binary** code if $\mathbb{F} = \mathbb{F}_2 := \mathbb{Z}/2$.

We shall for simplicity restrict to this case.

F. Catanese (✉)
Mathematisches Institut der Universität Bayreuth, Bayreuth, Germany
e-mail: Fabrizio.Catanese@uni-bayreuth.de

F. Catanese (ed.), *Varieties of Nodal Surfaces, Coding Theory and Discriminants of Cubic Hypersurfaces*, Lecture Notes of the Unione Matematica Italiana 31,
https://doi.org/10.1007/978-3-032-05899-7_1

It goes without saying that for a binary code any function is the characteristic function χ_A of a set A, and addition corresponds to the **symmetric difference**

$$(A \setminus B) \cup (B \setminus A) = (A \cup B) \setminus (A \cap B).$$

(ii) A bicoloured $\mathbb{F}$-linear code is an $\mathbb{F}$-vector subspace $\mathcal{K}'$ of the vector space $\mathbb{F}^{\mathcal{A} \cup \mathcal{B}}$, where $\mathcal{A} \cup \mathcal{B}$ is a disjoint union. Two such codes $\mathcal{K}'_1, \mathcal{K}'_2$ are equivalent if there are bijections $\mathcal{A}_1 \cong \mathcal{A}_2$, $\mathcal{B}_1 \cong \mathcal{B}_2$ carrying $\mathcal{K}'_1$ to $\mathcal{K}'_2$.

To $\mathcal{K}'$ we associate the code $\mathcal{K} := \mathcal{K}' \cap \mathbb{F}^{\mathcal{A}}$.

In this book $\mathcal{B}$ shall consist of a single element H (which we identify to its characteristic function), and, since we shall make the assumption that $H \notin \mathcal{K}'$, the bicoloured code $\mathcal{K}'$ will be determined by a pair of codes $\mathcal{K} \subset \mathcal{K}'' \subset \mathbb{F}^{\mathcal{A}}$, where $\mathcal{K}$ is as above and $\mathcal{K}''$ is the projection of $\mathcal{K}'$ into $\mathbb{F}^{\mathcal{A}}$ (see item (iii) below).

In particular $k' := \dim(\mathcal{K}') = \dim(\mathcal{K}'')$.

(iii) Given a code $\mathcal{K} \subset \mathbb{F}^{\mathcal{S}}$, and $\mathcal{N} \subset \mathcal{S}$, the **shortening** $\mathcal{K}_{\mathcal{N}}$ of $\mathcal{K}$ with respect to $\mathcal{N}$ is just the intersection $\mathcal{K} \cap \mathbb{F}^{\mathcal{N}}$ ($\mathcal{K}_{\mathcal{N}}$ is just the subspace of the functions in $\mathcal{K}$ with support in $\mathcal{N}$).

The **projection** of $\mathcal{K}$ determined by $\mathcal{N}$ is just the image of $\mathcal{K}$ via the restriction map $p_{\mathcal{N}} : \mathbb{F}^{\mathcal{S}} \to \mathbb{F}^{\mathcal{N}}$.

Hence the shortening with respect to $\mathcal{N}$ is just the kernel of the projection determined by $\mathcal{S} \setminus \mathcal{N}$.

(iv) The **effective length** n of $\mathcal{K}$ is just the cardinality of the union $\mathcal{S}' \subset \mathcal{S}$ of the supports of all the functions in $\mathcal{K}$. $\mathcal{K}$ is said to be a **spanning code** if $n = \nu$, where $\nu := |\mathcal{S}|$ is called the **length** of the code.

(v) For a vector $v \in \mathcal{K}$, the **weight** $w(v)$ (also called Hamming weight) is just the cardinality of the support of the function v. The **minimum weight** $w(\mathcal{K})$ is the minimum of the weights of non-zero vectors $v \in \mathcal{K}$.

The **weight enumerator function** of $\mathcal{K}$ is the function $w_{\mathcal{K}}(x, y) = \sum_i a_i x^{n-i} y^i$, where n is the effective length and a_i is the number of vectors of $\mathcal{K}$ with weight equal to i.

(vi) Dual to the injection $\mathcal{K} \subset \mathbb{F}^{\mathcal{S}}$ we have a surjection $\phi : \mathbb{F}^{\mathcal{S}} \to \mathcal{K}^{\vee}$. The **dual** code of $\mathcal{K}$ is just $\ker(\phi)$, that is, the perpendicular $\mathcal{K}^{\perp} \subset \mathbb{F}^{\mathcal{S}}$ to $\mathcal{K}$. Given a subset $\mathcal{N} \subset \mathcal{S}$, duality exchanges the projection with the shortening: i.e.,

$$(p_{\mathcal{N}}(\mathcal{K}))^{\perp} = (\mathcal{K}^{\perp})_{\mathcal{N}}.$$

$\mathcal{K}$ is said to be a **projective code** if the dual code admits no vector of weight equal to 2.

(vii) The **biduality code** is just the natural embedding $\mathcal{K} \to \mathbb{F}^{\mathcal{K}^{\vee} \setminus \{0\}}$. Its dual code is called the **Hamming code**, and the biduality code is also called the **simplex** code (some texts use the historically correct but cumbersome term: 'dual Hamming code').

(viii) The **MacWilliams identity** determines the weight enumerator of the dual code:

$$w_{\mathcal{K}^\perp}(x, y) = \frac{1}{2^k} w_{\mathcal{K}}(x + y, x - y). \tag{1.1}$$

(ix) The MacWilliams identity reads out as a series of identities, the first two being:

$$\sum_{i>0} a_i = 2^k - 1, \ \sum_{i>0} i a_i = 2^{k-1} n. \tag{1.2}$$

(x) Assume that we have a code $\mathcal{K}$ where all the weights are just 2^m, and let $k = dim(\mathcal{K})$, n be the effective length. In this case we get from the MacWilliams identity:

$$2^m(2^k - 1) = 2^{k-1} n \Rightarrow n = (2^k - 1)2^{(m-k+1)} \Rightarrow k \le m + 1.$$

If $k = m + 1$, we have the simplex code, in general a subcode of the simplex code.

(xi) Bonisoli [12] showed that each code with only one weight is gotten from the simplex code via a product structure: that is, $\mathcal{K} \subset \mathbb{F}_2^{X \times Y}$ via the composition of the inclusions $\mathcal{K} \subset \mathbb{F}_2^Y \subset \mathbb{F}_2^{X \times Y}$, in shorthand notation $v(x, y) = v(y)$.

In the next subsection we discuss the case of spanning codes, and Radon duality.

1.1.1 Codes and Radon Duality on Finite Projective Spaces

Consider now more generally a linear code over the finite field $\mathbb{F}_q$, that is, $\mathcal{K}$ is a vector subspace of $\mathbb{F}_q^\nu$ of dimension k and which is spanning (full), that is, it has effective length $n = \nu$.

Let $\mathbb{P}$ be the set of points of the projective space $\mathbb{P}_{\mathbb{F}_q}^{k-1}$.

Up to a permutation automorphism of $\mathbb{F}_q^n$, such a spanning $\mathbb{F}_q$-linear code $\mathcal{K}$ is determined by a multiset of n points in $\mathbb{P}$, given by the n rows of a generator matrix G (a multiset is a map $\mathcal{C} : \mathbb{P} \to \mathbb{Z}_{\geq 0}$): because no such row can be zero, else $\mathcal{K}$ is not spanning.

Two multisets yield the same code if and only if they are projectively equivalent (this amounts to a change of basis for $\mathcal{K}$). Moreover, since $\mathcal{K}$ has dimension k, the multiset, that is, the support of the function $\mathcal{C}$, must span $\mathbb{P}$.

For $A \subseteq \mathbb{P}$ we define the counting function $\mathcal{C}(A) := \sum_{x \in A} \mathcal{C}(x)$, which shall also be called the multiplicity of the set A.

To a codeword $\mathbf{v} = G\mathbf{x}$ corresponds the hyperplane $H_{\mathbf{x}} := \mathbf{x}^\perp$, and the weight $w(\mathbf{v})$ of the codeword equals $\mathcal{C}(\mathbb{P} \setminus H_{\mathbf{x}}) = n - \mathcal{C}(H_{\mathbf{x}})$. And two codewords lead to the same hyperplane if and only if the corresponding vectors $\mathbf{x}$ yield the same point of $\mathbb{P}$.

Before we show that the weights of code vectors determine the isomorphism class of a spanning code, we recall a known definition, introduced by Bolker, see for instance [11], [63].

Definition 1.1 The function on the dual projective space $\mathbb{P}^\vee$ (whose elements are the hyperplanes $H \subset \mathbb{P}$) defined by $\mathcal{C}(H) := \sum_{x \in H} \mathcal{C}(x)$ shall be called the Radon transform $R(\mathcal{C})$ of the function $\mathcal{C}(x)$.

The Radon transform $R(\mathcal{C})$ determines the function $\mathcal{C}(x)$ by what we call Radon duality.

Theorem 1.2 (Radon Duality) *We have the following Radon duality formula:*

$$\mathcal{C}(x) = \frac{1}{q^{k-2}}[R(R(\mathcal{C}))(x) - \frac{q^{k-2}-1}{q-1} \int_{\mathbb{P}} \mathcal{C})].$$

In particular, if the function $\mathcal{C}(x)$ has zero average, that is, $\int_{\mathbb{P}} \mathcal{C}(x) = 0$, then we have the simpler formula:

$$\mathcal{C}(x) = \frac{1}{q^{k-2}} R(R(\mathcal{C}))(x).$$

Finally, the Radon transform $R(\mathcal{C})$ determines the function $\mathcal{C}(x)$ as follows:

$$\mathcal{C}(x) = \frac{1}{q^{k-2}}[R(R(\mathcal{C}))(x) - \frac{(q^{k-2}-1)}{(q^{k-1}-1)} \int_{\mathbb{P}^\vee} R(\mathcal{C})].$$

Proof In fact, the double Radon transform is

$$R(R(\mathcal{C}))(x) =: F(x) = \sum_{x \in H} \mathcal{C}(H) = \frac{1}{q-1}[(q^{k-1}-1)\mathcal{C}(x)+(q^{k-2}-1)\sum_{y \neq x} \mathcal{C}(y)],$$

hence we get

$$F(x) = q^{k-2}\mathcal{C}(x)+\frac{1}{q-1}(q^{k-2}-1)\sum_{y} \mathcal{C}(y) = q^{k-2}\mathcal{C}(x)+\frac{1}{q-1}(q^{k-2}-1)\int_{\mathbb{P}} \mathcal{C},$$

the first assertion follows by dividing by q^{k-2}, while the second assertion follows right away from the first.

For the third assertion, observe that

$$\int_{\mathbb{P}^\vee} R(\mathcal{C}) = \sum_H R(\mathcal{C})(H) = \sum_H \sum_{x \in H} \mathcal{C}(x) =$$

$$= \frac{1}{q-1}[(q^{k-1}-1)\sum_x \mathcal{C}(x)] = \frac{1}{q-1}[(q^{k-1}-1)\int_{\mathbb{P}} \mathcal{C}].$$

Hence

$$F(x) = q^{k-2}\mathcal{C}(x) + \frac{1}{q-1}(q^{k-2}-1)\int_{\mathbb{P}} \mathcal{C} = q^{k-2}\mathcal{C}(x) + \frac{(q^{k-2}-1)}{(q^{k-1}-1)}\int_{\mathbb{P}^\vee} R(\mathcal{C}).$$

And the third assertion follows now. □

Corollary 1.3 *The isomorphism class of a spanning code $\mathcal{K}$ is completely determined by the weights of its code vectors.*

Proof Follows immediately from the previous Theorem 1.2 on Radon duality and from the fact that $\mathcal{C}(x)$ determines $\mathcal{K}$. □

1.1.2 Kummer Codes

To go back to our classification of the irreducible components of the variety of nodal surfaces of degree 4, recall that, for nodal surfaces of even degree d, we have a natural embedding

$$\mathcal{K} \subset \mathcal{K}',\ \mathcal{K} = \mathcal{K}' \cap \mathbb{F}_2^{\mathcal{S}},$$

where of course the codimension of $\mathcal{K}$ in $\mathcal{K}'$ is at most 1.

The Kummer codes are just the codes of the Kummer surfaces, the quartic surfaces with 16 nodes, and these codes are related to the so-called Reed-Muller codes. The strict Kummer code $\mathcal{K}_{Kum}$ is a first order Reed-Muller code, the (5-dimensional) space of affine functions on $\mathbb{F}_2^4$,

$$\mathcal{K}_{Kum} := Aff(\mathbb{F}_2^4, \mathbb{F}_2) \subset \mathbb{F}_2^{\mathbb{F}_2^4}$$

while the extended Kummer code $\mathcal{K}'_{Kum}$ is the (6-dimensional) code $\mathcal{K}'_{Kum} \subset (\mathbb{F}_2^{\mathbb{F}_2^4} \oplus \mathbb{F}_2 H)$ whose projection in $\mathbb{F}_2^{\mathbb{F}_2^4}$ is generated by the space of affine functions on $\mathbb{F}_2^4$, and by a rank two quadratic function $\beta = x_1 y_1 + x_2 y_2$.

When we speak of shortenings of $\mathcal{K}'_{Kum}$, we always refer to subsets $\mathcal{N} \subset \mathcal{S}$.

1.2 Normal Surfaces in $\mathbb{P}^3$, and Binary Codes of Nodal Surfaces

Let $Y \subset \mathbb{P}^3 := \mathbb{P}^3_{\mathbb{C}}$ be a normal surface, with singular set $\mathrm{Sing}(Y) = \{P_1, \ldots, P_\nu\}$.

Set $Y^* := Y \setminus \mathrm{Sing}(Y)$: the topological structure of Y^* is an important invariant of Y, which is not changed by equisingular deformations; in particular the fundamental group $\pi_1(Y^*)$ and its Abelianization, the first homology group $H_1(Y^*, \mathbb{Z})$, are two basic such invariants.

Given the singular point P_i, let (see [70]) U_i be a neighbourhood of P_i obtained by intersecting Y with a small Euclidean ball with centre P_i, and set $U_i^* := U_i \setminus \{P_i\}$; let further F_i be the Milnor fibre of the singularity (Y, P_i), and let Γ_i be the local fundamental group $\pi_1(U_i^*)$.

Proposition 1.4 *Let $Y \subset \mathbb{P}^3$ be a normal surface and let Y^* be its smooth locus: then $\pi_1(Y^*)$ is normally generated by a quotient of the free product $\Gamma_1 * \Gamma_2 * \cdots * \Gamma_\nu$ of the local fundamental groups at the singular points.*

In particular, $H_1(Y^*, \mathbb{Z})$ is a quotient of the direct product of the Abelianizations of the Γ_i's, and we have a surjection:

$$\Gamma_1^{ab} \times \Gamma_2^{ab} \times \cdots \times \Gamma_\nu^{ab} \twoheadrightarrow H_1(Y^*, \mathbb{Z}).$$

Proof We observe that Y deforms to a smooth hypersurface Y', which has $\pi_1(Y') = 1$ by Lefschetz' theorem.

Define F to be the union of the Milnor fibres F_i, together with a tree connecting the base point to each Milnor fibre with a respective segment (in particular $\pi_1(F) = 1$ by Milnor's theorem [70] stating that F_i is homotopically equivalent to a bouquet of spheres).

The first van Kampen's theorem says that $1 = \pi_1(Y')$ is the quotient of the free product $\pi_1(Y^*) * \pi_1(F) = \pi_1(Y^*)$ by the subgroup normally generated by $\pi_1(F^*)$, where $F^* := F \cap Y^*$. Now, by construction, $\pi_1(F^*)$ is the free product $\Gamma_1 * \Gamma_2 * \cdots * \Gamma_\nu$, hence the desired assertion follows.

When we abelianize, we obtain a surjection of $\Gamma_1 * \Gamma_2 * \cdots * \Gamma_\nu$ onto $H_1(Y^*, \mathbb{Z})$, which therefore factors through a surjection $\Gamma_1^{ab} \times \Gamma_2^{ab} \times \cdots \times \Gamma_\nu^{ab} \twoheadrightarrow H_1(Y^*, \mathbb{Z})$. □

We are especially interested in the simplest possible case, where Y is a nodal surface, that is, all singular points of Y are **nodes**, singularities of multiplicity two with nondegenerate Hessian (hence locally biholomorphic to the singularity $y_1 y_2 - y_3^2 = 0$ inside $\mathbb{C}^3$).

Corollary 1.5 *If Y is a nodal surface in $\mathbb{P}^3$, with $\mathrm{Sing}(Y) = \{P_1, \ldots, P_\nu\}$, then $H_1(Y^*, \mathbb{Z}) \cong (\mathbb{Z}/2)^k$, and there is a natural surjection $\oplus_1^\nu (\mathbb{Z}/2) \twoheadrightarrow H_1(Y^*, \mathbb{Z})$.*

Proof A node is the quotient $\mathbb{C}^2/(\pm 1)$, since, if u_1, u_2 are local coordinates on $\mathbb{C}^2$, the quotient is embedded by $y_i := u_i^2,\ i = 1, 2$ and by $y_3 := u_1 u_2$.

In particular, the local fundamental group is then $\mathbb{Z}/2$. □

1.2.1 Global Orbifold Abelian Covering

Remark 1.6 Proposition 1.4 and Corollary 1.5 (and other results below) hold more generally for:

(i) Y a normal surface whose singularities are rational double points (these have local homology groups Γ_i^{ab} which are finite) and
(ii) whose minimal resolution S is simply connected (since Y deforms differentiably to a smooth Y', which is diffeomorphic to S, see for instance [26]).

For such surfaces then $H_1(Y^*, \mathbb{Z})$ is finite.

Definition 1.7 Let Y be a normal surface such that, setting $Y^* = Y \setminus Sing(Y)$, $|H_1(Y^*, \mathbb{Z})| < \infty$.

Then there is a finite Galois covering $Z \to Y$, with Z normal, associated to the surjection $\pi_1(Y^*) \to H_1(Y^*, \mathbb{Z})$, and with group $G := H_1(Y^*, \mathbb{Z})$.

Z **is called the Global (orbifold) abelian covering of** Y.

The Galois group being abelian, we have a splitting

$$p_*(\mathcal{O}_Z) = \bigoplus_{\chi \in G^*} \mathcal{F}_\chi,$$

where, for χ a character of G, the eigensheaf $\mathcal{F}_\chi$ is torsion free of generic rank 1 and indeed reflexive.

The Global abelian covering has irregularity $q(Z) := h^1(\mathcal{O}_Z) = 0$ if and only if $q(Y) = 0$ and $h^1(\mathcal{F}_\chi) = 0$ for all characters χ.

Moreover, if Y has quotient singularities, Z is smooth if and only if the local fundamental group Γ_i injects into G. This criterion is particularly interesting for the case of singularities which are nodes: Z is smooth if and only if $\Gamma_i \cong \mathbb{Z}/2$ injects into $H_1(Y^*, \mathbb{Z})$. Dually, if and only if the dual map is surjective.

A geometric consequence of Corollary 1.5 is that every connected finite Galois covering $Z \to Y$, ramified only on the singular set of Y, and with abelian Galois group G, has $G = (\mathbb{Z}/2)^r$. Understanding the Galois theory of such covers reduces to the study of a binary code, defined as follows.

Definition 1.8 Let Y be a nodal surface in $\mathbb{P}^3$: then its binary code $\mathcal{K} \subset (\mathbb{Z}/2)^\nu$ is defined as the image of the dual map of the surjection

$$\bigoplus_1^\nu (\mathbb{Z}/2) \twoheadrightarrow H_1(Y^*, \mathbb{Z}).$$

We want now to show that the above definition is the same as the one given by Beauville in [9]. Recall for this purpose the following theorem [23]

Theorem 1.9 *Let X be a smooth compact complex manifold of complex dimension n, and let $D = D_1 \cup \cdots \cup D_\nu$ be a reduced divisor, where each D_i is irreducible.*

Then the surjection $H_1(X \setminus D, \mathbb{Z}) \twoheadrightarrow H_1(X, \mathbb{Z})$ *has kernel equal to the cokernel of* $\rho : H^{2n-2}(X, \mathbb{Z}) \to H^{2n-2}(D, \mathbb{Z}) = \oplus_1^\nu \mathbb{Z}[D_i]$.

For $n = 2$, $\rho(L) = \sum_i (L \cdot D_i)[D_i]$.

For Y a nodal surface in $\mathbb{P}^3$, let $\tilde{Y}$ the blow up at the singular points: then it is well known that $\tilde{Y}$ is smooth and the inverse image of P_i is a (-2)-curve, i.e. a smooth curve $E_i \cong \mathbb{P}^1$ such that $E_i^2 = -2$.

Corollary 1.10 *Let* Y *be a nodal surface in* $\mathbb{P}^3$*, and let* $\tilde{Y}$ *be the blow up of* Y *at the singular points: then its binary code* $\mathcal{K} \subset (\mathbb{Z}/2)^\nu$ *equals to the kernel of the map*

$$\bigoplus_{i=1}^{\nu} (\mathbb{Z}/2)[E_i] \to H^2(\tilde{Y}, \mathbb{Z}/2).$$

The vectors in $\mathcal{K}$ *correspond to subsets* $\mathcal{N}$ *of* $\operatorname{Sing}(Y)$ *such that* $\sum_{i \in \mathcal{N}} E_i$ *is linearly equivalent to* $2L$*, for some divisor class* L *on* $\tilde{Y}$*.*

The cardinality t *of such sets* $\mathcal{N}$*, i.e., the weights of the code vectors, are divisible by* 4*, and indeed by* 8 *if the degree* $d := \deg(Y)$ *is even.*

Finally, the dimension k *of* $\mathcal{K}$ *is at least* $\nu - [\frac{1}{2} b_2(\tilde{Y})]$.

Proof Since Y is nodal, by the theorem of Brieskorn and Tjurina [15] and [88], $\tilde{Y}$ is diffeomorphic to a smooth surface of degree $d = \deg(Y)$ in $\mathbb{P}^3$, which is simply connected, by Lefschetz' theorem. In particular, $H^2(\tilde{Y}, \mathbb{Z})$ is a free $\mathbb{Z}$-module, $H^2(\tilde{Y}, \mathbb{Z}/2) = H^2(\tilde{Y}, \mathbb{Z}) \otimes_{\mathbb{Z}} \mathbb{Z}/2$, and the second assertion follows immediately from the first.

By Theorem 1.9, applied to $X := \tilde{Y}$ with $D := E_1 \cup \cdots \cup E_\nu$, we get that

$$H_1(Y^*, \mathbb{Z}) \cong coker[H^2(\tilde{Y}, \mathbb{Z}) \to \bigoplus_{i=1}^{\nu} \mathbb{Z}[E_i]].$$

Since $E_i \cdot E_j = 0, i \neq j$, and $E_i^2 = -2$, we obtain the surjection

$$\bigoplus_{i=1}^{\nu} (\mathbb{Z}/2)[E_i] \to H_1(Y^*, \mathbb{Z}),$$

which we have already shown in another way.

Since we have an exact sequence

$$H^2(\tilde{Y}, \mathbb{Z}) \to \bigoplus_{i=1}^{\nu} \mathbb{Z}[E_i] \to H_1(Y^*, \mathbb{Z}) \to 0,$$

we tensor with $\mathbb{Z}/2$ and obtain an exact sequence

$$H^2(\tilde{Y}, \mathbb{Z}/2) \to \bigoplus_{i=1}^{\nu} \mathbb{Z}/2[E_i] \to H_1(Y^*, \mathbb{Z}) \to 0,$$

and taking the dual $\mathbb{Z}/2$-vector spaces, we obtain the exact sequence

$$0 \to \mathcal{K} \hookrightarrow \bigoplus_{i=1}^{\nu} (\mathbb{Z}/2)[E_i] \to H^2(\tilde{Y}, \mathbb{Z}/2).$$

The statement about $|\mathcal{N}|$ was proven in [21], Proposition 2.11. The final assertion follows since the dimension of the kernel equals ν minus the dimension of the image, which is at most 1/2 of the dimension $b_2(\tilde{Y})$ of $H^2(\tilde{Y}, \mathbb{Z}/2)$, since the image is an isotropic subspace for the intersection form, which is nondegenerate. □

In the case where the degree $d := \deg(Y)$ is even we have a bigger code containing $\mathcal{K}$.

Definition 1.11

(1) Let Y be a nodal surface in $\mathbb{P}^3$ of even degree d, and let H be a smooth hyperplane section of Y.
Then the **extended code** $\mathcal{K}'$ is defined as the kernel

$$\mathcal{K}' := \ker[\bigoplus_{i=1}^{\nu} (\mathbb{Z}/2)[E_i] \oplus (\mathbb{Z}/2)[H] \to H^2(\tilde{Y}, \mathbb{Z}/2)].$$

(2) The sets $\mathcal{N}$ which correspond to code vectors in $\mathcal{K}$ were defined in [21] and later in [20] as strongly even (even) sets of nodes, whereas those sets corresponding to code vectors in $\mathcal{K}' \setminus \mathcal{K}$ were called weakly even (later: half-even) sets of nodes.

The weakly even (half-even) sets of nodes, which are defined if $d = 2m$, have cardinality t such that $t \equiv m (mod\ 4)$.

Example 1.12 (The Quadric Cone) For $d = 2$ there is only one nodal surface, the quadric cone $Y := \{y_1 y_2 - y_3^2 = 0\}$.

Since a curve E with $E^2 = -2$ is not divisible by 2, it follows that $\mathcal{K} = 0$, while we claim that $\mathcal{K}'$ is spanned by $E + H$.

Y is in fact the quotient of $\mathbb{P}^2$, with coordinates (u_0, u_1, u_2), by the action of $\mathbb{Z}/2$ sending $(u_0, u_1, u_2) \mapsto (u_0, -u_1, -u_2)$. The quotient is embedded by

$$y_i := u_i^2,\ i = 0, 1, 2,\ y_3 := u_1 u_2,$$

hence

$$\mathbb{P}^2/(\mathbb{Z}/2) \cong Y := \{y_1 y_2 - y_3^2 = 0\},$$

and Y has a double covering ramified only on the singular point $y_1 = y_2 = y_3 = 0$ and the hyperplane section $H := Y \cap \{y_0 = 0\}$. Blowing up $\mathrm{Sing}(Y)$ we get a double covering of $\tilde{Y}$ isomorphic to the blow up of $\mathbb{P}^2$ at the point $u_1 = u_2 = 0$, and ramified exactly on $E \cup H$.

Hence the divisor $E + H$ is divisible by 2, and yields a class in $\mathcal{K}'$.

Also, $\pi_1(Y^* \setminus H) \cong \mathbb{Z}/2$ and $\mathbb{P}^2$ is the Global abelian covering of Y.

We have again another characterization for the extended code.

Corollary 1.13 *Let Y be a nodal surface in $\mathbb{P}^3$, let $\tilde{Y}$ be the blow up of Y at the singular points, and let Y' be the complement in $\tilde{Y}$ of the union of the exceptional curves E_i with a smooth hyperplane section H (not passing through any node).*

(i) Then the first homology group $H_1(Y', \mathbb{Z})$ is a quotient

$$\bigoplus_{i=1}^{\nu} (\mathbb{Z}/2)[E_i] \oplus (\mathbb{Z}/d)[H] \to H_1(Y', \mathbb{Z}),$$

and

(ii) moreover, its extended binary code $\mathcal{K}' \subset (\mathbb{Z}/2)^\nu \oplus (\mathbb{Z}/2)[H]$, by definition equal to the kernel of the map

$$\bigoplus_{i=1}^{\nu} (\mathbb{Z}/2)[E_i] \oplus (\mathbb{Z}/2)[H] \to H^2(\tilde{Y}, \mathbb{Z}/2),$$

determines the saturation of the lattice $\mathcal{L}$ generated by the curves E_i and by the hyperplane section H inside the lattice $\Lambda := H^2(\tilde{Y}, \mathbb{Z})$.

More precisely, the saturation $\mathcal{L}^{sat} := \Lambda \cap \mathbb{Q}\mathcal{L}$ consists of the vectors

$$\mathcal{L}^{sat} = \{\frac{1}{2}(\sum_i a_i E_i + bH) | \sum_i (a_i E_i + bH)(mod\ 2) \in \mathcal{K}'\}.$$

The same result holds for nodal surfaces Y whose minimal resolution $S = \tilde{Y}$ is simply connected and such that the hyperplane class H yields an indivisible element in $\mathrm{Pic}(S)$.

Proof Arguing as in Corollary 1.10, we see that the code $\mathcal{K}'$ is the cokernel

$$H_1(Y^* \setminus H, \mathbb{Z}) \cong coker(H^2(\tilde{Y}, \mathbb{Z}) \to \bigoplus_1^{\nu} \mathbb{Z}[E_i] \oplus \mathbb{Z}H),$$

of the dual map of the embedding inside the lattice $\Lambda := H^2(\tilde{Y}, \mathbb{Z})$ of the lattice $\mathcal{L}$ generated by the curves E_i and by the hyperplane section H.

Assertion (i) follows then since $E_i^2 = -2$, $E_i \cdot E_j = 0$, if $i \neq j$, $H^2 = d$, $H \cdot E_i = 0$.

In view of the exact sequence

$$0 \to \mathcal{L} \to \Lambda \to F \oplus T \to 0,$$

where T is a torsion group and F a free abelian group, dualizing we get (in view of the unimodularity of Λ)

$$0 \to F^\vee \to \Lambda \to \mathcal{L}^\vee \to Ext^1(T, \mathbb{Z}) \to 0,$$

showing that $T \cong H_1(Y^* \setminus H, \mathbb{Z})$.

Consider now the saturation $\mathcal{L}' := \mathcal{L}^{sat}$ of $\mathcal{L}$ inside Λ; $\mathcal{L}'$ is primitively embedded.

Since $\mathcal{L} \subset \mathcal{L}'$, and $\mathcal{L}'/\mathcal{L} \cong T$, it follows that $\mathcal{L}' = \mathcal{L}^{sat}$ is generated by $\mathcal{L}$ and by classes $\lambda \in \Lambda$ of the form

$$\lambda = \sum_i \frac{1}{2} a_i E_i + \frac{1}{d} bH, a_i, b \in \mathbb{Z}.$$

Since $2\lambda \in \Lambda$, we obtain that $\frac{2b}{d} H \in \Lambda$.

But H is indivisible (cf. for instance [21], page 464), hence d divides $2b$; if d is odd, this implies that d divides b, hence we may assume $b = 0$, if $d = 2m$ is even, then this means that m divides b, so that $\lambda = \sum_i \frac{1}{2} a_i E_i + \frac{1}{2} b' H, a_i, b' \in \mathbb{Z}$, hence the class of λ yields a divisor class L such that $2L = \sum_i a_i E_i + b' H, a_i, b' \in \mathbb{Z}$, and corresponds exactly to a codeword in $\mathcal{K}'$.

The same proof works in the more general case of a nodal surface Y satisfying the two properties that its minimal $S = \tilde{Y}$ is simply connected, and that the class of H is indivisible; it suffices to observe that the only facts that we have used are that $\mathrm{Pic}(\tilde{Y})$ is a free abelian group, because $\tilde{Y}$ is simply connected, and that the class of H is indivisible. □

Remark 1.14 In particular, the same result holds also for a nodal K3 surface Y, with minimal resolution S, such that class of the hyperplane divisor H on S is indivisible.

1.2.2 Examples

Example 1.15 (A Cubic with Maximal Milnor Number) A similar example to Example 1.12 is the quotient $\mathbb{P}^2/(\mathbb{Z}/3)$ for the action such that $(u_0, u_1, u_2) \mapsto (u_0, \epsilon u_1, \epsilon^2 u_2)$, where ϵ is a primitive third root of unity.

The quotient is embedded by $y_i := u_i^3,\ i = 0, 1, 2$ and by $y_3 := u_0 u_1 u_2$, so that

$$\mathbb{P}^2/(\mathbb{Z}/3) \cong Y := \{y_0 y_1 y_2 = y_3^3\}.$$

In this case the triple covering is only ramified in the three singular points $y_3 = y_i = y_j = 0$ (for $0 \leq i < j \leq 2$), hence we conclude that (since $\mathbb{P}^2$ minus three points is simply connected)

$$\pi_1(Y^*) \cong \mathbb{Z}/3,$$

and again we have a Global abelian covering.

This cubic surface is remarkable, since it has 3 singular points with Milnor number 2 (locally isomorphic to the singularity $x_1 x_2 = x_3^3$), and $3 \cdot 2 = 6$ is bigger than the maximum number of singular points that a normal cubic can have, which equals to 4.

For each normal cubic surface Y, which is not the cone over a smooth cubic curve, its singularities are rational double points, hence the sum $m := Milnor(Y)$ of the Milnor numbers of the singular points equals the number of the exceptional (-2)-curves appearing in the resolution $\tilde{Y}$; since $\tilde{Y}$ is the blow up of the plane in 6 points, the rank of $H^2(\tilde{Y}, \mathbb{Z})$ is at most 7, hence (these (-2)-curves being numerically independent, see [5]) $Milnor(3) \leq 6$, and equality is attained in this example.

Example 1.16 (The Cayley Cubic) A normal cubic can have at most 4 singular points, and if it does have 4 singular points, then it is projectively equivalent to the Cayley cubic $Y := \{\sigma_3(x_0, x_1, x_2, x_3) = 0\}$, where σ_3 is the third elementary symmetric function (the four singular points are the 4 coordinate points).

We give another proof of the following known theorem (see [41], and Theorem 34 of [27] which follows the same idea but contains more explicit details)

Theorem 1.17 *If Y is the Cayley cubic, then*

$$\pi_1(Y^*) \cong \mathbb{Z}/2.$$

Proof In fact, Y admits a double covering Z ramified only in the four singular points: this can be seen by representing Y as the determinant of a symmetric matrix of linear forms, or as a consequence of the B-inequality of Proposition 1.19 which shows that the code $\mathcal{K}$ of the Cayley cubic has dimension 1 and is generated by the vector with 4 coordinates equal to 1.

Z is a smooth Del Pezzo surface of degree 6, hence isomorphic to the blow up of $\mathbb{P}^2$ in three points. Since Z minus a finite number of points is simply connected, follows that $\pi_1(Y^*) \cong \mathbb{Z}/2$, and Z is the Global abelian covering. □

Instead for other nodal cubics the associated code is trivial, because (see [21]) the number of nonzero coordinates of a vector in $\mathcal{K}$ is always divisible by 4, and divisible by 8 if the degree of the nodal surface is even.

We have more (see Proposition 35 of [27] for a simple proof):

Proposition 1.18 *If Y is a nodal cubic with $\nu \leq 3$ nodes, then Y^* is simply connected.*

1.2.3 *Range of Non Triviality of the Binary Codes*

The important discovery by Beauville is that for surfaces of low degree with many nodes these codes are necessarily nontrivial.

We use the following standard notation which we have already introduced: $\mathcal{K}$ is a $\mathcal{C}_k^n$ binary code if $\mathcal{K} \subset \mathbb{F}_2^\nu$ has dimension k, and effective length n (n is the cardinality of the minimal subset $\mathcal{N} \subset \{1, \ldots, \nu\}$ such that $\mathcal{K} \subset \mathbb{F}_2^{\mathcal{N}}$); $\mathcal{K}$ is said to be a spanning code if $n = \nu$.

Beauville remarks that the second Betti number $b_2(d)$ of a smooth surface of degree d equals the one of $\tilde{Y}$ if Y has degree d, and by Noether's formula $b_2(d) = e - 2 = 12\chi - K^2 - 2 = 10 + 12\binom{d-1}{3} - d(d-4)^2$. Since the span of the curves E_i (and of H, if d is even) inside $H^2(\tilde{Y}, \mathbb{Z}/2)$ is an isotropic subspace, Beauville obtained the

Proposition 1.19 (B-inequalities = Beauville Inequalities)

$$k \geq \nu - \frac{1}{2}b_2(d) = \nu - (5 + (d-1)(d-2)(d-3)) + \lceil \frac{1}{2}d(d-4)^2 \rceil,$$

where we denote by $\lceil m \rceil$ *the smallest integer greater or equal to* m*, and*

$$k' \geq \nu + 1 - \frac{1}{2}b_2(d) = \nu + 1 - (5 + (d-1)(d-2)(d-3)) + \lceil \frac{1}{2}d(d-4)^2 \rceil,$$

Remark 1.20 We get the following inequalities for low degree d:

- $d = 3, k \geq \nu - 3,$
- $d = 4, k \geq \nu - 11, k' \geq \nu - 10,$
- $d = 5, k \geq \nu - 26,$
- $d = 6, k \geq \nu - 53, k' \geq \nu - 52.$

Since, by the Miyaoka inequality [71], it is known that $\nu \leq \frac{4}{9}d(d-1)^2$, the B-inequality is useful only if $(4/9)d(d-1)^2 > [b_2(d)/2]$.

I.e., only if

$$9d(d-4)^2 - 90 \geq (d-1)[10d^2 - 82d + 108],$$

equivalently

$$(20 - d)d^2 \geq 46d - 18,$$

which clearly does not hold for $d \geq 20$, fails also for $d = 18, 19$, and holds for $3 \leq d \leq 17$.

1.3 Code Shortenings and Unobstructed Nodal Surfaces

Definition 1.21

(1) Consider a linear code $\mathcal{K} \subset \mathbb{F}^{\mathcal{S}}$, and a subset $\mathcal{N} \subset \mathcal{S}$.

Then the **shortening** of the code $\mathcal{K}$ relative to the subset $\mathcal{N}$, denoted by $Res(\mathcal{K}, \mathcal{N})$, or more briefly by $\mathcal{K}_{\mathcal{N}}$, was defined as

$$\mathcal{K}_{\mathcal{N}} := Res(\mathcal{K}, \mathcal{N}) := \{f | f \in \mathcal{K}, \ \text{supp}(f) \subset \mathcal{N}\} \subset \mathbb{F}^{\mathcal{N}}.$$

In case of an extended code $\mathcal{K}' \subset \mathbb{F}^{\mathcal{S}} \oplus \mathbb{F}H$, we denote by shorthand notation

$$\mathcal{K}'_{\mathcal{N}} = Res(\mathcal{K}', \mathcal{N}) := Res(\mathcal{K}', \mathcal{N} \cup H).$$

(2) A binary code $\mathcal{K} \subset \mathbb{F}_2^{\mathcal{S}}$ is said to be d-realized if $\mathcal{K}$ is the code of a nodal surface of degree d in $\mathbb{P}^3$.

Definition 1.22 If Y is a nodal hypersurface of degree d in $\mathbb{P}^n$, Y is said to be **unobstructed** if its singular points impose linearly independent conditions on the vector space of homogeneous polynomials of degree d.

The geometrical meaning of unobstructedness is that the deformations of Y have a submersion onto the space of local deformations of the singularities, so that one can obtain independent smoothings of all the singular points.

In fact, (see for instance [98], [24], and [85]) for a normal surface Y in $\mathbb{P} := \mathbb{P}^n$, we have (by definition) the exact sequence

$$0 \to N_Y^* \to \Omega_{\mathbb{P}}^1 \otimes \mathcal{O}_Y \to \Omega_Y^1 \to 0,$$

where the conormal sheaf $N_Y^* \cong \mathcal{O}_Y(-d)$ is generated by the differential of F, where F is the polynomial equation (of degree d) of Y.

Dualizing with respect to $\mathcal{O}_Y$, we get an exact sequence

$$0 \to \Theta_Y := \mathcal{H}om(\Omega_Y^1, \mathcal{O}_Y) \to \Theta_{\mathbb{P}} \otimes \mathcal{O}_Y \to N_Y \cong \mathcal{O}_Y(d) \to \mathcal{E}xt^1(\Omega_Y^1, \mathcal{O}_Y) \to 0,$$

which can be split into two short exact sequences, the second one being:

$$0 \to N'_Y \to N_Y \cong \mathcal{O}_Y(d) \to \mathcal{E}xt^1(\Omega_Y^1, \mathcal{O}_Y) \to 0.$$

The sheaf $\mathcal{E}xt^1(\Omega_Y^1, \mathcal{O}_Y)$ is, for a normal surface, concentrated on the singular points, and there equal to the quotient Tjurina algebra

$$\mathcal{O}_{\mathbb{P}}/(F, \partial F/\partial x_i),$$

where the x_i are local coordinates.

Unobstructedness means, when we pass to the long exact cohomology sequence, in view of the surjection $H^0(\mathcal{O}_{\mathbb{P}}(d)) \to H^0(\mathcal{O}_Y(d))$, exactness of

$$0 \to H^0(N'_Y) \to H^0(\mathcal{O}_Y(d)) \to H^0(\mathcal{E}xt^1(\Omega^1_Y, \mathcal{O}_Y)) \to 0.$$

The main result is that then the projective space $\mathbb{P}(H^0(\mathcal{O}_{\mathbb{P}}(d)))$, which parametrizes surfaces of degree d, at the point corresponding to Y, has a submersion onto the manifold $\mathcal{L}ocDef(Y)$ of local deformations of the singularities of Y. Hence in the neighbourhood of Y we can find all possible local deformations of the singularities of Y.

This leads to a very effective criterion:

Theorem 1.23 *Let Y be an unobstructed nodal surface of degree d: then all shortenings of the code $\mathcal{K}$ (as well as those of $\mathcal{K}'$ if d is even) are d-realized, i.e. they are realized by some nodal surface of degree d.*

Proof Let $\mathcal{S} := \mathrm{Sing}(Y)$, and let $\mathcal{N} \subset \mathcal{S}$. Then there is a local deformation Y' of Y which keeps the nodes in $\mathcal{N}$ and does a smoothing for the others.

By the Brieskorn-Tyurina theorem [15] and [88], there is a smooth family whose fibres contain $\tilde{Y}$ and $\tilde{Y}'$, and $(Y')^*$ is diffeomorphic to the complement of the union of the curves E_i, for $i \in \mathcal{N}$, while Y^* is diffeomorphic to the complement of the union of all the curves E_i, for $i \in \mathcal{S}$.

Then from the surjective map $\oplus_{i \in \mathcal{S}} \mathbb{F}_2[E_i] \to H_1(Y^*, \mathbb{Z}) = \mathcal{K}_Y^\vee \to 0$, whose kernel we shall denote here by V, if we divide by the subspace $\oplus_{i \in \mathcal{S} \setminus \mathcal{N}} \mathbb{F}_2[E_i]$, we obtain a surjection $\oplus_{i \in \mathcal{N}} \mathbb{F}_2[E_i] \to H_1((Y')^*, \mathbb{Z}) = \mathcal{K}_{Y'}^\vee \to 0$, and get

$$\mathcal{K}_Y^\vee = \mathbb{F}_2^{\mathcal{S}}/(V), \quad 0 \to (V + \mathbb{F}_2^{\mathcal{S} \setminus \mathcal{N}}) \to \mathbb{F}_2^{\mathcal{S}} \to \mathcal{K}_{Y'}^\vee \to 0.$$

Dualizing the above exact sequence we obtain (using that $(A/B)^\vee = ker(A^\vee \to B^\vee)$):

$$0 \to \mathcal{K}_{Y'} \to \mathbb{F}_2^{\mathcal{S}} \to V^\vee \oplus \mathbb{F}_2^{\mathcal{S} \setminus \mathcal{N}}$$

whence

$$\mathcal{K}_{Y'} = \ker(\mathcal{K}_Y \to \mathbb{F}_2^{\mathcal{S} \setminus \mathcal{N}}),$$

which is what we wanted to prove. □

The next question is: which nodal surfaces are unobstructed?

Of course, one necessary condition is that the number of nodes ν be strictly smaller than $\frac{1}{6}(d+3)(d+2)(d+1)$. This condition is false in general, as shown by Beniamino Segre [84], since he proved that $\frac{\mu(2d)}{8d^3} > \frac{\mu(d)}{d^3}$. Hence, starting from the case of Kummer surfaces, where $d = 4$, $\mu(4) = 16$, we get

$$\limsup \left(\frac{\mu(d)}{d^3} \right)) > \frac{1}{4} = 0,25$$

and using the Barth sextic that

$$\limsup \left(\frac{\mu(d)}{d^3} \right) > \frac{65}{216} = 0,3009\ldots.$$

Indeed Beniamino Segre constructed also easy explicit examples:

Example 1.24 For even degree $d = 2m$, let $\lambda_1(x), \ldots, \lambda_d(x)$ be general linear forms, and let $B(x)$ be a general form of degree m.

Then the surface

$$Y := \{x | \lambda_1(x) \cdots \lambda_d(x) - B(x)^2 = 0\}$$

has $\nu(Y) = \frac{1}{4}d^2(d-1)$, its nodes being the points such that there are $i < j$ with $\lambda_i(x) = \lambda_j(x) = B(x) = 0$.

The above examples by B. Segre were used by Burns and Wahl [18] to produce smooth surfaces S (the minimal resolutions of the nodal surfaces X) whose local moduli space is singular.

It is not yet clear whether there are examples of Nodal Severi varieties $\mathcal{F}(d, \nu)$ which are not smooth (cf. Question 5 in the Introduction).

However, in small degree, the known hypersurfaces which yield $\nu = \mu(d)$ are unobstructed.

This is obvious for $d = 2, 3$; for instance, for $d = 3$, we have the Cayley cubic, whose singular points are the four coordinate points.

Theorem 1.25 *The following nodal hypersurfaces are unobstructed:*

(i) The 4-nodal Cayley cubic of equation $\sigma_3(x_0, \ldots, x_3) = 0$.
(ii) The 16-nodal Duke of Cefalú surface of equation

$$(\sum_0^3 x_i^2)^2 - 3\sum_0^3 x_i^4 = 0\,;$$

(iii) The 10-nodal Segre cubic hypersurface in $\mathbb{P}^4$ *of equations (in* $\mathbb{P}^5$*)*

$$s_1(x) := s_1(x_0, \ldots, x_5) = 0, \quad s_3(x) = 0.$$

(iv) The Goryunov 15-nodal cubic hypersurface in $\mathbb{P}^5$ *of equations (in* $\mathbb{P}^6$*)*

$$s_1(x) := s_1(x_0, \ldots, x_5) = 0, \quad 2s_3(x) - 3zs_2(x) + 12z^3 = 0.$$

(v) *The 35-nodal Segre cubic hypersurface in* $\mathbb{P}^6$ *of equations (in* $\mathbb{P}^7$*)*

$$s_1(x) := s_1(x_0, \ldots, x_7) = 0, \quad s_3(x) = 0.$$

(vi) *the 31-nodal Togliatti surface of equation*

$$2(x^5 - 5x^4w - 10x^3y^2 - 10x^2y^2w + 20x^2w^3 + 5xy^4 - 5y^4w$$
$$+ 20y^2w^3 - 16w^5) + 5(x^2 + y^2 + bz^2 + zw + dw^2)^2z = 0$$

(vii) *the 65-nodal Barth sextic of equation*

$$(\tau^2x^2 - y^2)(\tau^2y^2 - z^2)(\tau^2z^2 - x^2) - \frac{1}{4}(2\tau - 1)w^2(x^2 + y^2 + z^2 - w^2)^2 = 0,$$

where $\tau := \frac{1}{2}(1 + \sqrt{5})$ *is the inverse of the golden ratio.*

Proof

(i) is obvious, since the singular points are the 4 coordinate points e_i, and $x_i^3(e_j) = \delta_{i,j}$.

(ii) The 16 singular points are just the G-orbit of (0, 1, 1, 1), where G is the semidirect product of $(\mathbb{Z}/2)^3 \rtimes \mathfrak{S}_4$, and $(\mathbb{Z}/2)^3$ acts multiplying the coordinates by ± 1.

We write the homogeneous forms of degree 4 as

$$\sum_1^4 a_i z_i^4 + \sum_{\{i,j\}} d_{i,j} z_i^2 z_j^2 + \sum_{i \neq h} b_{i,h} z_i^3 z_h + \sum_{i \neq j} e_{i,j} z_i \frac{z_1 z_2 z_3 z_4}{z_j}.$$

Imposing the vanishing on the 4 points $(0, 1, \epsilon_3, \epsilon_4)$, for $\epsilon_3, \epsilon_4 \in \{1, -1\}$, leads to the equations

$$(\sum_i a_i) - a_1 + d_{2,3} + d_{2,4} + d_{3,4} = 0$$

$$b_{2,3} + b_{3,2} + e_{4,1} = 0$$

$$b_{2,4} + b_{4,2} + e_{3,1} = 0$$

$$b_{3,4} + b_{4,3} + e_{2,1} = 0.$$

Acting with $\mathfrak{S}_4$, we get 12 equations for $\{i, h, r, s\} = \{1, 2, 3, 4\}$:

$$b_{i,h} + b_{h,i} + e_{r,s} = 0$$

$$b_{i,h} + b_{h,i} + e_{s,r} = 0$$

implying that the 12 coefficients $e_{r,s}$'s are determined by the $b_{i,h}$'s.

And we get 4 equations

$$(\sum_i a_i) - a_j + \sum_{r,s \neq j} d_{r,s} = 0,$$

showing that the 4 a_j's are determined by the coefficients $d_{r,s}$.

(iii) In general the singular points of the Segre cubic are the points where the gradients of s_1, s_3 are proportional, hence $x_i^2 = x_j^2$ for all pairs i, j. The condition $s_1(x) = 0$, implies that half of the x_i are equal to 1, the other half to -1. For $\mathbb{P}^5$ we get 10 singular points, corresponding to the partitions of $\{0, 1, 2, 3, 4, 5\}$ into two subsets of 3 elements.

Consider the cubic polynomials of the form $(x_i + x_j)(x_a + x_b)(x_c - x_d)$ where the 6 indices are all different. Such a polynomial vanishes on a singular point unless $x_i = x_j$, $x_a = x_b$ whereas x_c and x_d have opposite sign.

So, it vanishes for all but 2 singular points, for instance $(1, 1, -1, -1, 1, -1)$ and $(1, 1, -1, -1, -1, 1)$ if the indices are in increasing order $1, 2, 3, 4, 5, 6$.

However, for any two such different singular points, since we are in projective space, they have four equal coordinates, two of them $= 1$, the other two $= -1$, and two opposite coordinates.

Hence there is a polynomial as above taking in both points the same value ± 8, and vanishing in the others. Hence we see that the image of the evaluation map contains a basis of $\mathbb{C}^{10}$.

(iv) The 15 singular points are the $\mathfrak{S}_6$-orbit of the point $(1, 1, 1, 1, -2, -2, -1)$, which corresponds, as a permutation representation, to the set of unordered pairs of elements in the set $\{0, 1, 2, 3, 4, 5\}$. The same holds for the subspace of cubic polynomials with basis $F_{i,j} := (x_i x_j)z$, for $i < j$.

The representation splits ([56] page 203) as the sum of the trivial representation, and two irreducible representations of respective dimensions 5, 9. The former, added to the trivial representation, yields the standard permutation representation on 6 letters, and has basis $f_i := x_i z \sum_{j \neq i} x_j$.

The kernel is therefore the sum of some of these three irreducible representations.

The invariant polynomial $\sum_{i<j} a_{i,j}(x_i x_j)z$ does not vanish on $(1, 1, 1, 1, -2, -2, -1)$, since its value is $-(6 - 2 \cdot 8 + 4) = 6$.

Also, if the kernel would contain the 5-dimensional irreducible representation, then $f_1 - f_4$ would be in the kernel, but its value on $(1, 1, 1, 1, -2, -2, -1)$ equals $-((3 - 2 - 2) - (-8 + 4)) = -3 \neq 0$.

The final possibility is that the kernel is either zero, or equals the 9-dimensional irreducible representation, but then the rank of the evaluation map would be 6, and this is a contradiction. Indeed, the condition that $\sum a_{i,j} F_{i,j}$ vanishes on $(1, 1, 1, 1, -2, -2, -1)$ is

$$\sum_{i<j\leq 3} a_{i,j} - 2\sum_{i\leq 3}(a_{i,4}+a_{i,5}) + 4\sum a_{4,5} = 0.$$

To show that these linear equations have rank at least 7, we reduce modulo 2, and it suffices then to verify that the functions with coefficients in $\mathbb{Z}/2$ associated to pairs of elements $\{x, y\}$ in the set $\{0, 1, 2, 3, 4, 5\}$, $\sum_{i<j,i,j\neq x,y} a_{i,j}$ generate a subspace of dimension at least 7, which is tedious but true (the pairs not containing 0 map to the full 5-dimensional space generated by the $a_{0,j}$, then the pairs containing 0 generate at least a 2-dimensional space).

(v) See Theorem 2.19.

(vi) and (vii) See the computer calculation contained in Sect. 2.5.

□

Remark 1.26 More generally, all nodal surfaces of degree 4 and nodal K3 surfaces can be shown to be unobstructed, see Theorem 1.37 of [18], especially Example 3.9.

The following surfaces:

$$Y_3 := \{w^3 = xyz\},\ Y_4 := \{w^4 = xyz(x+y+z)\}$$

are unobstructed, while $Y_d := \{w^d = \Pi_1^d L_i(x, y, z, w)\}$, where the L_i's are general linear forms, is obstructed for $d \geq 5$.

Nodal surfaces of degree $d \leq 7$ have been shown to be unobstructed by Dimca [42] and Kloosterman [61].

Corollary 1.27 *For degree $d \leq 6$, we have the 'no gaps' phenomenon for nodal surfaces, this means, for each ν with $1 \leq \nu \leq \mu(d)$ there exists a nodal surface of degree d in $\mathbb{P}^3$ with precisely ν nodes.*

More generally, for each $\nu \leq \binom{d+1}{3}$, there exists a nodal surface of degree d with exactly ν nodes.

Proof The second statement follows from Proposition 2.25 of [21], asserting that for a linearly symmetric set of nodes, the nodes impose independent conditions on the surfaces of degree $d-1$, hence a fortiori on the surfaces of degree d. Hence for all $\nu \leq \binom{d+1}{3}$ there is a partial smoothing yielding the desired number ν of nodes.

Another proof for the case of degree $d = 5$ is therefore also gotten as follows: for $\nu \leq 20$, we are done by the unobstructedness of a general quintic symmetroid; while we use, for $16 \leq \nu \leq 31$, Corollary 2.27.

Ditto for the case of degree $d = 6$ using (v). □

Remark 1.28

(i) The result is new for $d = 5$, while for $d = 6$ it follows from the explicit constructions by Catanese-Ceresa [28] ($\nu \leq 64$) and Barth [8] $\nu = 65$.

(ii) It is an interesting question to ask for which degrees d the 'no gaps' phenomenon holds true (this is a special case, for the case of surfaces, of Question 3 mentioned in the introduction).

1.3.1 An Instructive Extended Code: The Case of the Beniamino Segre Surfaces

We want here to give a simple illustration of the application of coding theory to geometry, giving a characterization of the Segre surfaces of Example 1.24 in low degree.

These have equation

$$Y := \{x|\lambda_1(x)\cdots\lambda_d(x) - B(x)^2 = 0\},$$

where $B(x)$ is a form of degree $m = \frac{1}{2}d$, and the points $\lambda_i(x) = \lambda_j(x) = B(x) = 0$, varying $1 \leq i < j \leq d$, are nodes.

The equation can be seen in many ways, for each permutation of $\{1, \ldots, d\}$ and for all $r < d$, as the determinant of a symmetric 2×2 matrix of forms, $F(x) = det A(x)$, where

$$A(x) := \begin{pmatrix} \lambda_{i_1}(x)\ldots\lambda_{i_r}(x) & B(x) \\ B(x) & \lambda_{i_{r+1}}(x)\ldots\lambda_{i_d}(x) \end{pmatrix}. \tag{1.3}$$

As we see from [21], such a determinantal expression produces an even set of nodes with cardinality $mr(d - r)$, and which is $\frac{1}{2}$-even for r odd, and strictly even for r even.

For $r = 1$ we obtain a codeword w_i of $\mathcal{K}''$ with weight $m(d-1) = \frac{1}{2}d(d-1)$, and it is easy to see that these vectors (codewords) generate a subcode $\mathcal{K}''_S$ of dimension $d - 1$, since the sum $w_1 + \cdots + w_d = 0$ of these vectors is zero; we shall call this code the Segre code.

This code is derived from the 2-partition code $\mathcal{C}(2, \Sigma)$ on the set $\Sigma := \{1, \ldots, d\}$, which is defined as follows.

Let $\mathcal{M}$ be the family of subsets $\mathcal{D}$ of Σ with 2 elements, and associate to a 2-partition of Σ (Σ is the disjoint union $\mathcal{A} \cup \mathcal{B}$), the function $\phi_{\mathcal{A},\mathcal{B}}$ such $\phi_{\mathcal{A},\mathcal{B}}(\mathcal{D}) = 1$ if and only if $\mathcal{D}$ contains an element of $\mathcal{A}$ and an element of $\mathcal{B}$ (these functions shall be called 2-partition functions).

Then

$$\mathcal{C}(2, \Sigma) \subset \mathbb{F}_2^{\mathcal{M}},\ \mathcal{C}(2, \Sigma) := \{\phi_{\mathcal{A},\mathcal{B}}|\Sigma = \mathcal{A} \cup \mathcal{B}, \mathcal{A} \cap \mathcal{B} = \emptyset.\}$$

$\mathcal{C}(2, \Sigma)$ has 2^{d-1} elements, equal to $\frac{1}{2}$ the cardinality of the power set $\mathcal{P}(\Sigma)$, hence $\mathcal{C}(2, \Sigma)$ has dimension $(d - 1)$.[1] The sum of elements $\phi_{\mathcal{A},\mathcal{B}} + \phi_{\mathcal{A}',\mathcal{B}'}$ corresponds to the symmetric difference $\mathcal{A}\Delta\mathcal{A}'$, that is, to the 2-partition $((\mathcal{A}\cap\mathcal{A}') \cup (\mathcal{B}\cap\mathcal{B}')) \cup ((\mathcal{A} \cap \mathcal{B}') \cup (\mathcal{A}' \cap \mathcal{B}))$.

[1] One can define in a similar way the k-partition functions for all $k \leq d$: these do not in general form a code, they only generate a code which we call $\mathcal{C}(k, \Sigma)$.

We see that $\mathcal{K}''_S$ is an m-times repeated $\mathcal{C}(2, \Sigma)$ code, i.e. obtained via the product structure $\mathcal{N} := \mathcal{M} \times \{1, \dots, m\}$.

We have then the following characterization

Theorem 1.29 *Let Y be a nodal surface of degree $d = 2m$ in $\mathbb{P}^3$, and assume that the code $\mathcal{K}''$ admits a shortening which is (isomorphic to) the Segre code $\mathcal{K}''_S$ in degree d, i.e. $\mathcal{K}''_S$ is an m-times repeated $\mathcal{C}(2, \Sigma)$ code.*

Then, for $d \leq 8$, Y admits an equation of the form

$$Y := \{x | \lambda_1(x) \cdots \lambda_d(x) - B(x)^2 = 0\},$$

where the $\lambda_i(x)$ are linear forms.

Proof Let $\mathcal{N}_i$ be the $\frac{1}{2}$-even set corresponding to the codeword w_i, so that $|\mathcal{N}_i| = m(d-1)$: we have a linear equivalence on the minimal resolution S of Y,

$$2L_i \equiv - \sum_{P \in \mathcal{N}_i} E_P + H.$$

Using Theorem 1.10, part (ii) of Endraß [44], we infer that L_i is effective for $d \leq 8$, hence there is a linear form $\lambda_i(x)$ such that its divisor on S is of the form $2C_i + \sum_{P \in \mathcal{N}_i} E_P$, where C_i contains no exceptional divisor (the multiplicity of a linear form at a node being either 1 or 0).

Consider now the product $\Lambda(x) := \lambda_1(x) \dots \lambda_d(x)$. Its divisor on S is an effective divisor of the form $2D$, because the sum of the vectors w_i equals zero.

Then, since there is no torsion in $\mathrm{Pic}(S)$, D is cut by a form $B(x)$ of degree $m = \frac{1}{2}d$, and we have that $\lambda_1(x) \dots \lambda_d(x) - B(x)^2 = 0$ is zero on Y. But it is not identically zero because the linear forms are distinct, hence we obtained the equation of Y. □

Remark 1.30 It was conjectured by Endraß [44] on the basis of a result by Gallarati [47] that for half-even sets of nodes $\mathcal{N}$ the cardinality is always at least $\geq \frac{1}{2}d(d-1)$, equality holding if and only if $\mathcal{N}$ is symmetric and L is effective, that is, it corresponds to writing the equation F of Y as $\lambda(x)\phi(x) - B(x)^2$, with $\lambda(x)$ a linear form.

If the conjecture is true, then the above characterization of the B. Segre surfaces of Example 1.24 holds for all (even) degrees.

1.4 Codes and 'Fundamental Groups' of Nodal Quartic Surfaces

A nodal quartic surface can have at most 16 nodes.

This follows from the cruder estimate that the dual surface $Y^\vee$ of a nodal surface Y with ν nodes has degree $\delta = d(d-1)^2 - 2\nu$, and the observation that the degree

δ of $Y^\vee$ is at least d (for $d \geq 4$ the argument is that, since Y and $Y^\vee$ are birational, they have the same geometric genus, while, for $d \leq 3$, this follows by biduality $(Y^\vee)^\vee = Y$).

Hence, for $d = 3$, we get $\nu \leq 4$, as we already discussed, while for $d = 4$ $\nu \leq 16$, equality holding if and only if $Y^\vee$ is also a quartic surface.

We shall now discuss the well known fact that if $\nu = 16$, then Y is the quotient of a principally polarized Abelian surface A by the group $\{\pm 1\}$, embedded by the theta functions of second order on A (see for instance [27], which shows also that, in suitable coordinates, $Y = Y^\vee$).

This can be shown in several ways, but we choose here the coding theoretic approach, pioneered by Nikulin [73]: the main idea is that, knowing that the strict code $\mathcal{K}$ associated to a nodal quartic has weights only 8 or 16, there exists a codeword of weight 16, hence a double covering ramified exactly at the 16 nodes.

We begin with a rather elementary result in coding theory, for which we give a detailed proof as a preparation for other precise results (a shorter proof could be obtained invoking Bonisoli's theorem [12]).

Proposition 1.31 *Let $\mathcal{K} \subset \mathbb{F}_2^\nu$ be a code with $\nu \leq 16$ and weights only 8, 16. Then $\mathcal{K}$ is contained in the Kummer code $\mathcal{K}_{Kum} := Aff(\mathbb{F}_2^4, \mathbb{F}_2)$, a first order Reed-Muller code, consisting of the vector space of affine functions inside $\mathbb{F}_2^{\mathbb{F}_2^4}$.*

Moreover, if the weights are never equal to 16, it is a shortening of the Kummer code.

Proof We observe that $\mathcal{K}$ is contained in $\mathbb{F}_2^{16}$, where there is a unique vector e of weight 16.

We take $V := \mathcal{K}$ if $\mathcal{K}$ contains e, otherwise we let V be the span of $\mathbb{F}_2 e$ and of $\mathcal{K}$.

Notice that e corresponds to the constant function 1.

If there is another non zero vector $v_1 \in V$, we can assume by a permutation of coordinates that $v_1 = \sum_1^8 e_i$.

If V has dimension strictly bigger than two, take v_2 linearly independent of v_1, e. We can write $v_2 = v_2' + v_2''$, where $v_2' = \sum_{i \in A} e_i$, $v_2'' = \sum_{i \in B} e_i$, and $A \subset \{1, 2, \ldots, 8\}$, $B \subset \{9, 10, \ldots, 16\}$.

The condition that the weights of v_2 and $v_1 + v_2$ are equal to 8 (since they are both different from 0, e) amount to

$$|A| + |B| = 8, \ 8 - |A| + |B| = 8 \leftrightarrow |A| = |B| = 4.$$

By a permutation of the first 8 and of the second 8 coordinates we achieve that $v_2 = \sum_1^4 e_i + \sum_9^{12} e_i$.

If V has dimension at least 4, take v_3 linearly independent of e, v_1, v_2.

Then v_3 and $v_2 + v_3$ are different from 0, e and, writing as before $v_3 = v_3' + v_3''$, $v_3' = \sum_{i \in A'} e_i$, $v_3'' = \sum_{i \in B'} e_i$, and $A' \subset \{1, 2, \ldots, 8\}$, $B' \subset \{9, 10, \ldots, 16\}$, we find that $|A'| = 4$ and that the symmetric difference $(A \cup A') \setminus (A \cap A')$ has cardinality 4. Similarly for B, B'.

Hence we can permute the coordinates, leaving the partition determined by the integer part of $\frac{i}{4}$ invariant, and assume that $v_3 = \sum_{i \equiv 1,2 (mod\ 4)} e_i$.

A similar argument shows that , if $k := dim V \geq 5$, then if v_4 is linearly independent of e, v_1, v_2, v_3 then we can permute the coordinates, leaving the partition determined by the integer part of $\frac{i}{2}$ invariant, and assume $v_4 = \sum_{i \equiv 1 (mod\ 2)} e_i$.

That V has at most dimension 5 follows from a similar argument, or from the MacWilliams identity (1.2) applied to a supplement W of $\mathbb{F}_2 e$ inside V.

In fact we have then, if (W) is a C^n_h code, that is, a code of effective length n and dimension h, and a_8 is the number of vectors of weight 8,

$$a_8 = 2^h - 1, \quad 8 a_8 = n 2^{k-1} \Rightarrow n = (2^h - 1) 2^{4-h},$$

implying that $h \leq 4$.

We construct now a bijection of $\{1, \dots, 16\}$ with $\mathbb{F}_2^4$, by using the functions v_1, v_2, v_3, v_4 (we identify e_i with its characteristic function).

Hence it is proven that $\mathcal{K}$ is contained in the Kummer code.

For the second assertion, observe that, if $\mathcal{K}$ does not contain e, there are the four cases, where by a permutation of the coordinates we may assume that a basis of $\mathcal{K}$ is given by $v_1, \dots, v_k$.

For $k = 1$ we get the shortening associated to any subset of $\{1, 2, \dots, 11\}$ which contains the subset $\{1, 2, \dots, 8\}$, for $k = 2$ the shortening associated to the subsets $\{1, 2, \dots, 12\}$ and $\{1, 2, \dots, 13\}$, for $k = 3$ the shortening associated to the subset $\{1, 2, \dots, 14\}$ and for $k = 4$ the shortening associated to the subset $\{1, 2, \dots, 15\}$.

Indeed, when we take the shortening of the Kummer code by removing one point in affine space, this point may be chosen as the origin in a vector space, and we have the 4-dimensional space $\mathcal{K}$ of linear functions. Removing one, respectively two vectors, the dimension of $\mathcal{K}$ drops to 3, respectively 2. Removing a third vector, there are two cases, depending on whether the three vectors are linearly independent or not; the dimension drops to 1, respectively 2.

Once we remove at least four vectors plus the origin, three of them are independent and generate a 3-dimensional linear subspace U, and if the dimension of the shortening $\mathcal{K}$ stays 1, this means that the fourth vector we are removing is another one of the four remaining nonzero vectors in U.

Finally, the condition that $\mathcal{K}$ does not contain e is necessary, since $\mathbb{F}_2 e$ is not a shortening of the Kummer code. □

Remark 1.32 In fact, as done in [25] for generalized codes of this type, it is convenient to have the following combinatorial representation of the Kummer code as a chessboard code.

We represent $\{1, \dots, 16\} \cong \mathbb{F}_2^4$ as $W_1 \times W_2$, where $W_i := V_i^\vee$ and V_i is a 2-dimensional vector space over $\mathbb{F}_2$.

Then $V_1 \otimes V_2$ is the space of linear functions on $W_1 \times W_2$, and the space $\mathcal{K}$ of affine functions on $W_1 \times W_2$ is represented as the set of functions f such that, $\forall i, j$,

(i) $f(i, j) = f(0, 0) + f(0, j) + f(i, 0)$,

(ii) $\sum_j f(i, j) = 0$,
(iii) $\sum_i f(i, j) = 0$.

In fact, then, if $\{e_1, e_2\}$ is a basis for W_1, and $\{\epsilon_1, \epsilon_2\}$ is a basis for W_2, then $F := f + f(0, 0)$ satisfies the three equations

$$F(i, j) = F(0, j) + F(i, 0), \tag{1.4}$$

$$F(0, \epsilon_1) + F(0, \epsilon_2) + F(0, \epsilon_1 + \epsilon_2) = 0, \tag{1.5}$$

$$F(e_1, 0) + F(e_2, 0) + F(e_1 + e_2, 0) = 0, \tag{1.6}$$

hence F is linear.

In this representation, v_1 has entries equal to 1 exactly on the first two rows, v_2 on the first and third row, v_3 on the first two columns, v_4 on the first and third column.

Proposition 1.33 *Let $\mathcal{K}_{Kum}$ be the Kummer code $Aff(W_1 \times W_2, \mathbb{F}_2) \subset \mathbb{F}_2^{W_1 \times W_2}$, where W_1, W_2 are 2-dimensional vector spaces over $\mathbb{F}_2$, and let $\mathcal{K}_{Kum} \subset \mathcal{K}' \subset \mathbb{F}_2^{W_1 \times W_2} \oplus \mathbb{F}_2 H$ be a code such that $\mathcal{K}_{Kum} \neq \mathcal{K}'$, $\mathcal{K}_{Kum}$ is the $W_1 \times W_2$-shortening of $\mathcal{K}'$, and where the weights of vectors in $\mathcal{K}' \setminus \mathcal{K}$ are either 7 or 11.*

Then there is only one isomorphism class for the bicoloured code $\mathcal{K}'$, and $\mathcal{K}'$ is generated by $\mathcal{K}_{Kum}$ and by the pair (β, H), where β is the bilinear function $\beta(x, y) := x_1 y_1 + x_2 y_2$. We call this bicoloured code the ***extended Kummer code*** *$\mathcal{K}'_{Kum}$.*

Proof We use the notation introduced in the previous remark. $\mathcal{K}'$ is generated by $\mathcal{K}$ and by the pair (g, H), where the function g has the property

$$\text{the weight of } (g + f) \text{ equals 6 or 10}, \forall f \in \mathcal{K}. \tag{1.7}$$

By replacing g by $g+1$, we may assume that the weight of g equals 6. Conversely, if property (1.7) is satisfied, then necessarily $\mathcal{K}'$, the code generated by $\mathcal{K}$ and by the pair (g, H), fulfills the required properties.

Letting

$$a_i := \sum_j g(i, j), \ b_j := \sum_i g(i, j)$$

be the respective sums of the rows and of the columns of $g(i, j)$, we have

$$a_1 + a_2 + a_3 + a_4 = 6, \ b_1 + b_2 + b_3 + b_4 = 6.$$

Adding suitable linear combinations of $v_1, v_2, 1$ we may add to g two rows having all entries equal to 1. For instance, if we add the first two rows, we obtain the condition that

$$2 + 2(a_3 + a_4) = (4 - a_1) + (4 - a_2) + a_3 + a_4 = 6 \text{ or } 10 \Leftrightarrow a_3 + a_4 = 2 \text{ or } 4.$$

Playing the same game for the other pairs of rows, we conclude that for each partition into two pairs, one pair gives sum equal to 2, and the other gives sum equal to 4.

As a consequence, we observe that, if for instance $a_1 + a_2 = 4$, then necessarily $a_1 \leq 3$, and $a_2 \geq 1$.

It follows that, for all i, we have $a_i \leq 3$, else there is a pair with $a_i + a_j \leq 1$. If there is a j with $a_j = 3$, then all others are odd, and equal to 1; otherwise, $\forall i,\ a_i \leq 2$, and since there is a j with $a_j = 2$, we conclude that all the a_i's are even.

Hence either the four numbers a_i are, after a permutation, $(2, 2, 2, 0)$, or they are $(3, 1, 1, 1)$.

The same must also be true for the column sums.

To simplify the argument, we use the fact that the group of affinities of W_i equals the full symmetric group, permuting rows for $i = 1$ (respectively: columns for $i = 2$) and that we can use the symmetries in $Aff(W_1) \times Aff(W_2)$, and also exchange rows and columns (this amounts to exchanging W_1 with W_2): we shall consider two functions equivalent if obtained one from the other via a sequence of these operations.

Case $(3, 1, 1, 1)$ **for the Row Sums**

Here there is a row with three entries equal to 1: hence, using the above symmetries we can assume that it is the first row, and that its last entry equals 0.

Assume that also the column sums are of type $(2, 2, 2, 0)$: then the last column is equal to zero, and the first three columns contain exactly one more 1, according to a permutation of the three indices. Using the above symmetries, we can assume that the minor obtained deleting first row and last column is the identity matrix, and we have the function G corresponding to the matrix

$$G := \begin{pmatrix} 1 & 1 & 1 & 0 \\ 1 & 0 & 0 & 0 \\ 0 & 1 & 0 & 0 \\ 0 & 0 & 1 & 0 \end{pmatrix}.$$

Assume now that also the column sums are of type $(3, 1, 1, 1)$: then either the last column has all entries equal to 1, and we have the function g corresponding to the matrix

$$g := \begin{pmatrix} 1 & 1 & 1 & 0 \\ 0 & 0 & 0 & 1 \\ 0 & 0 & 0 & 1 \\ 0 & 0 & 1 & 1 \end{pmatrix}.$$

or the last column contains just one entry equal to 1, which we can assume to be the one in the last row. Then we get the function g' corresponding to the matrix

$$g' := \begin{pmatrix} 1 & 1 & 1 & 0 \\ 1 & 0 & 0 & 0 \\ 1 & 0 & 0 & 0 \\ 0 & 0 & 0 & 1 \end{pmatrix}.$$

Case $(2, 2, 2, 0)$ **for the Row Sums**

The case where the column sums are of type $(3, 1, 1, 1)$ is equivalent to the case of the function G, so we are left with the case where also the column sums are of type $(2, 2, 2, 0)$.

In this case we may assume that the first row and the first column are zero, and then since all other rows and columns contain exactly one entry equal to zero, we may assume that the entry zero occurs exactly for the diagonal elements, hence we get the function h corresponding to the matrix

$$h := \begin{pmatrix} 0 & 0 & 0 & 0 \\ 0 & 0 & 1 & 1 \\ 0 & 1 & 0 & 1 \\ 0 & 1 & 1 & 0 \end{pmatrix}.$$

Up to the above symmetry, we have then four cases, corresponding to the functions g, g', G, h such that

- g has support contained in the first row and last column, and vanishes for $(1, 4)$
- g' has support consisting of the first three elements of the first row and column, and of $(4, 4)$
- G has value 1 for the first three elements of the first row, and for the three elements $(i + 1, i)$
- h vanishes on the first row and column, and on the elements (i, i).

We show now the equivalence of these four cases by using other symmetries, given by other affinities of $W_1 \times W_2$, such that for instance

$$(i, j) \in W_1 \times W_2 \mapsto A(i, j) = (i, j + \phi(i)),$$

where $\phi : W_1 \to W_2$ is an affine map (in particular, we may get any permutation!).

These above transformations preserve the rows, and we have similar ones preserving the columns.

For instance, $j \mapsto j + e_1$ transposes the first two columns and the last two ones. Define $e_3 = e_1 + e_2, \epsilon_3 = \epsilon_1 + \epsilon_2$ and for each sequence of four points $[v_0, v_1, v_2, v_3]$ of W_2, take ϕ such that

$$0 \mapsto v_0, \ \epsilon_1 \mapsto v_1, \ \epsilon_2 \mapsto v_2, \ \epsilon_3 \mapsto v_3.$$

Then:

(i) for $[0, e_1, e_2, e_3]$, ϕ sends g' to a matrix of type $(3, 1, 1, 1) \times (2, 2, 2, 0)$ (equivalent to G)
(ii) for $[e_1, e_3, e_2, 0]$, ϕ sends h to a matrix of type $(3, 1, 1, 1) \times (2, 2, 2, 0)$ (equivalent to G)
(iii) for $[e_3, e_2, 0, e_1]$, ϕ sends g to a matrix of type $(3, 1, 1, 1) \times (2, 2, 2, 0)$ (equivalent to G)

Hence there is only one case, and the proof is accomplished, if we observe that

$$h \sim \beta := x_1 y_1 + x_2 y_2,$$

since h is the matrix of the form $x_1 y_2 + x_2 y_1$. □

Remark 1.34 The proof could also be accomplished (as we did in the first version) by showing that in each of the four cases the extra function belongs to the extended Kummer code.

For instance, $\sim$ denoting equivalence,

$$1+g \sim q := x_1x_2+x_1+x_2+y_1y_2 = \beta+x_1+x_2, \ 1+G \sim q(x_1, x_2, x_1+y_1, x_2+y_2).$$

We opted here for a conceptually clearer proof.

Proposition 1.35 *Let $\mathcal{K}' \subset \mathbb{F}_2^\nu \oplus \mathbb{F}_2 H$ be a bicoloured code such that*

(i) $\nu \leq 16$,
(ii) $\mathcal{K} := \mathcal{K}' \cap \mathbb{F}_2^\nu$ *has only weights* 8, 16 *while*
(iii) the projection $\mathcal{K}''$ of $\mathcal{K}'$ to $\mathbb{F}_2^\nu$ has the property that the vectors in $\mathcal{K}'' \setminus \mathcal{K}$ have only weights 6 *or* 10.

Then $\mathcal{K}'$ is contained in the extended Kummer code $\mathcal{K}'_{Kum}$.

In particular, if $\mathcal{K}'$ satisfies the B-inequality $k' := dim(\mathcal{K}') \geq \nu - 10$, $\mathcal{K}'$ is a shortening of the extended Kummer code.

Proof Let us prove the first assertion, namely the inclusion $\mathcal{K}' \subset \mathcal{K}'_{Kum}$.

If $\mathcal{K}' = \mathcal{K}$, this is the content of Proposition 1.31, hence we may now assume that $\mathcal{K}' \neq \mathcal{K}$.

We can clearly also assume that $\mathcal{K}''$ contains the vector e, that is, the function 1. Observe that the assertion is true if the dimension of $\mathcal{K}$ equals 1 or 5 (by virtue of Proposition 1.33).

Let us use the same notation as in Propositions 1.31 and 1.33, and consider a function $g \in \mathcal{K}''$ of weight 6.

If the dimension k of $\mathcal{K}$ equals 2, we have that $g = g' + g''$, where $g' = \sum_{i \in A} e_i$, $g'' = \sum_{i \in B} e_i$, and since

$$|A| + |B| = 6, \ 8 - |A| + |B| \in \{6, 10\}, \ 8 + |A| - |B| \in \{6, 10\}$$

we obtain

$$2(|A| - |B|) \in \{0, -4, 4\}.$$

Since $|A| = |B| = 3$ is impossible, we may assume, possibly exchanging A with B, that

$$|A| = 4, |B| = 2.$$

By a suitable permutation of the set $\mathcal{S}$, leaving invariant the partition $\mathrm{supp}(v_1), \mathrm{supp}(v_1 + e)$, we achieve that the function g equals the one of Proposition 1.33, and we are done for $k = 2$.

If the dimension $k = 3$, then we may assume that $\mathcal{K}$ is generated by v_1, v_2, e, and we have that the row sums a_i satisfy $a_1 + a_2 + a_3 + a_4 = 6$. As in Proposition 1.33 we conclude that these numbers are either $(2, 2, 2, 0)$ or $(3, 1, 1, 1)$. Hence, by a suitable permutation of the set $\mathcal{S}$, leaving invariant the partition given by the rows, we achieve that the function g is either the function g or the function h of Proposition 1.33, hence we are done for $k = 3$.

For the final case where the dimension of $\mathcal{K}$ equals 4 the argument is similar: the four rows split into 8 blocks, first and second half, and we may assume without loss of generality that the sum of the first pair of columns is equal to 2.

For row sums $(3, 1, 1, 1)$ there are two possibilities, namely

$$\begin{pmatrix} 2 \text{ or } 1 & 1 \text{ or } 2 \\ 0 \text{ or } 1 & 1 \text{ or } 0 \\ 0 & 1 \\ 0 & 1 \end{pmatrix}.$$

For row sums $(2, 2, 2, 0)$ there are also a priori two possibilities, namely

$$\begin{pmatrix} 0 & 2 \\ 1 \text{ or } 0 & 1 \text{ or } 2 \\ 1 \text{ or } 2 & 1 \text{ or } 0 \\ 0 & 0 \end{pmatrix},$$

but the latter option cannot occur, since, adding v_2 and then v_4, we obtain a vector of weight 2, which is not possible.

It is easy now to verify that the three possibilities are given by the respective functions g, G with first and last columns exchanged, h with first and last rows exchanged.

We pass now to proving the last assertion, that if $\mathcal{K}' \subset \mathcal{K}'_{Kum}$ and $k' \geq \nu - 10$, then $\mathcal{K}'$ is a shortening of $\mathcal{K}'_{Kum}$.

We argue on $\mathcal{K}'' \subset \mathcal{K}''_{Kum}$.

If $k' = 6$ we have equality, hence as well if $\nu = 16$ by the B-inequality $k' \geq \nu - 10$.

Let

$$\mathcal{A} := \mathrm{supp}(\mathcal{K}''_{Kum}) \supset \mathcal{N}' := \mathrm{supp}(\mathcal{K}'').$$

Let $\mathcal{C} \supset \mathcal{K}'$ be the $\mathcal{N}'$-shortening of the extended Kummer code.

If $|\mathcal{N}'| = 15, 14, 13$ then $\mathcal{C}$ has dimension d equal respectively to $d = 5, 4, 3$, while k' is at least respectively (since $k' \geq \nu - 10$) 5, 4, 3, hence $d = k'$ and $\mathcal{K}'' = \mathcal{C}$.

If $|\mathcal{N}'| = 12$, our assertion is proven in the same way if $d = 2$; and $d = 3$ only if, looking at the list given in Sect. 1.8.3 for the shortenings $\mathcal{C}$, we have for $\mathcal{C}$ case (1) with $\nu = 12$, and $d = 3$.

Obviously, if $d = 1$, we conclude since by assumption $k' \geq 1$.

Observe also that the case where $\mathcal{K}'' = \mathcal{K}$, $k' \leq 2$, is evidently a shortening of the extended Kummer code (cases (II) and (a)).

In the critical case (1) for $\mathcal{C}$, with $|\mathcal{N}'| = 12$, and $d = 3$, $\mathcal{K}'$ is a subspace of $\mathcal{C}$, which is spanned by $y_1, y_2, x_1y_1 + x_2y_2$; since we may assume $\mathcal{K} \neq \mathcal{K}''$, if $k' = 2$, we find that $\mathcal{K}''$ is either isomorphic to the subspace of the shortening (III-2), or to the one generated by $y_2, x_1y_1 + x_2y_2 + y_1 = x_1y_1 + x_2y_2 + y_1^2 = (x_1 + y_1)y_1 + x_2y_2$, which is equivalent to the shortening (III-2), and we are done.

If $|\mathcal{N}'| = 11$, then $d \leq 2$, hence we are done unless $k' = 1$, $k = 0$: then we use the normal form of quadratic forms to conclude that we have one of the two shortenings (b) or (c). □

Corollary 1.36 *Let Y be a nodal surface of degree $d = 4$ in $\mathbb{P}^3$, and with $\nu = 16$ nodes. Then Y is a Kummer surface, $Y = A/\pm 1$, where A is a principally polarized Abelian surface.*

In particular, $\pi_1(Y^*)$ is the semidirect product $\mathbb{Z}^4 \rtimes \mathbb{Z}/2$ corresponding to multiplication by ± 1.

Its extended code $\mathcal{K}'$ is the extended Kummer code $\mathcal{K}'_{Kum}$, that is, the direct sum $\mathcal{K}_{Kum} \oplus \mathbb{F}_2\beta$ of Proposition 1.33, where β is the bilinear form $\beta := x_1y_1 + x_2y_2$.

More generally, let Y be a nodal surface of degree d and with $\nu = 16$ nodes. Then d is divisible by 4, $d = 4m$, and Y is a Kummer surface, $Y = A/\pm 1$, where A is an Abelian surface with a polarization of type $(1, m)$.

Again $\pi_1(Y^*)$ is the semidirect product $\mathbb{Z}^4 \rtimes \mathbb{Z}/2$ corresponding to multiplication by ± 1.

Its extended code $\mathcal{K}'$ is the extended Kummer code for m odd, and for m even the space generated by the affine functions and the characteristic function f_π of an affine plane π.

Proof By the B-inequality the code $\mathcal{K}$ of Y has dimension at least 5, hence the previous Proposition 1.31 tells that we have exactly the Kummer code, in particular there exists a double covering $A \to Y$ branched exactly on the 16 nodes (and where A is smooth).

We have that K_A is a trivial divisor, in particular A is minimal.

By Proposition 2.11 of [21] we have that $\chi(\mathcal{O}_A) = 0$, hence, by the classification theorem of algebraic surfaces, A is an Abelian surface. Since there is an involution ι on A which has fixed points, we can assume, up to a translation on A, that ι is multiplication by ± 1.

Finally, the pull back of the hyperplane divisor H on Y is a divisor D with $D^2 = 8$ and with $h^0(A, \mathcal{O}(D)) \geq 4$. Hence D is a polarization of type (d_1, d_2) with $d_1 d_2 = 4$. Since however all the sections of $H^0(A, \mathcal{O}_A(D))$ are even functions, it follows that $d_1 = d_2 = 2$ and D is twice a principal polarization.

Writing $A = \mathbb{C}^2/\Lambda$, where $\Lambda \cong \mathbb{Z}^4$, Y is the quotient of $\mathbb{C}^2$ by the group Γ of transformations $z \mapsto \pm z + \lambda$, for $\lambda \in \Lambda$.

Γ is then isomorphic to the semidirect product $\mathbb{Z}^4 \rtimes \mathbb{Z}/2$ corresponding to multiplication by ± 1.

We calculate its abelianization, set $\Lambda := \mathbb{Z}^4$ and let g be the generator of $\mathbb{Z}/2$: since $g^2 = 1$, $g\lambda g = -\lambda$, in the abelianization $\mathcal{K}$ we have

$$2[g] = 0, \ \forall \lambda \in \Lambda, \ [\lambda] = -[\lambda] \Leftrightarrow 2[\lambda] = 0,$$

hence $\mathcal{K} \cong (\mathbb{Z}/2)^5$.

Since $\mathbb{C}^2 \setminus \Lambda$ is simply connected, and $Y^* = (\mathbb{C}^2 \setminus \Lambda)/\Gamma$, where the action is free, we obtain that $\pi_1(Y^*) \cong \Gamma$.

The extended code $\mathcal{K}'$ is as claimed by virtue of Proposition 1.33, but one can also use geometry as follows. $\mathcal{K}'$ is obtained adding a linear equation saying that there is a divisor L on $\tilde{Y}$ such that $H \equiv 2L + \sum_{i \in \mathcal{N}} E_i$.

Take a symmetric theta divisor D, so that D is ι-invariant, and it descends to a Weil divisor D' on Y. Observe that D passes exactly through the six odd half periods, i.e., those for which $\beta(x_1, x_2, y_1, y_2)$ equals 1, and with multiplicity one.

The desired divisor L is the proper transform of D': and since $2D'$ is a hyperplane section, we have the desired linear equivalence $H \equiv 2L + \sum_{i \in \mathcal{N}} E_i$, where $\mathcal{N}$ is the set of six nodes corresponding to set of the odd half periods.

In the more general case where Y has 16 nodes and degree d, we have to slightly change some part of the argumentation as follows.

As before $Y = A/\pm 1$.

The extended code $\mathcal{K}'(Y)$ has dimension 6, hence there is a divisor L on $\tilde{Y}$ such that $H \equiv 2L + \sum_{i \in \mathcal{N}} E_i$.

Since, for some $h \in \mathbb{N}$, $L + hH$ is effective, for the pull back D of H we have that $(2h + 1)D \equiv 2\Delta$, where the divisor Δ is effective. Hence D is 2 divisible, $D = 2D'$ and D^2 is divisible by 8, hence $d = H^2$ is divisible by 4, $d = 4m$. And then D' is of type (a, b) with $ab = m$.

Recall that $a|b$, hence, if $a \neq 1$, then $D' = aD''$ and the divisor $D = 2aD''$ is the pull-back of H, and since $2D''$ is the pull back of a Cartier divisor H'' on Y, we get $H = aH''$, contradicting that H is indivisible. Hence $a = 1, b = m$.

The cardinality of the half-even set $\mathcal{N}$ (which must be $\neq 0$) is the number of half periods where the divisor D' has odd multiplicity. We can take D' the pull back of a principal polarization under an isogeny with kernel $\mathbb{Z}/m$.

More precisely, we consider the lattice

$$\Lambda' := (\mathbb{Z} \oplus m\mathbb{Z}) \oplus \tau\mathbb{Z}^2 \subset \Lambda := (\mathbb{Z} \oplus \mathbb{Z}) \oplus \tau\mathbb{Z}^2 \subset \mathbb{C}^2,$$

for τ in the Siegel domain, and calculate the Riemann theta function $\theta(\tau, z)$ at the half periods.

The group of half periods for Λ' maps, for m odd, isomorphically to the group of half periods for Λ, hence for m odd we get the points where the multiplicity is odd, these correspond to the solutions of $\beta(x_1, x_2, y_1, y_2) = 1$.

While, for m even, x_2 maps to $2x_2$, hence we get the solutions of $x_1 y_1 = 1$: these are the points of the affine plane $x_1 = y_1 = 1$. □

We want now to give a first sharpening of a result shown in [30], theorem B and in [19], asserting the finiteness of the fundamental group of the smooth locus; we shall extend this result to other nodal K3 surfaces in the sequel (Theorem 1.74).

Theorem 1.37 *Let Y be a nodal quartic surface in $\mathbb{P}^3$, with ν nodes, where $\nu \leq 15$: then the fundamental group $\pi_1(Y^*)$ of its smooth locus is finite and is isomorphic to its binary code $\mathcal{K}$, which is in turn determined by its dimension k, $0 \leq k \leq 4$.*

The subset of the Nodal Severi variety $\mathcal{F}(4, \nu)$ corresponding to nodal quartic surfaces with fixed number $\nu \leq 16$ of nodes and fixed binary code $\mathcal{K}'$ is smooth and connected, hence irreducible.

The possible codes $\mathcal{K}'$ appearing are in particular all the shortenings of the extended Kummer code of Corollary 1.36, and they are listed in Sect. 1.8.3 (cases (0), (00), (i), (1), (2) , (II), (III-1), (III-2), (a), (b), (c), (d), hence we have 11 nontrivial codes in their minimal embedding).

The Nodal Severi variety $\mathcal{F}(4, \nu)$ has exactly one irreducible component for $\nu = 0, 1, 2, 3, 4, 5, 14, 15, 16$, exactly two irreducible components for $\nu = 6, 7, 13$, exactly three irreducible components for $\nu = 8, 9$, exactly four irreducible components for $\nu = 11, 12$, exactly five irreducible components for $\nu = 10$. Hence its stratification is made of 34 strata, where each stratum is in the boundary of another if and only if the corresponding extended code admits the other as a shortening, as in the following Genealogy tree = Kummer nontrivial genealogy for nodal quartic surfaces (Fig. 1.1).

Remark 1.38 The full list of the connected components of $\mathcal{F}(4, \nu)$ is given in the Table of Sect. 1.8 (another description of the possible codes was given in [45]).

Since $\mathcal{F}(4, \nu)$ is smooth, the connected components are also irreducible components.

1.5 Potential Nodal K3 Lattices

Before we begin with the proof of Theorem 6, we establish some general results valid for all nodal K3 surfaces. For this purpose we introduce the following definition (condition ii) being dictated by Proposition 1.47).

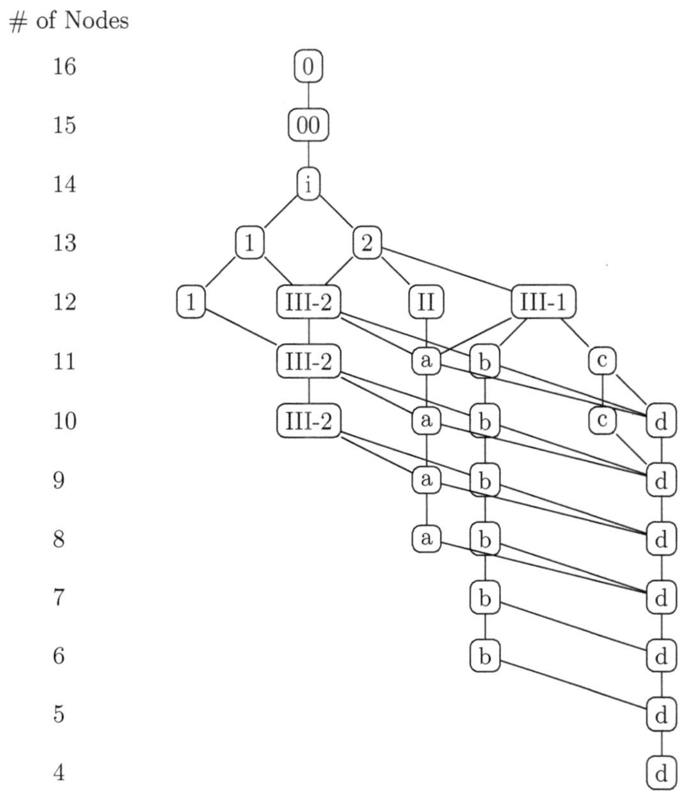

Fig. 1.1 Kummer nontrivial genealogy for nodal quartic surfaces

Definition 1.39 Consider a lattice

$$\mathcal{L} := \bigoplus_{1}^{\nu} \mathbb{Z}E_i \oplus \mathbb{Z}H =: \mathcal{L}^0 \oplus \mathbb{Z}H,$$

of rank $\nu + 1 \leq 17$, with integral symmetric bilinear form defined by:

$$E_i \cdot E_j = -2\delta_{ij},\ H \cdot E_i = 0,\ H^2 = d := 2d' > 0.$$

($\mathcal{L}$ generated by a basis which in the case of a nodal algebraic K3 surface corresponds to the (-2)-exceptional curves E_i and the hyperplane section H).

We define further a lattice $\mathcal{L}'$ which is candidate to be primitively embedded inside Λ, the 22-dimensional K3 lattice.

$\mathcal{L}'$ consists of the vectors

$$\mathcal{L}' := \{\frac{1}{2}(\sum_i a_i E_i + bH) | a_i, b \in \mathbb{Z},\ \sum_i (a_i E_i) + bH (mod\ 2) \in \mathcal{K}'\},$$

where $\mathcal{K}'$ is a bicoloured binary code

$$\mathcal{K}' \subset (\bigoplus_1^{\nu}(\mathbb{Z}/2)E_i \oplus (\mathbb{Z}/2)H) =: V^0 \oplus (\mathbb{Z}/2)H =: V.$$

We say that $\mathcal{L}'$ is a **potential nodal K3 lattice** if $\mathcal{K}'$ is a **potential nodal K3 code**, which by definition amounts to the requirements:

(i) letting $\mathcal{K} := \mathcal{K}' \cap (\bigoplus_1^{\nu}(\mathbb{Z}/2)E_i)$, then the weights $w(v)$ of the vectors in $\mathcal{K}$ are divisible by 8,

(ii) the weights $w(v)$ of the vectors in $\mathcal{K}' \setminus \mathcal{K}$ satisfy, once we set $w(v) =: t + 1$,

$$4|(t - d').$$

(iii) the B-inequality $dim(\mathcal{K}') \geq \nu - 10$ is satisfied,

(iv) the vector $H \notin \mathcal{K}'$ (hence, $\mathcal{K}'$ is isomorphic to its projection $\mathcal{K}''$ inside V^0).

Let us finally define $\mathcal{L}^*$ by the property that $\mathcal{L}^*/\mathcal{L} \cong \mathcal{K}$, so that $\mathcal{L}^* = \hat{\mathcal{L}} \oplus \mathbb{Z}H \supset \mathcal{L}^0 \oplus \mathbb{Z}H = \mathcal{L}$ and $\hat{\mathcal{L}}/\mathcal{L}^0 \cong \mathcal{K}$.

Remark 1.40 Recall the definition of discriminant group of $\mathcal{L}'$, defined (see Nikulin [74]) by

$$Disc(\mathcal{L}') := \mathcal{L}'^{\vee}/\mathcal{L}', \ \mathcal{L}'^{\vee} := Hom(\mathcal{L}', \mathbb{Z}).$$

Since $\mathcal{L} \subset \mathcal{L}'$, and $\mathcal{L}'/\mathcal{L} \cong \mathcal{K}'$, we have the series of (finite index) inclusions:

$$0 \to \mathcal{L} \to \mathcal{L}' \to \mathcal{L}'^{\vee} \to \mathcal{L}^{\vee},$$

where $Disc(\mathcal{L}) = \oplus_1^{\nu}(\mathbb{Z}/2)E_i \oplus (\mathbb{Z}/2d')H$, and the successive quotients for the series of inclusions are respectively isomorphic to $\mathcal{K}'$, $Disc(\mathcal{L}')$, $V/((\mathcal{K}')^{\perp})$.

The last assertion follows since $\mathcal{L}'^{\vee} \subset \mathcal{L}^{\vee} = \mathcal{L}^{0\vee} \oplus \mathbb{Z}(\frac{1}{2d'}H)$ is defined by the condition

$$\mathcal{L}'^{\vee} = \{u := \frac{1}{2}(\frac{a}{d'}H + \sum_i b_i E_i) \,|u \cdot \mathcal{L}' \in \mathbb{Z}\}.$$

$$\Leftrightarrow \mathcal{L}'^{\vee} = \{u | a\delta - \sum_{i \in \mathcal{A}} b_i \equiv 0 \ (mod\ 2), \forall\, v' := \frac{1}{2}(\delta H + \sum_{i \in \mathcal{A}} E_i) \in \mathcal{K}'\}.$$

Indeed, by the exact sequence, $0 \to \mathcal{L} \to \mathcal{L}' \to \mathcal{K}' \to 0$ (from which we deduce, by dualizing, the other exact sequence: $0 \to \mathcal{L}'^{\vee} \to \mathcal{L}^{\vee} \to Ext^1(\mathcal{K}', \mathbb{Z}) \cong \mathcal{K}'$) it suffices to observe that $u \in \mathcal{L}'^{\vee}$ if and only if

$$a\delta - \sum_{i \in \mathcal{A}} b_i \equiv 0 \ (mod\ 2), \forall\, v' := \frac{1}{2}(\delta H + \sum_{i \in \mathcal{A}} E_i) \in \mathcal{K}' \Leftrightarrow aH + \sum_i b_i E_i \in (\mathcal{K}')^{\perp}.$$

Proposition 1.41

(1) *The discriminant group of* $\mathcal{L}'$, $Disc(\mathcal{L}') = \mathcal{L}'^{\vee}/\mathcal{L}'$, *admits an orthogonal decomposition*

$$D(\mathcal{K}') \oplus (\mathbb{Z}/2)^{\nu-n},$$

where n is the effective length of the code $\mathcal{K}''$, *projection of* $\mathcal{K}'$ *in* $V^0 := (\mathbb{Z}/2)^{\nu}$.

(2) *In the case where* $\mathcal{K}' = \mathcal{K}$ *we have an orthogonal direct sum:*

$$Disc(\mathcal{L}') = \mathbb{Z}/d \oplus Disc(\hat{\mathcal{L}}).$$

(3) *Moreover if* $\mathcal{K}' = \mathcal{K}$ $Disc(\hat{\mathcal{L}}) \cong (\mathbb{Z}/2)^{\nu-2k'}$, $k' = k = dim\mathcal{K}$.

(4) *Also if* $\mathcal{K}' \neq \mathcal{K}$ *we have a group decomposition*

$$Disc(\mathcal{L}') = \mathbb{Z}/d \oplus (\mathbb{Z}/2)^{\nu-2k'},$$

unless possibly if d' *is odd, and* $\nu = 2k'$.

(5) *The binary part of the finite Abelian group* $Disc(\mathcal{L}')$ *is then of the form* $Disc(\mathcal{L}')_2 \cong \mathbb{Z}/2^r \oplus (\mathbb{Z}/2)^{\nu-2k'}$, *where* 2^r *is the maximal power of* 2 *which divides d.*

Proof First of all, we consider the support $\mathcal{S}'$ of $\mathcal{K}''$, which has n elements, and we view $\mathcal{L}'$ as the direct sum of the lattice $\mathcal{M}$ corresponding to the $\mathcal{S}' \cup \{H\}$-shortening of $\mathcal{K}'$, and of the lattice generated by $\oplus_{i \notin \mathcal{S}'} E_i$. Then our first assertion follows right away, if we set $D(\mathcal{K}') := Disc(\mathcal{M})$.

The second assertion is clear since by assumption

$$\mathcal{L}' = \mathcal{L}^* = \hat{\mathcal{L}} \oplus^{\perp} \mathbb{Z}[H].$$

The third assertion follows because of the series of (finite index) inclusions:

$$0 \rightarrow \mathcal{L}^0 \rightarrow \hat{\mathcal{L}} \rightarrow (\hat{\mathcal{L}})^{\vee} \rightarrow (\mathcal{L}^0)^{\vee},$$

where $Disc(\mathcal{L}^0) = \oplus_1^{\nu}(\mathbb{Z}/2)[E_i] =: V^0$, and the successive quotients are respectively isomorphic to $\mathcal{K}$, $Disc(\hat{\mathcal{L}})$, $V^0/(\mathcal{K})^{\perp} \cong Ext^1(\mathcal{K}, \mathbb{Z})$.

For the fourth assertion, we first claim that we can find a vector $v \in \mathcal{K}'^{\perp}$ such that the coefficient a of H is odd. (that is, $|\,\mathrm{supp}(v) \cap \mathcal{A}|$ is odd for all $v' \in \mathcal{K}'$ written in the above form). Otherwise, each vector in $\mathcal{K}'^{\perp}$ is contained in V^0, hence $\mathcal{K}'$ contains the vector H, contradicting assumption (iv).

Since $\frac{1}{d'}H \in \mathcal{L}'^{\vee}$, there is an element $v'' \in \mathcal{L}'^{\vee}$ with $a = 1$, and this element gives an element of the discriminant of order $d = 2d'$ if d' is even: because $d'v'' = \frac{1}{2}H \notin \mathcal{K}'$.

If instead d' is odd, and $d'v'' \in \mathcal{K}'$, then the vector $v \in \mathcal{K}'$, and we are done if we show that we may choose v not to be in $\mathcal{K}'$. Otherwise, $\mathcal{K}'^{\perp} \subset \mathcal{K}'$; since d' is odd, then t is odd, hence the vectors in $\mathcal{K}'$ have even weight, hence $\mathcal{K}' \subset \mathcal{K}'^{\perp}$, hence we have equality $\mathcal{K}' = \mathcal{K}'^{\perp}$, and $k' = \nu - k'$.

Since we have an element of order d of the discriminant group, and $Disc(\mathcal{L}) \cong \mathbb{Z}/d \oplus (\mathbb{Z}/2)^{nu}$, and as we saw $|Disc(\mathcal{L}')| = d \cdot 2^{\nu - 2k'}$, follows the desired assertion. □

Corollary 1.42 *The length $l(\mathcal{L}')$ (that is, the minimal number of generators of the discriminant group $Disc(\mathcal{L}')$) satisfies the inequality*

$$l(\mathcal{L}') \leq \text{rank}(\Lambda) - \text{rank}(\mathcal{L}') = 21 - \nu.$$

Moreover equality holds if and only if $\nu = 10 + k'$.

Proof If $\mathcal{K} = \mathcal{K}'$, then $Disc(\mathcal{L}') = \mathbb{Z}/d \oplus (\mathbb{Z}/2)^{\nu - 2k'}$, hence we are done by the B-inequality ($k' = dim(\mathcal{K}') \geq \nu - 10$), since the length

$$l(\mathcal{L}') = 1 + \nu - 2k' \leq 1 + 20 - \nu = 22 - (\nu + 1) = \text{rank}(\Lambda) - \text{rank}(\mathcal{L}').$$

If instead $\mathcal{K} \neq \mathcal{K}'$ and $Disc(\mathcal{L}^*) = \mathbb{Z}/d \oplus (\mathbb{Z}/2)^{\nu - 2k'}$, we get the same inequality.

Finally, if $\nu = 2k'$, then $l(\mathcal{L}') \leq 2 + \nu - 2k' = 2 < 4 \leq 21 - \nu$. □

We want now to show that if a potential nodal K3 lattice $\mathcal{L}'$ admits an isometric embedding into the K3 lattice Λ, then this embedding is unique (up to isometries of Λ).

To this purpose we shall apply Nikulin's theorem 1.14.4 of [74], asserting that under three (sufficient) conditions there exists a unique primitive embedding. This assertion must be taken with a grain of salt, since what really Nikulin means is just that there exists at most one such embedding (it is unique, if it exists).

The conditions are:

(1) Positivity of (Λ) > Positivity of $(\mathcal{L}')$, Negativity of (Λ) > Negativity of $(\mathcal{L}')$,
(2) $\text{rank}(\Lambda) \geq \text{rank}(\mathcal{L}') + 2 + l(Disc(\mathcal{L}')_p)$ for all primes $p \neq 2$,
(3) If $\text{rank}(\Lambda) = \text{rank}(\mathcal{L}') + l(Disc(\mathcal{L}')_2)$, then the quadratic form q of $\mathcal{L}'$ splits a summand either of the form $u_+^{(2)}(2)$ or of the form $v_+^{(2)}(2)$;
(4) Condition (3) is automatically verified if $Disc(\mathcal{L}') \cong A' \oplus (\mathbb{Z}/2)^3$ (see remark 1.14.5 of [74]).

To explain the above conditions, recall that, if M is an integral lattice, $A_M := Disc(M) = M^{\vee}/M$ inherits a bilinear form with values in $\mathbb{Q}/\mathbb{Z}$, and an associated *finite* quadratic form $q_M : A_M \rightarrow \mathbb{Q}/2\mathbb{Z}$.

Writing the discriminant group A_M as the direct sum of its p-primary components $(A_M)_p$, we can write $q_M = \oplus_p (q_M)_p$.

The p-adic part of the finite quadratic form can be computed by writing the normal form for the p-adic lattice $M_p := M \otimes_{\mathbb{Z}} \mathbb{Z}_p$, since $(q_M)_p = q_{M_p}$.

Now, see also [37], M_p splits as an orthogonal direct sum of lattices of rank ≤ 2, which in Nikulin's notation are called

(i) $K_\theta^{(p)}(p^h)$, corresponding to the 1×1-matrix (θp^h), where θ is a unit, uniquely defined modulo $(\mathbb{Z}_p^*)^2$ (hence we have four possibilities $\theta = 1, 3, 5, 7$); the associated discriminant group is $\cong \mathbb{Z}/p^h$, and the value of the finite quadratic form on the generator $e' := p^{-h}e$ is $q_\theta^{(p)}(p^h)(e') = \theta \cdot p^{-h}$.

(ii) $U^{(2)}(2^h)$, corresponding to the 2×2-matrix

$$\begin{pmatrix} 0 & 2^h \\ 2^h & 0 \end{pmatrix}, \tag{1.8}$$

its associated quadratic form is denoted $u_+^{(2)}(2^h)$; the associated discriminant group is $\cong (\mathbb{Z}/2^h)^2$ and the value of the finite quadratic form on the vectors e_1', e_2' is $2^{-(h-1)}$.

(iii) $V^{(2)}(2^h)$, corresponding to the 2×2-matrix

$$\begin{pmatrix} 2^{h+1} & 2^h \\ 2^h & 2^{h+1} \end{pmatrix}, \tag{1.9}$$

its associated quadratic form is denoted $v_+^{(2)}(2^h)$, the associated discriminant group is $\cong (\mathbb{Z}/2^h)^2$ and the value of the finite quadratic form on the vector $e_1' + e_2' = 2^{-h}(e_1 + e_2)$ is $2^{-(h-1)}$.

Observe that always $h \geq 1$, else one gets a trivial summand to the discriminant group.

Theorem 1.43 *Given a potential nodal K3 lattice, if it admits an isometric embedding into the K3 lattice Λ, then this embedding is unique.*

Proof Condition (1) is verified, since Λ has positivity index 3, negativity 19, these for $\mathcal{L}'$ are respectively 1, $\nu \leq 16$.

Condition (2) is verified since, for a prime $p \neq 3$, then $l(Disc(\mathcal{L}')_p) \leq 1$, since $Disc(\mathcal{L}')_p$ is cyclic (there is an element of order d'), hence we are done, since $22 > 20 \geq \nu + 4$.

Regarding the prime $p = 2$, equality holds if and only if $\nu = 10 + k'$ by Corollary 1.42.

Condition (3) is verified, indeed we shall now explain Nikulin's condition (4) concerning the prime $p = 2$.

First of all, recall that $Disc(\mathcal{L}')_2 \cong \mathbb{Z}/2^r \oplus (\mathbb{Z}/2)^{\nu - 2k'}$, hence we only have one summand of type $K_\theta^{(2)}(2^r)$, and all other summands have $h = 1$.

Hence the quadratic form q_2 of $\mathcal{L}'$ can only split summands of type $u_+^{(2)}(2)$, $v_+^{(2)}(2)$, $q_\theta^{(2)}(2)$, plus only one summand of the form $q_\theta^{(2)}(2^r)$.

Either condition (3) is verified, or we have at least $\nu - 2k'$ copies of summands of the type $q_\theta^{(2)}(2)$.

We use now Prop. 1.8.2 of [74], formula (d). The possible θ's are distinguished by their congruence class (modulo 8), namely, 1, 3, 5, 7. However, by formula i) ibidem, $q_\theta^{(2)}(2) \cong q_{5\theta}^{(2)}(2)$, hence for the forms $q_\theta^{(2)}(2)$ there are only 2 isomorphism classes.

If we have three summands, then two must give the same class, so we may suppose that they have the same θ. If the other is with $\theta' \equiv -\theta(mod 4)$, we split off $u_+^{(2)}(2)$, if instead $\theta' \equiv \theta(mod 4)$ we split off $v_+^{(2)}(2)$.

Condition (3) is therefore verified as soon as $\nu - 2k' \geq 3$. Condition (3) pertains just to the case where equality in claim of Corollary 1.42 is attained. Hence we should have $\nu = 10 + k'$, and we are done unless $\nu \leq 2k' + 2$. We only need to observe that $k' \leq 6$, hence $10 + k' \leq 2k' + 2$ is impossible, so we are finished. □

The conclusion of the above theorem is that $\mathcal{L}'$ and its embedding in Λ is uniquely defined by the code $\mathcal{K}'$ and by the integer d.

The next question is when does a potential nodal K3 lattice admit an isometric embedding into the K3 lattice Λ.

Theorem 1.12.2 of [74] answers this question giving necessary and sufficient conditions.

These conditions are:

(1') Positivity of $(\Lambda) \geq$ Positivity of $(\mathcal{L}')$, Negativity of $(\Lambda) \geq$ Negativity of $(\mathcal{L}')$,
(2') $\text{rank}(\Lambda) \geq \text{rank}(\mathcal{L}') + l(Disc(\mathcal{L}'))$,
(3') a certain congruence condition for the primes $p \geq 3$ such that $\text{rank}(\Lambda) = \text{rank}(\mathcal{L}') + l(Disc(\mathcal{L}')_p)$,
(4') a certain congruence condition between 2-adic units (here $K(q_2)$ is the 2-adic lattice having associated quadratic form q_2):

$$|Disc(\mathcal{L}')| \equiv \pm discr(K(q_2)) \; (mod \; (\mathbb{Z}_2^*)^2),$$

which must be verified in the case where

$$\text{rank}(\Lambda) = \text{rank}(\mathcal{L}') + l(Disc(\mathcal{L}')_2),$$

and where moreover q_2 does not split off a summand $q_\theta^{(2)}(2)$.

Remark 1.44

(i) As we shall see, the last condition (4') will be the only one which is not always verified, and which, fixed the code $\mathcal{K}'$, will lead to a distinction depending on the congruence class of d modulo (16). This distinction is important, since we shall classify the potential nodal K3 codes, and show that their classification depends on the congruence class of d modulo (8).
(ii) In many cases however we shall show the existence of these embeddings by purely geometric methods (smoothings or projections).

1.5.1 Proof of Theorem 6 = Theorem 1.37

Proof Consider a Kummer surface Y_0, which is unobstructed (cf. instance the Duke of Cefalú surface illustrated in Sect. 1.3): then if Y is a small deformation of Y, then Y^* is obtained from $\tilde{Y}_0$ by removing $16 - \nu$ (-2)-curves E_i.

We have seen that the fundamental group of Y_0^* is the affine group of transformations of $\mathbb{R}^4$

$$x \mapsto \pm x + \lambda, \lambda \in \mathbb{Z}^4,$$

and it is generated by simple small loops γ_i around the curves E_i, which correspond to the symmetry around a half period $\frac{1}{2}\lambda_i$ of $\mathbb{R}^4/\mathbb{Z}^4$, $\gamma_i(x) = -x + \lambda_i$.

Without loss of generality we may assume that the half period 0 corresponds to one of the $16 - \nu$ (-2)-curves E_i which we stick in again to get Y^* from Y_0^*. Hence the fundamental group of Y^* is a quotient of Γ by the normal subgroup generated by $x \mapsto -x$, and is therefore a quotient of $(\mathbb{Z}/2)^4$, hence it is Abelian, and equal to $H_1(Y^*, \mathbb{Z})$. The latter is in turn isomorphic to $\mathcal{K}$.

That $\mathcal{K}$ is determined by its dimension was shown in the course of the proof of Proposition 1.31.

Clearly $\nu, \mathcal{K} \subset \mathcal{K}'$, are deformation invariant for equisingular deformations, and we can take different shortenings of the Kummer code yielding the same number ν, but different codes: this shows that fixing only ν we get a disconnected set.

On the other hand, the code $\mathcal{K}'$, as in Corollary 1.13, is shown to be equal to the cokernel

$$H_1(Y^* \setminus H, \mathbb{Z}) \cong coker[H^2(\tilde{Y}, \mathbb{Z}) \to \oplus_1^{\nu} \mathbb{Z}[E_i] \oplus \mathbb{Z}H].$$

of the dual map of the embedding inside the 22-dimensional K3 lattice $\Lambda := H^2(\tilde{Y}, \mathbb{Z})$ of the lattice $\mathcal{L}$ generated by the curves E_i and by the hyperplane section H.

We have seen (ibidem) that the saturation $\mathcal{L}' := \mathcal{L}^{sat} := \Lambda \cap \mathbb{Q}\mathcal{L}$ consists of the vectors

$$\mathcal{L}' = \mathcal{L}^{sat} = \{\frac{1}{2}(\sum_i a_i E_i + bH) | \sum_i (a_i E_i + bH)(mod\ 2) \in \mathcal{K}'\}.$$

$\mathcal{L}'$ is primitively embedded inside Λ.

We have proven that for each nodal quartic surface $\mathcal{K}'$ is a shortening of the extended Kummer code $\mathcal{K}'_{Kum}$ (Proposition 1.35), indeed the proof shows that every potential nodal K3 code for $d = 4$ occurs as such a shortening.

Hence, for $d = 4$, all the potential nodal K3 lattices $\mathcal{L}'$ come from small deformations of a Kummer quartic surface, hence are all the lattices associated to a nodal quartic surface.

Theorem 1.43 asserts that any such lattice $\mathcal{L}'$ admits a unique primitive embedding in the K3 lattice Λ, and we observe that it possesses one such embedding, induced by the inclusion $\mathcal{L}' \subset \mathcal{L}'_{Kum}$.

Therefore, we can take a marking of the second cohomology (a lattice isomorphism of $H^2(\tilde{Y}, \mathbb{Z})$ with the K3 lattice Λ^0 of a Kummer surface), such that the subgroup $\mathcal{L}' \subset \mathcal{L}'_{Kum}$ maps onto a fixed subgroup $\mathcal{L}'^0 \subset \mathcal{L}'_{Kum}$. Denote by e_i the image of the class of E_i, for $i = 1, \ldots, 16$ and by e_{20} the image of the class of H.

We can complete to an orthogonal basis, $e_1, \ldots, e_{22}$ of $\Lambda^0 \otimes_{\mathbb{Z}} \mathbb{R}$, such that the intersection form is negative definite on the span of the first 19 vectors, and positive definite on the span of the last 3 vectors.

The period domain is

$$\mathcal{D} := \{[\omega] \in \mathbb{P}(\Lambda^0 \otimes_{\mathbb{Z}} \mathbb{C}) | \omega \wedge \omega = 0, \ \omega \wedge \overline{\omega} > 0\},$$

and if $\omega = (z_1, \ldots, z_{19}, w_{20}, w_{21}, w_{22},)$ the above equations boil down to:

$$\sum_1^{19} z_i^2 = \sum_1^3 w_j^2, \ |w| > |z|.$$

$\mathcal{D}$ contains the submanifold

$$\mathcal{M} = \{[\omega] | \omega \wedge e_i = 0, i = 1, \ldots, \nu, i = 20\} = \mathcal{D} \cap \{z_1 = \ldots z_\nu = w_{20} = 0\},$$

which we shall show to have two connected components.

Indeed, we write $Z := (z_{\nu+1}, \ldots, z_{19}), \ u := (y, v) := (w_{21} + i w_{22}, w_{21} - i w_{22})$.

Then we have the subset $\mathcal{M}$ of projective space consisting of the vectors $(Z, u) = (Z, (y, v))$ such that

$$(**) \ \sum_j Z_j^2 = yv, \ 2|Z|^2 < |y|^2 + |v|^2.$$

Hence $u = (y, v) \neq (0, 0)$ and by the second inequality either $y \neq 0$, or $v \neq 0$. Therefore $\mathcal{M}$ has a holomorphic map to the projective line $\mathbb{P}^1$ via (y, v).

By the Cauchy inequality we have $|\sum_j Z_j^2| \leq |Z|^2$, hence if the inequality (**) holds we must have, setting $r := |Z|^2$:

$$2|yv| \leq 2r < |y|^2 + |v|^2 \Rightarrow |y|^2 - |v|^2 \neq 0 \Leftrightarrow |y| \neq |v|.$$

Hence we have two disjoint open sets, defined by $|y| < |v|$, respectively by $|v| < |y|$, exchanged by complex conjugation.

Let us show that these open sets are connected; without loss of generality let us stick to the case $|v| < |y|$: then $y \neq 0$ and, since we are in projective space, we set $y = 1$, and we get $y = 1, |v| < 1$.

Since we have $v = \sum_j Z_j^2$, we get the domain

$$\mathcal{D} := \{Z := (z_{v+1}, \dots, z_{19}) || \sum_j Z_j^2| < 1, 2|Z|^2 < 1 + |\sum_j Z_j^2|^2\}.$$

This domain $\mathcal{D}$ is connected (indeed, it is known to be a symmetric bounded domain of type (IV), as we now shortly prove.

In fact, the inequalitites imply $|Z| < 1$, $\mathcal{D}$ contains the origin in its interior and for every complex line L through the origin (we assume now $|Z| = 1$ for the direction of the line $\{\lambda Z | \lambda \in \mathbb{C}\}$ and set $b := |\sum_j Z_j^2|$) the intersection $\mathcal{D} \cap L$ is the following open set in $\mathbb{C}$:

$$\{\lambda \in \mathbb{C} | \; -2|\lambda|^2 + |\lambda|^4 b^2 + 1 > 0 \, , \; |\lambda|^2 b < 1\},$$

where $0 \le b \le 1$. This last open set is connected since for $b = 0$ we get, setting $r := |\lambda|^2$, $r < \frac{1}{2}$; whereas, for $0 < b \le 1$ we get

$$rb < 1, \; b^2(b^2 r^2 - 2r + 1) > 0 \Leftrightarrow rb < 1,$$

$$1 - rb^2 > \sqrt{1 - b^2} \Leftrightarrow rb < \frac{1}{b}(1 - \sqrt{1 - b^2}) \le 1.$$

We end the proof by invoking firstly the following tools:

(1) the surjectivity of the period map for K3 surfaces and
(2) the Torelli theorem for K3 surfaces (see for instance [76] and [10]) implying that the given family of nodal quartic surfaces with v nodes, and with fixed extended code $\mathcal{K}'$ is a holomorphic image of $\mathcal{M}$, and has therefore at most two connected components.
(3) Moreover, since the deformations of a Kummer surface are unobstructed (it suffices that this holds for just one of them, even if this is true in general) each such shortening $\mathcal{K}'$ is realized, see Sect. 1.8, especially Corollary 1.77 and the list preceding it; and
(4) each family of such nodal quartic surfaces contains small deformations of a Kummer surface, since these correspond in the period domain to the subspace

$$\mathcal{M}_{Kum} = \{[\omega] | \omega \wedge e_i = 0, i = 1, \dots, 16, i = 20\} = \mathcal{D} \cap \{z_1 = \dots z_{16} = w_{20} = 0\}.$$

(5) we need an argument to see that there is only one connected component: for this, we observe that if we take the complex conjugate surface $\overline{Y}$ of Y, then we have an orientation preserving diffeomorphism among them, which however sends the holomorphic 2-form η to $\overline{\eta}$. Hence the period point of $\overline{Y}$ is obtained by exchanging the two coordinates y, v, since $\eta, \overline{\eta}$ span the positivite definite subspace $H^{2,0} \oplus H^{0,2}$, which is orthogonal to H.

Hence Y and $\overline{Y}$ have period points which, with the given marking, belong to the two connected components of $\mathcal{D}$. The proof is then completed if we show that there is, for each choice of $\mathcal{K}'$, a real surface Y ($Y = \overline{Y}$).

This can be proved via the following observation: take the Duke of Cefalu' Kummer quartic surface Y_0 (see Theorem 1.25, (ii)). Y_0 has a polynomial equation with coefficients in $\mathbb{Z}$, and also its singular points have coordinates $\in \{0, 1\} \subset \mathbb{Z}$; hence all the local components of the smoothing of Y_0 are defined over $\mathbb{R}$ and contain real surfaces. We are done in view of point (4) above. □

1.6 Cubic 4-Folds and Nodal K3 Surfaces of Degree 6

The present short section gives some motivation for the study of nodal K3 surfaces of degree 6, which shall be useful also for the study of nodal quintic surfaces.

We begin by recalling a well known elementary result, which is however crucial for our description

Lemma 1.45 *Let P be a point of multiplicity 2 of a cubic hypersurface $X \subset \mathbb{P}^N$. Take coordinates $(x, z) := (x_0, \ldots, x_n, z)$ where $N = n + 1$ and $P = (0, 1)$; consider the Taylor development at P of the equation F of X, $F = Q(x)z + G(x)$.*

Define $Y \subset \mathbb{P}^n$ as the complete intersection $Y = \{x | Q(x) = G(x) = 0\} \subset \mathbb{P}^n$. Then

(i) $\mathrm{Sing}(X) \subset \{(x, z) | Q(x) = G(x) = 0\} = (P * Y)$: *the singular set of X is contained in the cone over Y with vertex P.*
(ii) If $x \in Y$, $x \notin \mathrm{Sing}(Q)$, then the line $L = P * x$ (contained in X), contains at most another singular point (x, z) of X ($x \neq 0$), and this happens if and only if $x \in \mathrm{Sing}(Y)$.
(iii) In the previous situation, x is a node of Y if and only if (x, z) is a node of X.
(iv) If $x \in \mathrm{Sing}(Q) \cap Y$ ($\subset \mathrm{Sing}(Y)$) , then the line $L = P * x$ contains another singular point of X (different from P) if and only if $L \subset \mathrm{Sing}(X)$; this happens if and only if $x \in \mathrm{Sing}(G)$.

Proof

(i)

$$\{(x, z) \in \mathrm{Sing}(X) | x \neq 0\} =$$

$$= \{(x, z) | x \neq 0,\ Q(x) = 0, G(x) = 0, z(\partial_i Q) + \partial_i G = 0,\ \forall i = 0, \ldots, n\}.$$

(ii) if there is an i such that $\partial_i Q(x) \neq 0$, then z is uniquely determined, and it exists if and only if the gradient of G is proportional to the one of Q.
(iii) if the gradient of Q vanishes at x, there is another singular point if and only if also the gradient of G vanishes at x.

(iv) observe that we have a node of X if and only if the ideal $(Q, z(\partial_i Q) + \partial_i G)$ is locally the maximal ideal of the point; but, if $\partial_0 Q \neq 0$, this ideal is the ideal

$$(Q, -(\partial_0 G)(\partial_i Q) + (\partial_i G)(\partial_0 Q)).$$

We may take Q as a local coordinate y, and then in local coordinates (u_j, y) we get the ideal $(y, \partial_j G)$, which is the Jacobian ideal of Y. □

Corollary 1.46 *X is nodal if and only if*

(I) Q is smooth
(II) Y is nodal.

Moreover, in this case the number γ of nodes of X equals the number of nodes ν of Y plus 1.

Conversely, if Y is nodal, then the corank of Q is at most 2, and if the corank equals 2, then X has $\nu - 3$ nodes plus the other singular point P, which is not a node.

Proof For the first assertion, (I) is a necessary condition, since P is a node and Q is the quadratic part of F at P. If Q is smooth, then the equivalence of X and Y being nodal follows from (iii) of Lemma 1.45.

For the second assertion, since Y has isolated singularities, $\mathrm{Sing}(Q)$ has projective dimension at most 1, and if equality holds, then G intersects this singular line in exactly 3 distinct points, which are nodes for Y. □

1.7 Nodal K3 Surfaces of Arbitrary Degree

The following proposition motivates the definition, given earlier, of potential nodal K3 codes $\mathcal{K}'$.

Proposition 1.47 *Let Y be*

- *a nodal K3 surface in $\mathbb{P}^4$ (a complete intersection of type $(2, 3)$), respectively*
- *a nodal K3 surface of degree $d = 4m + 2$ (where the ample divisor H is indivisible).*

Then the number ν of nodes of Y is at most 15, the weights of the code $\mathcal{K}$ are equal to 8, while

- *the weights of $\mathcal{K}'$ are divisible by 4 in the first case, respectively*
- *the weights of $\mathcal{K}'$ are divisible by 4 for m odd, and congruent to 2 modulo 4 if m is even.*

Moreover, $dim(\mathcal{K}) \geq dim(\mathcal{K}') - 1$, $dim(\mathcal{K}') \geq \nu - 10$ (hence $dim(\mathcal{K}) \geq \nu - 11$).

Finally, if Y is a nodal K3 surface of degree $d = 4m$, the weights of the code $\mathcal{K}$ are either 8 or 16, while the weights of vectors in $\mathcal{K}' \setminus \mathcal{K}$, when written as $w = t+1$, satisfy $t \equiv 2m \, mod(4)$ (that is, for m odd $t \equiv 2 \, mod(4)$, for m even $t \equiv 0 \, mod(4)$).

Proof The determination of the weights follows from the theorem of Riemann Roch: if $2L = \sum_1^t E_i + \delta H$ on the resolution of singularities of Y, with $\delta \in \{0, 1\}$, then $\chi(-L) = \chi(\mathcal{O}) + \frac{1}{2}L^2$, hence L^2 is an even number.

But $4L^2 = -2t + d\delta$ hence 4 divides $\frac{d}{2}\delta - t$. For $\delta = 1$ we get that $t \equiv \frac{d}{2} \, mod(4)$; for $\delta = 0$ we use the usual fact that the first cohomology group carries a non-degenerate alternating form, hence L^2 is divisible by 4 and t is divisible by 8.

For the other assertion, we know in any case that $\nu \leq 16$: if equality holds, then by the B-inequalities the dimension of $\mathcal{K}$ is equal to $16 - 11 = 5$, hence $\mathcal{K}$ is the strict Kummer code and $\mathcal{K}' \neq \mathcal{K}$. Therefore, for m odd, there is a vector in $\mathcal{K}'$ with $t \equiv 3$ (*mod* 4), and adding the vector in $\mathcal{K}$ with all coordinates equal to 1, we get a vector in $\mathcal{K}$ with $t' = 16 - t \equiv 1$ (*mod* 4), a contradiction; similarly for the case m even. □

Remark 1.48 If the degree d is 2, then Y is a finite double cover of the plane branched on a plane curve B with only nodes as singularities, which correspond to the nodes of Y.

The codewords in $\mathcal{K}' \setminus \mathcal{K}$ correspond to components Δ of B of odd degree, with t equal to the number of nodes where Δ has odd multiplicity ($t = 5$ if Δ is a line, or has degree 5, $t = 9$ if Δ is a cubic).

The codewords in $\mathcal{K}$ correspond to divisors $\Delta \leq B$ of even degree, with t equal to the number of nodes where Δ has odd multiplicity.

Then we have an easy picture: the maximum number $\nu = 15$ is obtained when B consists of 6 lines intersecting in 15 points (and we get the projection of a Kummer surface in $\mathbb{P}^3$ if there is a smooth conic tangent to all the six lines in points different from the 15 nodes).

In this case each line in the branch curve B produces a codeword with weight 6 in $\mathcal{K}'$, and ($t = 5$) a codeword of weight 5 in the projected code $\mathcal{K}''$, namely

$$v_i := \sum_{i \neq j} E_{i,j}.$$

The sum of these 6 vectors v_i is the zero vector, and these vectors span $\mathcal{K}'$, which has dimension 5.

By our above remarks on the weights of vectors in $\mathcal{K}'$, $\mathcal{K}'$ is contained in the projection of the extended Kummer code (contained in the second Reed-Muller code for the affine space $\mathbb{F}_2^4$). If we take the hyperplane point H as corresponding to the origin of the affine space, we get the following description (see Sect. 1.8).

(d=2) $\mathcal{K}'$ consists of the subspace generated by the linear forms and the quadratic function $q'(x, y) := x_1 y_1 + x_2 y_2 + 1$.

All the other configurations are obtained by smoothings of the curve which is the union of six lines, and the corresponding codes are shortenings of the above code.

In view of Proposition 1.35 we record the conclusion of the above discussion.

Proposition 1.49 *Let $\mathcal{K}' \subset \mathbb{F}_2^\nu \oplus \mathbb{F}_2 H$ be a bicoloured code such that $\nu \leq 15$, such that the weight of vectors of $\mathcal{K}$ is 8, and the other weights of $\mathcal{K}'$ are 6 or 10.*

Then $\mathcal{K}'$ is a shortening of the subspace of the second order Reed-Muller code which is generated by the linear forms and the quadratic function $q'(x, y) := x_1 y_1 + x_2 y_2 + 1$.

It is a fun exercise to classify geometrically all the possible cases (we proceed by smoothing one node at each step). Recall that $k' := dim(\mathcal{K}') \geq k := dim(\mathcal{K})$, and observe that in the following list $\mathcal{K}' = 0$ if B is irreducible.

(i) $\nu = 14$: conic plus 4 lines ($k = 3, k' = 4$)
(ii) $\nu = 13$: two conics plus 2 lines ($k = 2, k' = 3$) or nodal cubic plus 3 lines ($k = 2, k' = 3$);
(iii) $\nu = 12$: three conics ($k = k' = 2$) or smooth cubic plus 3 lines ($k = 2, k' = 3$) or 3-nodal quartic plus two lines ($k = 1, k' = 2$) or nodal cubic plus conic plus line ($k = 1, k' = 2$)
(iv) $\nu = 11$: two nodal cubics ($k = 0, k' = 1$) or smooth cubic plus conic plus line ($k = 1, k' = 2$) or 2-nodal quartic plus two lines ($k = 1, k' = 2$) or 6-nodal quintic plus line ($k = 0, k' = 1$) or conic plus three-nodal quartic ($k = k' = 1$)
(v) $\nu = 10$: nodal cubic plus smooth cubic ($k = 0, k' = 1$) or conic plus two-nodal quartic ($k = k' = 1$) or 1-nodal quartic plus two lines ($k = 1, k' = 2$) or irreducible 5-nodal quintic plus line ($k = 0, k' = 1$) or irreducible 10-nodal sextic
(vi) $\nu = 9$: two smooth cubics ($k = 0, k' = 1$) or conic plus one-nodal quartic ($k = k' = 1$) or smooth quartic plus two lines ($k = 1, k' = 2$) or irreducible 4-nodal quintic plus line ($k = 0, k' = 1$) or irreducible 9-nodal sextic
(vii) $\nu = 8$: conic plus smooth quartic ($k = k' = 1$) or irreducible 3-nodal quintic plus line ($k = 0, k' = 1$) or irreducible 8-nodal sextic
(viii) $5 \leq \nu \leq 7$: irreducible $(\nu - 5)$-nodal quintic plus line ($k = 0, k' = 1$);
(ix) $0 \leq \nu \leq 7$: irreducible ν-nodal sextic.

All these configurations can be realized over the reals, and we get in this way, as we shall further see, the classification of the nodal Severi varieties of K3 surfaces of degree 2.

We mention here a result of [59], theorem 5.6, where, following ideas of Togliatti, he constructed degree six K3 surfaces with 15 nodes (we do not reproduce the full proof since we reobtain this result as a consequence of Nikulin's theorems on quadratic forms).

Proposition 1.50 *There exists a configuration in the plane, of a nodal curve B of degree six consisting of two smooth conics $\{Q_1 = 0\}, \{Q_2 = 0\}$, plus 2 lines $\{Q_3 = 0\}$, and of two smooth conics $C_1 = \{q_1 = 0\}$, $C_2 = \{q_1 = 0\}$, both not passing through the nodes of B, and everywhere tangent to B in distinct points, indeed such that*

$$Q_1 Q_2 Q_3 = r^2 + 4 q_0 q_1 q_2,$$

where r is a cubic form, and q_0 is the square x_0^2 of a linear form.

Let $p : Y \to \mathbb{P}^2$ be the double cover branched over B, yielding a nodal degree two K3 surface Y, which is birational to the degree six K3 surface $Y' \subset \mathbb{P}^4$,

$$y_0 y_1 + q_0(x_0, x_1, x_2) = 0, \quad y_0 q_1(x_0, x_1, x_2) + y_1 q_2(x_0, x_1, x_2) + r(x_0, x_1, x_2) = 0.$$

Y' has 15 nodes.

Proof *(Idea)* The points e_0, e_1 are nodes for Y', and projection from the line $\{x_0 = x_1 = x_2 = 0\}$ joining them yields a double cover of the plane branched on B: actually projection from e_1 yields the nodal quartic surface of equation

$$y_0^2 q_1(x) - q_0(x) q_2(x) + r(x) y_0 = 0,$$

Since B has 13 nodes, we see that Y' has $13 + 2 = 15$ nodes. □

For our application, we need however a stronger result, namely a K3 surface which is defined over $\mathbb{R}$, and with all the nodes being real points.

This was achieved in [27], where the following theorem was proven:

Theorem 1.51 *For each degree $d = 4m$ there is a nodal K3 surface X' of degree d with 16 nodes which is defined over $\mathbb{R}$, and such that all its singular points are defined over $\mathbb{R}$.*

Similarly, for each degree $d = 4m - 2$ there is a nodal K3 surface X'' of degree d with 15 nodes which is defined over $\mathbb{R}$, and such that all its singular points are defined over $\mathbb{R}$.

1.7.1 Codes for K3 Surfaces of Degree $d \equiv 6 \pmod 8$

We shall now classify all the pairs of codes as in Proposition 1.47 with m odd; later we shall state analogues of the theorem concerning nodal quartic surfaces in $\mathbb{P}^3$.

The following theorem is due to collaboration with Michael Kiermaier.

Theorem 1.52 *Let $\mathcal{S}$ be a set of cardinality $\nu \le 15$, and consider a non-zero bicoloured code $\mathcal{K}' \subset \mathbb{F}_2^{\mathcal{S}} \oplus \mathbb{F}_2 H$, such that*

(i) the code $\mathcal{K} := \mathcal{K}' \cap \mathbb{F}_2^{\mathcal{S}}$ admits only the weight 8,
(ii) the weights of vectors in $\mathcal{K}'$ are divisible by 4,
(iii) $dim(\mathcal{K}) \ge \nu - 11$, $dim(\mathcal{K}') \ge \nu - 10$, $dim(\mathcal{K}') \le dim(\mathcal{K}) + 1$.

Then $\mathcal{K}$ is a shortening of the simplex code, in particular its dimension $k \le 4$, and we can view $\mathcal{S} \subset \mathbb{P} := \mathbb{P}(\mathbb{F}_2^4)$, and the non-zero elements of $\mathcal{K}$ as the characteristic *functions of complements of hyperplanes* $H \subset \mathbb{P}$ *with the property that* $\mathcal{T} := (\mathbb{P} \setminus \mathcal{S}) \subset H$.

We have several cases:

(1) $\mathcal{K} = \mathcal{K}'$: $k = k' = 1, 2$, *accordingly* $\nu \in \{8, 9, 10, 11\}$, *resp.* $\nu = 12$;
(2) $\mathcal{K} \neq \mathcal{K}'$, *the weight* 16 *occurs in* $\mathcal{K}'$: *then* $\nu = 15$ *and* $\mathcal{K}'$ *is the strict Kummer code;*
(3) $\mathcal{K} \neq \mathcal{K}'$, *all the weights are* 8: *then* $k' = 1, 2, 3, 4$, $\nu \in \{7, 8, 9, 10, 11\}$, *resp.*$\nu = 11, 12$, *resp.* $\nu = 13$, *resp.* $\nu = 14$.
(4) $\mathcal{K} \neq \mathcal{K}'$, *there exists a weight* 4 *vector (if* $k > 0$ *this condition is implied by the condition that there exists a weight* 12 *vector): then* $k = 0, 1, 2, 3, 4$, *and respectively* $\nu \in \{4, 5, 6, 7, 8, 9, 10, 11\}$, *resp.* $\nu \in \{9, 10, 11, 12\}$, *resp.* $\nu \in \{12, 13\}$, *resp.* $\nu = 14$, *resp.* $\nu = 15$. *The bicoloured code* $\mathcal{K}'$ *in this case (4) is unique for all pairs* k, ν *except for* $k = 1, \nu \geq 11$, *when we have two cases, and also for* $k = 2$, $\nu = 13$, *where we have two cases.*
(5) $\mathcal{K} \neq \mathcal{K}'$, $k' = 1$, $\nu = 11$, *there is only the weight* 12 *in* $\mathcal{K}'$.

Proof In the first case, as we already saw, $\mathcal{K} = \mathcal{K}'$ is determined by its dimension $k = k' \in \{1, 2, 3, 4\}$, and then its length n equals, respectively, 8, 12, 14, 15. Since we have $n \leq \nu \leq 10 + k$, only the cases $k = 1, 2$ are possible.

In the second case obviously $\nu = 15$, hence $k' \geq 5$, so we have the strict Kummer code.

In the third case again $\mathcal{K} \subset \mathcal{K}'$ are determined by their dimensions $k = k' - 1$, $k' \in \{1, 2, 3, 4\}$ and again we have that $n \leq \nu + 1 \leq 11 + k'$, hence all the cases $k' = 1, 2, 3, 4$ are possible.

The remaining cases are more interesting, with exception of the subcases $k = 0$, which are straightforward to discuss: there is only a non zero weight, and this can be either 4 or 12, but in the latter case we have $\nu \leq 11, \nu \geq 11 \Rightarrow \nu = 11$, and we are in case (5).

Assume that $k > 0$: then let u be a vector with $w(u) = 4$. We denote as usual by v_1 the vector corresponding to the first two columns (here H corresponds to the entry in the last row and last column).

Let a be the number of non-zero coordinates of u in the first two columns, and b the number of non-zero coordinates in the last two columns, and not the coordinate of H (that is, we are looking at the code $\mathcal{K}''$, projection from H of the code $\mathcal{K}'$); then $a + b = 3$, and adding v_1 we must have $(8 - a) + b = 5 + 2b \equiv 3 \pmod 4$, which means that $b = 1, 3$. We have two cases: $(a, b) = (2, 1)$ or $(a, b) = (0, 3)$, distinguished by the non occurrence - occurrence of the weight 12 in $\mathcal{K}'$.

Conversely, if the weight 12 occurs, and $k = 1$, then $a+b = 11$ and $19-2a \equiv 3 \pmod 4$, hence a is even. However, by the B-inequality, $\nu \leq 12$, hence $b \leq 4$, hence $b = 3$ (and we have the previous case).

For the respective dimensions $k = 4, 3, 2$ we must have $\nu = 15$, resp. $\nu = 14$, resp. $\nu \in \{12, 13\}$.

These are the respective cases where we remove from $\mathbb{P}$ a set $\mathcal{T}$ of cardinality 0, 1, or just two or three points contained in two hyperplanes, hence two or three collinear points.

$\mathcal{K}''$ is spanned by the characteristic function of a subset $\mathcal{A} \subset \mathcal{S} = \mathbb{P} \setminus \mathcal{T}$ which has 3 or 11 elements. Moreover, for each hyperplane H containing $\mathcal{T}$, if $\mathcal{A}$ contains b points in H, and a points not in H, then we must have that $b+8-a \equiv 3\ (mod\ 4)$. For $a+b=3$ this means as before that $b=1,3$. For $a+b=11$ this means that b is odd, and since $a \leq (\nu-8) \leq 7$, $b=5,7$ (then respectively $a=6,4$).

Assume that there is a vector of $\mathcal{K}'$ corresponding to such a set $\mathcal{A}$ of cardinality 3. Each plane H must contain either all the 3 elements of $\mathcal{A}$, or just one. Said in other words, every plane containing $\mathcal{T}$ intersects $\mathcal{A}$ in only one point, or contains $\mathcal{A}$.

If $\mathcal{T}$ has 0 elements, this means that $\mathcal{A}$ consists of three collinear points.

If $\mathcal{T}$ has 1 element T, let $\{A_1, A_2, A_3\} = \mathcal{A}$: then two points in $\mathcal{A}$ cannot be collinear with T, else there are planes H containing two point of $\mathcal{A}$. There is a plane H' containing $\mathcal{A}$ and $\mathcal{T}$, and this case is always combinatorially equivalent to the one where the three points in $\mathcal{A}$ are collinear, and the line Λ they span does not contain T.

If $\mathcal{T}$ has 3 elements, which form a line L, there are a priori two possibilities: projection from L to a skew line Λ maps the three points A_i to three different points (this case is combinatorially equivalent to the one where the three points in $\mathcal{A}$ are collinear, and the line Λ they span is skew to L), or there exists a plane H' containing L and $\mathcal{A}$ (combinatorially, there is only one case).

However, the second case is not possible: since adding to the characteristic function of $\mathcal{A}$ the characteristic function of the complement of a plane H containing L and different from H' we would get the characteristic function of a set $\mathcal{B}$ of cardinality $8-3=5$, which is a contradiction.

Let us look at the case where $\mathcal{T}$ consists of two points $T_1, T_2 \in L$, and let us denote by T_3 the third point of L.

If $T_3 \in \mathcal{A}$, then each plane H containing L contains T_3, and if $T_3 = A_3$, there is a plane containing A_1 and A_2, which is combinatorially equivalent to the case where the points of $\mathcal{A}$ are collinear, but L and $\mathcal{A}$ are lines incident in $T_3 = A_3$.

If $\mathcal{A} \subset \mathbb{P} \setminus L$, again we have either three collinear points, spanning a line $\mathcal{A}$ skew to L, or there is a plane H' containing L and $\mathcal{A}$, but this gives a contradiction as before.

These are exactly three different cases.

Finally, if there is no such set $\mathcal{A}$ of cardinality 3, we have that elements of $\mathcal{K}'' \setminus \mathcal{K}$ correspond to subsets $\mathcal{B}$ of cardinality 11. Then each plane H (containing $\mathcal{T}$) must contain exactly 5 or 7 points of $\mathcal{B}$.

The second alternative is not possible if $\mathcal{T}$ is non-empty, since $\mathcal{T} \subset H, \mathcal{B} \cap \mathcal{T} = \emptyset$.

If $\mathcal{T}$ is empty, then $\mathcal{B}$ consists of a plane H_1 and 4 other points. Hence there are two planes H_1, H_2 such that either $\mathcal{B} = H_1 \cup H_2$, or there are points P, P' such that $\mathcal{B} = ((H_1 \cup H_2) \setminus \{P\}) \cup \{P'\}$.

Take a plane H containing the line $H_1 \cap H_2$ and different from these two planes: then adding the characteristic function of the complement of H we get the characteristic function of a set $\mathcal{A}$ with either 3 or 4 points: in both cases we reach a contradiction.

We can therefore assume that $\mathcal{B}$ is such that every hyperplane H (containing $\mathcal{T}$) must contain exactly 5 points of $\mathcal{B}$. Then $\mathcal{T}$ has at most two elements, and if it has two elements, for each plane H containing $\mathcal{T}$ we have

$$(H \setminus \mathcal{T}) \subset \mathcal{B} \Rightarrow \mathcal{B} = \mathbb{P} \setminus \mathcal{T},$$

a contradiction.

If $\mathcal{T}$ is one point, take two points P, P' not in $\mathcal{B}$, and a plane H containing $\mathcal{T}$, P, P': then H contains less than 4 points of $\mathcal{B}$, a contradiction. Similarly if $\mathcal{T}$ is empty, take a plane H containing three points in the complement of $\mathcal{B}$. □

Corollary 1.53

(I) We have two codes as above for $\nu = 15$, two also for $\nu = 14$, three for $\nu = 13$, five for $\nu = 12$, seven for $\nu = 11$, four for $\nu = 9, 10$, three for $\nu = 8$, two for $\nu = 7$, and just one for $\nu \in \{4, 5, 6\}$. For a total of 35 nontrivial bicoloured codes $\mathcal{K}'$.
(II) All the above bicoloured codes are shortenings of those two with $\nu = 15$.
(III) Case (3) with $\nu = 14$ is a shortening of case (4) with $\nu = 15$.

Proof Assertion (I) is a direct consequence of the previous theorem, assertion (II) is a consequence of its proof.

Indeed, cases (1) and (3) are clearly shortenings of case (2), while cases (4) contemplate always the case where $\mathcal{A}$ is a line, hence they shortenings of the case (4) with $\nu = 15$. If we take the shortening of case (4) with $\nu = 15$ corresponding to a projective basis of $\mathbb{P}^3$, then we get case (5).

Concerning assertion (III), recall that $\mathcal{S}$ is here the projective space $\mathbb{P}^3_{\mathbb{F}_2}$, the code $\mathcal{K}'$ is generated by the characteristic functions of complements of planes π, and by the characteristic function $\chi_{\mathcal{A}}$, where $\mathcal{A}$ is a line.

The vectors in $\mathcal{K}'' \setminus \mathcal{K}$ correspond to $\chi_{\mathcal{A}}$ and $\chi_{\mathcal{A}} + 1 + \chi_{\pi}$, and if $\pi \supset \mathcal{A}$ we get the characteristic function of a set of cardinality 11 containing $\mathcal{A}$. If $\mathcal{A} \cap \pi = \{P\}$, then we get the characteristic function of a set of cardinality 7 containing exactly the point P of $\mathcal{A}$.

Hence, if we take the shortening corresponding to the complement of a point of $\mathcal{A}$, we get a code whose weights are just 7, 8. □

We borrow the two following results from [59], and we provide proofs for them since the Thesis [59] is not so easily accessible (even if for the first result we shall give another proof in Theorem 1.62).

Proposition 1.54 *Let Y be a nodal complete intersection of type $(2, 3)$ in $\mathbb{P}^4$ and let $\mathcal{N}$ be an half-even set of nodes on S. Then we have*

(a) its cardinality $t := |\mathcal{N}|$ satisfies $t \leq 11$,
(b) its associated sheaf is Cohen-Macaulay,
(c) the associated double cover is a regular surface.

In particular the extended code $\mathcal{K}'$ of Y cannot be the strict Kummer code (case (2) of the previous classification in Theorem 1.52).

Proof We use a similar notation to the one used in the section on half-even sets of nodes on sextics: Y is a nodal complete intersection of type $(2, 3)$ in $\mathbb{P}^4$, $S := \tilde{Y}$ is its minimal resolution, $\mathcal{N}$ is an half-even set of nodes on S, and $t := |\mathcal{N}|$.

We denote by $E_1, \ldots, E_\nu$ the exceptional (-2)-curves, therefore there exists a divisor L on S such that

$$(*) \sum_{i \in \mathcal{N}} E_i + H \equiv 2L.$$

We denote by $\Delta := \sum_{i \in \mathcal{N}} E_i = \sum_1^t E_i$.

In this situation we attach to $(*)$ a double covering $f : Z \to S$ ramified on $\Delta + H$, where H is here any smooth hyperplane section, and such that

$$f_*(\mathcal{O}_Z) = \mathcal{O}_S \oplus \mathcal{O}_S(-L).$$

The inverse images of the (-2)-curves $E_1, \ldots, E_t$ are (-1) curves, and can be contracted to smooth points, yielding $p : Z \to Z'$ and a finite double covering $f' : Z' \to Y$.

We have

$$K_Z = f^*(K_{\tilde{Y}} + L) = f^*(L)$$

and, since by Kodaira-Ramanujam vanishing $H^1(Z, \mathcal{O}_Z(K_Z + nf^*H)) = 0$ for $n \geq 1$, we get

$$0 = H^1(Z, \mathcal{O}_Z(f^*(nH + L)) = H^1(S, \mathcal{O}_S(nH + L)) \oplus H^1(S, \mathcal{O}_S(nH)).$$

We obtain $H^1(S, \mathcal{O}_S(mH + L)) = 0$ for $m \geq 1$, and using again

(i) Serre duality on S:

$$\begin{aligned} h^1(S, \mathcal{O}_S(mH + L)) &= h^1(S, \mathcal{O}_S((-m)H - L)) \\ &= h^1(S, \mathcal{O}_S(-(m+1)H + L - \Delta)) \end{aligned}$$

(ii) the exact cohomology sequence associated to the exact sequence

$$0 \to \mathcal{O}_S(aH + L - \Delta) \to \mathcal{O}_S(aH + L) \to \mathcal{O}_\Delta(L) \to 0,$$

(we have vanishing of all cohomology groups of $\mathcal{O}_\Delta(L)$ since $h^j(\mathcal{O}_{\mathbb{P}^1}(-1) = 0)$, yielding

$$(**)\ h^i(\mathcal{O}_S((a+1)H - L)) = h^i(\mathcal{O}_S(aH + L - \Delta)) = h^i(\mathcal{O}_S(aH + L))\ \forall i,$$

we conclude that

$$h^1(S, \mathcal{O}_S(mH + L)) = h^1(S, \mathcal{O}_S((-m)H - L)) = h^1(S, \mathcal{O}_S((-m-1)H + L),$$

hence we have that the first cohomology groups $h^1(S, \mathcal{O}_S(mH + L))$ and $h^1(S, \mathcal{O}_S(mH - L))$ are zero for all m if and only if $h^1(S, \mathcal{O}_S(mH + L)) = 0$ for $m = 0$ or for $m = -1$, or, equivalently, Z is a regular surface.

We use now, following [59], that Z' is a canonical surface of general type, hence $H^1(\mathcal{O}_{Z'}(2K_{Z'})) = 0$.

Write $f^*(H) = 2H'$ on Z, then, since H' is disjoint from the (-1) curves contracted by p, the Leray spectral sequence implies

$$H^1(\mathcal{O}_Z(f^*(H))) = H^1(\mathcal{O}_Z(2H')) = H^1(p_*\mathcal{O}_Z(2H')) = H^1(\mathcal{O}_{Z'}(2K_{Z'})) = 0.$$

Hence $h^1(S, \mathcal{O}_S(H - L)) = 0$, and $h^1(S, \mathcal{O}_S(L)) = 0$, so that all cohomology groups vanish.

In particular, we have that

$$0 \le \chi(S, \mathcal{O}_S(L)) = \chi(\mathcal{O}_S) + \frac{1}{2}L^2 = 2 + \frac{1}{4}(-t + 3) \Leftrightarrow t \le 11.$$

□

The following is Lemma 3.13 of [59]

Proposition 1.55 *If Y is a nodal complete intersection of type $(2, 3)$ in $\mathbb{P}^4$, $\mathcal{K}'$ contains a vector of weight $w = 4$ if and only if the unique quadric Q containing Y has corank $= 2$.*

Proof One direction is easy, because a quadric of rank 3 is the join of its vertex, a line L, with a smooth plane conic C: and the join of L with the tangent lines of C yield a hyperplane H which is everywhere tangent to Q, hence to Y, and passes in general (simply) through the three nodes of Y, intersection of L with the cubic G such that $Y = \{Q = G = 0\}$.

In the other direction, as in the previous proposition, since $t = 3$, we get

$$\chi(S, \mathcal{O}_S(L)) = \chi(S, \mathcal{O}_S(-L)) = 2.$$

Since $2L \equiv H + \Delta = H + E_1 + E_2 + E_3$, then L is effective, $-L$ is not, and $|L|$ has dimension 1. Since $L \cdot E_i = -1$, we have that $|L| = |D| + \sum_1^3 E_i$, where $D^2 = 0$, since $L^2 = 0$. Since D is nef, we get a base point free pencil $|D|$ on our K3 surface, consisting of elliptic curves D with $D \cdot H = 3$, corresponding to plane cubic curves on Y. Each of these planes is therefore contained in Q, moreover, these planes contain the three nodes of $\mathcal{N}$. Hence the three nodes are collinear (since we have infinitely many planes), they belong to a line L contained in these planes and in Q. If Q had rank =5, then it would not contain any plane; if instead Q had rank =4, then it would be the join of its vertex point V with a rank 4 quadric Q' in $\mathbb{P}^3$, hence a family of planes contained in Q would be the join of V with a family of lines in Q'. But then these planes would intersect just in the vertex V.

We already know that the corank of Q is at most 2 (since Y has isolated singularities), hence Q has corank $= 2$. □

Remark 1.56 The above result shows that the condition that $Y = Q \cap G$, where $\text{corank}(Q) = 2$, is a topological property of the nodal surface Y. Hence, the nodal Severi varieties are split into two disjoint open sets by the condition $\text{corank}(Q) = 2$, or $\text{corank}(Q) \leq 1$.

Theorem 1.57 *The irreducible components of the Nodal Severi variety of nodal K3 surfaces Y of degree 6 in $\mathbb{P}^4$ are in bijection with those isomorphism classes of the extended codes $\mathcal{K}'$ of Theorem 1.52, which are different from case (2), the strict Kummer code.*

There is only one irreducible component for $\nu = 15$, and for it the unique quadric Q containing Y satisfies $\text{corank}(Q) = 2$; while there are exactly two irreducible components for $\nu = 14$, distinguished by the condition $\text{corank}(Q) = 2$ or $\text{corank}(Q) < 2$. All other components are gotten from small deformations of these two components, and belong to two disjoint open sets (cf. (i) of Remark 1.56).

In particular, the Nodal Severi variety $\mathcal{F}_4(3, 15)$ of Togliatti cubics (the cubic fourfolds with 15 nodes) is irreducible.

Proof The last assertion follows easily from the previous ones, since by Corollary 1.46 we have that giving the pair of a Togliatti cubic $X \subset \mathbb{P}^5$ and a node $P \in X$ is equivalent, up to projectivities, to giving a nodal K3 surface $Y \subset \mathbb{P}^4$ of degree 6 with 14 nodes, and such that the unique quadric Q containing it has $\text{corank}(Q) \leq 1$.

We are going to prove the first assertion exactly in the same way as in Theorem 6, using Nikulin's primitive embedding theorems, and the Torelli theorem for K3 surfaces.

The first step is to observe that, by Theorem 1.43, if the extended codes $\mathcal{K}'$ of Theorem 1.52, different from case (2), admit a primitive embedding in the K3 lattice Λ, then this embedding is unique.

Later on, to prove existence, we shall use for each of them the embedding induced by being a shortening of the two codes $\mathcal{K}'$ with $\nu = 14$.

Then the Torelli theorem shall prove the second assertion: showing, as in Theorem 6, that these components are irreducible, and that there are exactly two irreducible components for $\nu = 14$, distinguished by $\text{corank}(Q) = 2$, $\text{corank}(Q) < 2$, and all other components are gotten from small deformations of these two components.

The third assertion follows right away from the fact that for $\nu = 14$, $\text{corank}(Q) < 2$, there is only one irreducible component.

In order to prove existence, because all these codes are shortenings of two fixed ones, and since case (2) cannot occur by Proposition 1.54, it suffices to show existence for these two:

(A) case (4) with $\nu = 15$, and
(B) case (3) with $\nu = 14$.

We can use geometry in case (B), since we can take the Goryunov cubic X, and a node $P \in X$ to obtain, projecting with centre P, a complete intersection of type $(2, 3)$ possessing 14 nodes by virtue of Corollary 1.46. And then use the unobstructedness of X (Theorem 1.25).

Using the Goryunov cubic is important, since again we obtain K3 surfaces defined over the reals, and with all the singular points defined over the reals, so that all the local smoothings are defined over the reals and have real points.

The same can be done for case (A), where existence and the reality conditions are guaranteed by Theorem 1.51.

More generally, we can also apply Nikulin's existence theorems.

In the case that $\nu + 1 \leq 11$ we are indeed done by theorem 1.12.4 of [74], else we shall use theorem 1.12.2 of [74], part (d).

In fact Λ satisfies that the signature is divisible by 8, and we have already shown that the inequalities (2) ibidem hold true. Property (3) is satisfied since for every odd prime p (here, just $p = 3$) the length of the p-primary part of the discriminant group is $1 < 22 - (1 + \nu)$.

We remain with property (4), concerning the prime $p = 2$, where we have to verify that, if $\nu = 10 + k'$, then

(I) either the quadratic form q_2 splits off a rank 1 term with diagonal element $\theta \cdot 2$, (θ a 2-adic unit) or
(II) the following congruence is valid:

$$|A_q| \equiv \pm\, disc(K(q_2))\ (mod(\mathbb{Z}_2^*)^2),$$

where A_q is the discriminant group, $K(q_2)$ is the unique 2-adic lattice of rank $l(A(q_2))$ with discriminant form q_2.

Once more, we associate to a code vector $v = \sum_i a_i E_i + aH$, $a = 0, 1$ the lattice element $\frac{1}{2}(\sum_i a_i E_i + aH)$, which we still denote v.

We have a basis of $\mathcal{L}'$ provided by

$$v_1, \dots, v_k, v', E_1, \dots, E_{10},$$

where $v_1, \dots, v_k$ correspond to a basis of $\mathcal{K}$, v' to an element in $\mathcal{K}' \setminus \mathcal{K}$. Set for commodity $v_{k+1} = v'$.

The elements $v_1, \dots, v_k, v_{k+1}$ generate an isotropic subspace for the quadratic form, while $v_i E_j = 0$ or $= -1$ according to the property that j is not or is on the support of v_i, while as usual $E_i^2 = -2,\ E_i E_j = 0,\ i \neq j,\ i, j = 1, \dots, 10$.

Since $k' = k + 1 \leq 5$, we can pair each v_j with some E_j, without loss of generality we assume that $v_j E_j = -1$, for $j = 1, \dots, k + 1$.

We can reduce the intersection matrix modulo 4, since the quadratic form takes vales in $\frac{1}{2}\mathbb{Z}/2\mathbb{Z}$, and then with column operations followed by the same operations for the rows, we end up with the direct sum of -2 times the identity (6×6)- matrix with $k + 1 = 5$ copies of the matrix A, equivalent to the matrix U:

$$A := \begin{pmatrix} 0 & -1 \\ -1 & -2 \end{pmatrix} \cong U := \begin{pmatrix} 0 & -1 \\ -1 & 0 \end{pmatrix}. \tag{1.10}$$

and we are done concerning the existence of a primitive isometric embedding, since we have verified that property (I) holds. □

1.7.2 Nodal K3 Surfaces of All Degrees

Our next task shall be to extend our previous results concerning nodal K3 surfaces of degree $d = 2, 4, 6$ to the case of all degrees $d = 2d'$.

First of all one has to determine the a priori possible bicoloured codes $\mathcal{K}'$ occurring, equivalently, the pair of codes $\mathcal{K} \subset \mathcal{K}'' \subset \mathbb{F}_2^{\nu}$, keeping in mind the inequalities $k' \geq \nu - 10$, $k' - 1 \leq k \leq k'$ (and later see whether the corresponding lattices admit a primitive embedding in the K3 lattice).

We shall do this according to the congruence class of d (mod 8).

In all the cases $d \equiv 2 \ mod(8)$, $d \equiv 4 \ mod(8)$, $d \equiv 6 \ mod(8)$ we have a first description of the possible codes under the assumption that the weights t of vectors in $\mathcal{K}'' \setminus \mathcal{K}$, equivalently the weights $t + 1$ of vectors in $\mathcal{K}' \setminus \mathcal{K}$, are the same as in the respective cases $d = 2, 4, 6$.

If $d \equiv 4 \ mod(8)$ we have seen in Proposition 1.47 that $t \in \{2, 6, 10, 14\}$, while for $d = 4$ only $t = 6, 10$ occur.

Not only the case $t = 2$, but also the case $t = 14$ cannot be excluded code-theoretically.

Example 1.58 For instance, we have the case $t = \nu = 14$, $k' = 4$, where, for each $v \in \mathcal{K}$, setting $u := \sum_1^{14} e_i \in \mathcal{K}''$, $u + v$ has weight 6; in this case t can only be 6, 14.

Lemma 1.59 *If the weight $t = 14$ occurs for $\mathcal{K}''$, then $\nu = 14$ and the code $\mathcal{K}''$ is unique, as in the previous example.*

Proof We show now that, if $t = 14$ occurs, then necessarily $\nu = 14$: $\nu \geq 14$ is obvious, and if $\nu = 16$, then $\mathcal{K}$ is the strict Kummer code, hence, if $\mathcal{A}$ has cardinality 14, then there is an affine function x_W vanishing in exactly one point of the complement of $\mathcal{A}$; hence $\mathcal{K}''$ would admit a vector with $t = 8$, a contradiction. Similarly if $\nu = 15$, we can find a linear function x_W not vanishing in the complement of $\mathcal{A}$, so that we would again get $t = 8$, a contradiction.

Then there is $u := \sum_1^{14} e_i \in \mathcal{K}''$ and $\mathcal{K}$ is generated, up to a permutation, by $\sum_1^8 e_i$, $\sum_5^{12} e_i$, $\sum_0^3 (e_{1+4j} + e_{2+4j})$. □

We shall discuss later the case where $t = 2$ occurs.

If $d \equiv 0 \ mod(8)$ we are done: we have seen in Proposition 1.47 that $t \in \{4, 8, 12, 16\}$, but indeed we must have $t \in \{4, 8, 12\}$.

In fact, the case $t = 16$ can be here excluded right away, since then $\nu = 16$, $\dim \mathcal{K} \geq 5$, hence $\mathcal{K}$ is the Kummer code, therefore follows that H is 2-divisible, a contradiction.

For $d = 8m$, $\nu = 16$, indeed all $t \in \{4, 8, 12\}$ do in fact occur, as can be immediately seen: if $\nu = 16$, $\mathcal{K}$ is the Kummer code, and $t = 4$ occurs if and only $t = 12$ does, since $\mathcal{K}$ contains the codeword with weight 16. Hence $t = 4$ and $t = 12$ must occur because of a pure code-theoretical argument: since the case where only $t = 8$ occurs would lead to $\mathcal{K}'' = \mathcal{K}$, a contradiction to $dim(\mathcal{K}'') \geq 5$.

We shall classify the corresponding codes in Theorem 1.83, which shows in particular that they are all the shortenings of a single code, which we call the **K8 code**, see Corollary 1.36 and Definition 1.81.

If $d \equiv 2 \; mod(8)$ we have seen in Proposition 1.47 that $t \in \{1, 5, 9, 13\}$, and, for $d = 2$, only $t = 5, 9$ occur.

Lemma 1.60 *The case $t = 13$ can occur only if $\nu = 13, k = 2$. The code $\mathcal{K}''$ is then unique.*

Proof If $t = 13$ occurs, then $\nu \geq 13$, and then recall that $k' \geq \nu - 10$.

For $\nu = 15$ we have that $\mathcal{K}$ corresponds to hyperplane complements in $\mathbb{P}^3_{\mathbb{F}_2}$, and we would have the characteristic function of a set $\mathcal{B}$ of cardinality 13, the complement of two points P_1, P_2. We have planes π containing the two points, respectively one of them, respectively none. Then we get vectors $u \in \mathcal{K}''$ with respective weights $5, 7, 9$, a contradiction.

For $\nu = 14$, we are taking the shortening corresponding to P_1, hence complements of planes π containing P_1. Again, take a plane not containing P_2 but containing P_1, and again we get $t = 7$, a contradiction.

For $\nu = 13$, we have no code theoretical restriction, the other half-even sets have then cardinality 5.

We conclude that $t = 13$ can occur only if $\nu = 13, k = 2$ and the code $\mathcal{K}''$ is unique. □

We shall later discuss the case $t = 1$.

We are now going to settle the existence question for the case $t = 13$.

Theorem 1.61 *The case $t = 13$ (hence also $\nu = 13, k = 2$) occurs exactly for $d \equiv 10 \; (mod \; 16)$.*

Moreover, the corresponding primitive embedding is unique.

Proof We have seen that $t = 13$ can only occur for $d = H^2 = 2 + 8h$. Set $d' := d/2 = 1 + 4h$. We know that here $\nu = 13$, hence we have the lattices

$$\mathcal{L} = \bigoplus_1^{13} \mathbb{Z}E_i \oplus \mathbb{Z}H,$$

$$\mathcal{L}' = \langle \mathcal{L}, v_1 := \frac{1}{2}\sum_1^8 E_i, v_2 := \frac{1}{2}\sum_5^{12} E_i, v_3 := \frac{1}{2}(H + \sum_9^{13} E_i\rangle.$$

In particular, a basis of $\mathcal{L}'$ is given by

$$v_1, v_2, v_3, E_2, \dots E_{11}, E_{13}.$$

The dual lattice $(\mathcal{L}')^\vee$ can be explicitly described (requiring integrality of the values on v_1, v_2, v_3) as

$$(\mathcal{L}')^\vee = \{\frac{1}{2}(\sum_i b_i E_i + \frac{a}{d'} H) | \sum_1^8 b_i \equiv 0(2), \sum_5^{12} b_i \equiv 0(2), a + \sum_9^{13} b_i \equiv 0(2)\}.$$

Set $u_3 := \frac{1}{2}(E_{13} + \frac{1}{d'} H)$. Then u_3 has order d in the Discriminant group

$$\mathcal{A}_{\mathcal{L}'} := (\mathcal{L}')^\vee / \mathcal{L}',$$

since $d' u_3 \equiv \frac{1}{2}(E_{13} + H) \neq 0 \ (mod \mathcal{L}')$.

Hence, defining $b' := \sum_1^4 b_i, b'' := \sum_5^8 b_i, b''' := \sum_9^{12} b_i$, and

$$\hat{\mathcal{L}} := \{\frac{1}{2}(\sum_i b_i E_i) | b' \equiv b'' \equiv b''' \equiv b_{13} \ (mod 2)\},$$

$$\mathcal{A}_{\mathcal{L}'} \cong (\mathbb{Z}/d) u_3 \oplus \hat{\mathcal{L}}/\mathcal{L}' \cong (\mathbb{Z}/d) \oplus (\mathbb{Z}/2)^7.$$

Hence the length (minimal number of generators) $\ell(\mathcal{A}_{\mathcal{L}'}) = 8$, and also the length of the 2-primary part $= 8$.

Since

$$8 = \ell((\mathcal{A}_{\mathcal{L}'})_2) \leq 22 - rk(\mathcal{L}') = 22 - 14,$$

by Theorem 1.12.2 of [74] there is an embedding of $\mathcal{L}'$ in the K3 lattice if and only if condition 4) holds. This requires to calculate the 2-adic discriminant finite quadratic form.

For this, we define $w := \frac{1}{2}(E_1 + E_5 + E_9 + E_{13})$, and we complete u_3, w to a basis of $(\mathcal{L}')^\vee$ choosing a basis of

$$\tilde{\mathcal{L}} := \{\frac{1}{2}(\sum_1^{12} b_i E_i) | b' \equiv b'' \equiv b''' \equiv 0 \ (mod 2)\}.$$

We note that $\tilde{\mathcal{L}}$ is isomorphic to the direct sum of three copies of $\{\frac{1}{2}(\sum_1^4 b_i E_i) | \sum_i b_i \equiv 0 \ (mod 2)\}$, with basis

$$\frac{1}{2}(E_1 + E_2), \frac{1}{2}(E_3 + E_4), \frac{1}{2}(E_2 + E_3), E_1.$$

It suffices, to find the whole basis of $\tilde{\mathcal{L}}$, to take the previous 4 vectors and add 4, respectively 8, to the lower indices.

We observe that the three summands described above are orthogonal to each other, and are orthogonal to u_3. While w has the property that the scalar product of w with each summand has values in $\mathbb{Z}$.

Hence, if we take as fourth summand $\mathbb{Z}u_3 \oplus \mathbb{Z}w$, we have given an orthogonal direct sum for the quadradic form q_2 of the discriminant group.

The form of the summand $\mathbb{Z}u_3 \oplus \mathbb{Z}w$ is

$$-q(au_3 + bw) = a^2 \frac{2h}{4h+1} + ab + 2b^2,$$

hence its binary (2-primary) part has matrix

$$U := \frac{1}{2} \begin{pmatrix} 0 & 1 \\ 1 & 0 \end{pmatrix}.$$

Instead, for the summand with basis $\frac{1}{2}(E_1 + E_2)$, $\frac{1}{2}(E_3 + E_4)$, $\frac{1}{2}(E_2 + E_3)$, E_1, we get the matrix

$$\frac{1}{2} \begin{pmatrix} 2 & 0 & 1 & 2 \\ 0 & 2 & 1 & 0 \\ 1 & 1 & 2 & 0 \\ 2 & 0 & 0 & 4 \end{pmatrix}$$

which, modulo $2\mathbb{Z}$, boils down to the same matrix where last row and column are deleted.

Subtracting second row to the first, and doing the same for the columns, we get modulo $2\mathbb{Z}$ the matrix

$$V := \frac{1}{2} \begin{pmatrix} 2 & 1 \\ 1 & 2 \end{pmatrix}.$$

The corresponding 2-adic quadratic forms, in Nikulin's notation, have respective matrices

$$u := \begin{pmatrix} 0 & 2 \\ 2 & 0 \end{pmatrix} v := \begin{pmatrix} 4 & 2 \\ 2 & 4 \end{pmatrix}.$$

By condition (4) of Nikulin's theorem 1.12.2 of [74] (the others are trivially satisfied) there exists a primitive embedding of $\mathcal{L}'$ into the K3 lattice if and only if

$$\pm\, disc(q_2) = \pm 3^3 2^8 \equiv |\mathcal{A}_{\mathcal{L}'}| = 2^8(4h+1) mod (\mathbb{Z}_2^*)^2,$$

$$\Leftrightarrow \pm 3^3 \equiv 4h + 1 (mod 8) \Leftrightarrow \pm 3 \equiv 4h + 1 (mod 8) \Leftrightarrow h \equiv 1 (mod 2).$$

Unicity of the primitive embedding follows by the general result of Theorem 1.43, in this case it follows also directly from the above computations and applying theorem 1.14.4 of [74]: condition 3) is satisfied, since q_2 is a direct sum where both $u = u_+^{(2)}(2)$, $v = v_+^{(2)}(2)$ occur by our calculation (the other conditions are also trivially satisfied). □

If $d \equiv 6\ mod(8)$ we have seen in proposition 1.47 that $t \in \{3, 7, 11, 15\}$ and the question (in view of the case $d = 6$, that we have seen) is when does the case $t = 15$ occur. This occurence is equivalent to $\nu = 15$, $k' = 5$, and $\mathcal{K}'$ is the Kummer code.

In this case we shall show

Theorem 1.62 *The case $t = 15$ (hence also $\nu = 15, k' = 5$, $d \equiv 6\ (mod\ 8)$ and $\mathcal{K}'$ is the Kummer code) occurs exactly for $d \equiv 14\ (mod\ 16)$.*

Moreover, the corresponding primitive embedding is unique.

Proof The proof is similar in spirit to the one of Theorem 1.61, but the calculations quite different.

A quick proof is the following: $d = H^2 \equiv -2\ (mod 8)$, hence $H^2 = 2(-1+4h)$, and the finite quadratic form is the same as the code $\mathcal{K}$ of a Kummer surface.

What varies in condition (4) of Theorem 1.12.2 of [74] is the size of the discriminant group, which has order $|4h - 1|2^6$. We claim that the discriminant for $\mathcal{K}$ is 2^6, hence there exists a primitive embedding in the K3 lattice if and only if $4h - 1 \equiv \pm 1\ (mod 8)$, this means that h is even, hence $d \equiv -2\ (mod 16)$.

A slick argument is that we have embedding for the Kummer code $\mathcal{K}$. A more precise and interesting calculation is the following, where $M := \{0, 1, 2, 3\}$, and $\mathcal{L} = (\mathbb{Z})^{M \times M}$:

$$\mathcal{K} = \{(k_{ij}) \in (\mathbb{Z}/2)^{M \times M} | k_{00} + k_{0j} + k_{i0} + k_{ij} = \sum_j k_{0j} = \sum_i k_{i0} = 0 \in \mathbb{Z}/2\}.$$

Just by the above description

$$\mathcal{K}^{\perp} = \langle e_{00} + e_{0j} + e_{i0} + e_{ij}, \forall 1 \leq i, j \leq 3, \sum_j e_{0j}, \sum_i e_{i0} \rangle.$$

$dim(\mathcal{K}) = 5$, $dim(\mathcal{K}^{\perp}) = 11$, and

$$\mathcal{L}' = \langle \mathcal{L}, \{\frac{1}{2} w\rangle_{w \in \mathcal{K}}, \ (\mathcal{L}')^{\vee} = \langle \mathcal{L}, \{\frac{1}{2} w\rangle_{w \in \mathcal{K}^{\perp}}.$$

A basis of $(\mathcal{L}')^{\vee}$ is given by

$$U_{ij} := \frac{1}{2}(e_{00} + e_{0j} + e_{i0} + e_{ij}),\, U_{0*} := \frac{1}{2}\sum_j e_{0j},\, U_{*0} := \frac{1}{2}\sum_i e_{i0},$$

$$e_{00}, e_{01}, e_{02}, e_{10}, e_{20}.$$

q_2 is defined on

$$\mathcal{A}_{\mathcal{L}'} = (\mathcal{L}')^\vee/\mathcal{L}' = \mathcal{K}^\perp/\mathcal{K} \cong (\mathbb{Z}/2)^6.$$

To calculate the associate bilinear form with values in $\mathbb{Q}/\mathbb{Z}$, we can forget about the basis elements in $\mathcal{L}$.

We see that the negatives of the intersection numbers are:

$$-U_{ij}^2 = -U_{0*}^2 = -U_{*0}^2 = 2,$$

$$-(U_{ij}, U_{hk}) = \frac{1}{2} \text{ for } i \neq h,\, j \neq k, \text{ else } = 1,\ -(U_{0*}, U_{*0}) = \frac{1}{2}.$$

An easy calculation shows, using twice that a (block) matrix (with $A = {}^t A$)

$$\begin{pmatrix} 0 & A & A \\ A & 0 & A \\ A & A & 0 \end{pmatrix}$$

is equivalent modulo $2\mathbb{Z}$ to

$$\begin{pmatrix} 0 & A \\ A & 0 \end{pmatrix}$$

that the binary quadratic form is given by three copies of

$$U := \frac{1}{2}\begin{pmatrix} 0 & 1 \\ 1 & 0 \end{pmatrix},$$

one occurring for the span of U_{0*}, U_{*0}; hence q_2 is a direct sum of three copies of $u = u_+^{(2)}(2)$. Follows that its discriminant equals 2^6 as claimed.

Note that the occurrence of a summand of the type $u = u_+^{(2)}(2)$ confirms the proven (Theorem 1.43) unicity of the primitive embedding. □

1.7.3 One-Node Projection

We shall now resort to geometry, namely, projecting the surface Y from a node of Y, to obtain more information, and in this way we shall introduce a new operation on these bicoloured codes $\mathcal{K}'$, that we shall call **one-node projection**.

In the next proposition we use an old result by Saint Donat [81], to obtain that for a general surface in a nodal Severi variety of ν-nodal K3 surfaces of degree d projection from a node gives an embedding of the blow up of the surface at the nodal point.

Proposition 1.63 (Saint Donat) *Let Y be a nodal K3 surface of degree $d = 2g - 2 \geq 10$ embedded in $\mathbb{P}^g$, let P_1 be a node of Y, and let $S := \tilde{Y}$ be the minimal resolution of Y.*

Then projection from the node P_1 is a morphism of the blow up Y' of Y at P_1, and induces a morphism $\psi : S = \tilde{Y} \to \mathbb{P}^{g-1}$.

The image $Z := \psi(S)$ is either

- *(0) Isomorphic to Y', or*
- *(I) a K3 surface with Rational Double points, obtained from Y' contracting the strict transforms C of lines $L \subset Y$ passing through P_1 ($C \cdot D = 0$, $D := H - E_1$), or*
- *(II) ψ is of degree 2 onto a rational surface Σ of degree $g - 2$, and this happens precisely when either*
- *(II a) there is an irreducible curve $\mathcal{E}$ of arithmetic genus 1 ($\mathcal{E}^2 = 0$) with $D \cdot \mathcal{E} = 2$, or*
- *(II b) $D = 2B$, B of arithmetic genus 2, in particular $B^2 = 2$, $H^2 = 10$, and Σ is the Veronese surface in $\mathbb{P}^5$.*

In all the cases different from (0) we have that Y is not a general surface in the Nodal Severi variety of ν-nodal K3 surfaces of degree d: in cases (I), (II a) because the Picard number $\rho(S) > \nu + 1$.

In the case where Y' is embedded, it is primitively embedded unless possibly in the case $d \equiv 2 \bmod(8)$ if $H - E_1$ is 2-divisible in Pic(S).

Proof In case (I), if there is such a line, then S contains at least $\nu + 2$ curves, and we claim that these are numerically independent.

Otherwise, a numerical equivalence $C = aH - \sum_i b_i E_i$ with rational coefficients implies, in view of $C \cdot E_1 = 1 \Leftrightarrow b_1 = \frac{1}{2}$, and of the fact that $2C$ is in the span of the curves $H, E_1, \ldots, E_\nu$, that $2C \equiv aH - \sum_i b_i E_i$, with $a, b_1, \ldots, b_\nu$ integers.

But then $CH = 1 \Rightarrow 2 = 2C \cdot H = ad \geq d$ is a contradiction always, since if Y is embedded we have $d \geq 4$.

In case (II) since $D \cdot E_1 = 2$, and $E_1^2 = -2$, E_1 maps to a conic.

In case (II a), we claim that the curves $H, E_1, \ldots, E_\nu, \mathcal{E}$ are numerically independent. Consider in fact the associated $(\nu + 2) \times (\nu + 2)$ intersection matrix, where we observe that $a_i := E_i \cdot \mathcal{E}$ is a non-negative integer. Linear dependence of the columns (since the first $(\nu + 1)$ ones are linearly independent) boils down to

$$\frac{1}{2}\sum_{1}^{\nu+1} a_j^2 - \frac{4}{d} = 0, \Leftrightarrow \sum_{1}^{\nu+1} a_j^2 = \frac{8}{d}.$$

But the left hand side is a positive integer, the right hand side is strictly smaller than 1, a contradiction.

In case (II b) $2B = H - E_1$, and E_1 maps to a line in the plane $\mathbb{P}^2$. This line must be everywhere tangent to the branch curve Δ. This condition determines in general a family of K3 surfaces with 17 moduli: but since the blow up of Y in P_1 has $\nu - 1$ nodes, the same holds for Δ, and we get a family of dimension $19 - (2 + \nu - 1)$ hence of codimension $\nu + 1$ and not ν.

For the last assertion, the question is whether $H' := H - E_1$ can be divisible in the Picard group (in homology).

Since $H' \cdot E_1 = 2$, the question is whether $H - E_1 = 2L$, that is, P_1 is a half-even set nodes. But $t = 1$ implies that $d \equiv 2 \, mod(8)$. □

Proposition 1.64 *Assume that Y is a nodal K3 of degree d, and let $P_1, \dots, P_\nu$ be the nodes of Y.*

Let $\mathcal{N}$ be a half-even set of nodes on Y, and assume that there is a node of $\mathcal{N}$, say P_1, such that $H^* := H - E_1$ yields an embedding of the blow up Y^* of Y at P_1. Then the code $\mathcal{K}(Y^*)$ is the shortening of $\mathcal{K}(Y)$ with respect to the complement of P_1.

While the coset $\mathcal{K}'(Y^*) \setminus \mathcal{K}(Y^*)$ is the projection of the subset of the coset $\mathcal{K}'(Y) \setminus \mathcal{K}(Y)$ consisting of the words where P_1 has coefficient 1.

In particular, all half-even sets of nodes on Y have cardinality $t + 1$, where t is the cardinality of some half-even set of nodes on a K3 surface Y^* of degree $d - 2$.

Proof Since Y and Y^* have the same minimal resolution, we have to see when

$$\sum_{2}^{\nu} a_i E_i + bH' \in 2\,\mathrm{Pic}(\tilde{Y}).$$

But this means

$$bE_1 + \sum_{2}^{\nu} a_i E_i + bH \in 2\,\mathrm{Pic}(\tilde{Y}),$$

and the cases $b = 0$, respectively $b = 1$ yield the desired assertion.

In particular, if for all half-even sets of nodes $\mathcal{N}$ on Y there is a node of $\mathcal{N}$, say P_1, such that $H^* := H - E_1$ yields an embedding of the blow up Y^* of Y at P_1, $\mathcal{N}$ has cardinality $t + 1$, where t is the cardinality of some half-even set of nodes on a K3 surface Y^* of degree $d - 2$.

Thanks to Proposition 1.63, this holds for a general surface in the component of the nodal Severi variety of ν-nodal K3 surfaces of degree d containing Y; and we use the fact that $\mathcal{K}'$ is an invariant of this component. □

Corollary 1.65 *If the weight $t = 15$ occurs for $d \equiv 6 \, mod(8)$, $d \geq 14$, then also the weight $t = 14$ occurs for $d - 2 \equiv 4 \, mod(8)$, $d \geq 12$, and the weight $t = 13$ occurs for $d - 4 \equiv 2 \, mod(8)$, $d \geq 10$.*

We give here a different, code theoretic proof, of a result which was already shown earlier in Corollary 1.36 (see also Theorem 1.83).

Corollary 1.66 *Let Y be a 16-nodal K3 surface of degree d divisible by 8: then all the numbers 4, 8, 12 occur as weights t for vectors in $\mathcal{K}'' \setminus \mathcal{K}$.*

The code $\mathcal{K}''$ is generated by the Kummer code $\mathcal{K}$ and by the characteristic function f_π of an affine plane π.

Proof Projecting from a node yields, if Y is general, a sextic K3 surface with 15 nodes: hence with all the numbers $t = 3, 7, 11$ as weights t for vectors in $\mathcal{K}'' \setminus \mathcal{K}$, in view of our explicit classification in the proof of Theorem 1.52.

Apply now the previous Proposition 1.64.

For the second assertion, observe that if ϕ is affine and not constant, the weight of $f_\pi + \phi$ equals $12 - 2a$, where $a := |\pi \cap \{\phi = 1\}|$, and this last cardinality a can be $4, 2, 0$.

Unicity of the code $\mathcal{K}''$ follows for instance from the recipe given in Proposition 1.64 and the classification for $d = 6$. □

The next result is contained in Sect. 1.9, Theorem 1.83:

Corollary 1.67 *Let Y be a nodal K3 surface of degree d divisible by 8: then its code $\mathcal{K}'$ is a shortening of the code of a 16-nodal K3 surface Z of degree $d = 8$ (and all these do occur).*

We make now the useful remark, the analog of the observation that deformation and projection are operations which commute with each other:

Lemma 1.68 *The two operations of one-node-shortening and node-projection, both reducing ν by 1, commute with each other.*

Proof We can assume that the node-shortening corresponds to the first element of $\mathcal{S}$, the node-projection to the second element of $\mathcal{S}$.

For the strict code $\mathcal{K}$ we get, independently of the order chosen for the composition of the two operations, the shortening with respect to the complement of the set $\{1, 2\}$.

For $\mathcal{K}''$ the assertion is clear if there is no vector $u \in \mathcal{K}'' \setminus \mathcal{K}$ with $u_2 = 1$, or if there is such a vector with $u_2 = 1$ and with $u_1 = 0$. There remains the case where each such a vector with $u_2 = 1$ has $u_1 = 1$. This implies that the shortening corresponding to 1 does not change $\mathcal{K}$, and makes $\mathcal{K}''$ become equal to $\mathcal{K}$, so that then the node-projection just affects the shortening of $\mathcal{K}$ corresponding to 2. In the other direction, we get the shortening of $\mathcal{K}$ corresponding to 2, and we get the projection u' of u, but $u'_1 = 1$ holds for each vector in $\mathcal{K}'' \setminus \mathcal{K}$, hence the next shortening corresponding to 1 reduces $\mathcal{K}''$ to $\mathcal{K}$. □

We observe now, in particular, that one-node projection shows the existence of the cases $t = 1$ and $t = 2$.

Corollary 1.69 *For $d \equiv 0 \, mod(8)$ we get all the possible shortenings of the K8 code, and for $d \equiv 6 \, mod(8)$ we get all the shortenings of the one-node projection of the K8 code; the latter are all the codes $\mathcal{K}'$ satisfying the B-inequality and with cardinalities $t \in \{3, 7, 11\}$ of the half-even sets of nodes, which are all the possible codes of K3 surfaces of degree $d = 6$.*

By further projection we get codes with $t = 2$ for $d \equiv 4 \, mod(8)$, with $t = 1$ for $d \equiv 2 \, mod(8)$.

Example 1.70 Let Z be a quartic K3 surface with ν nodes, with a line L contained in the smooth locus of Z.

Let D be the hyperplane section of Z, and consider the linear system $H := 2D + L$ on the resolution S of Z, which contains the exceptional (-2)-curves $E_1, \dots, E_\nu$.

Then $H^2 = 18$ and $H \cdot L = 0$, hence, since $|H|$ yields a birational morphism in view of the exact sequence

$$0 \to \mathcal{O}_S(2D) \to \mathcal{O}_S(H) \to \mathcal{O}_L(H) \cong \mathcal{O}_L \to 0,$$

the image of S is a K3 surface Y with $\nu + 1$ nodes in $\mathbb{P}^{10}$, and with $t = 1$. Observe that we may take here $\nu = 0$, so Y has only one node.

Assume now that we have a nodal K3 surface Y with a half-even set $\{P_1\}$ of cardinality $t = 1$: then, since every other half-even set has cardinality $t' \in \{1, 5, 9, 13\}$, every (strictly) even set $\mathcal{N}$ must be disjoint from P_1, in particular $\mathcal{S} \setminus \{P_1\}$ has at most 14 elements, but we have indeed more:

Proposition 1.71 *If Y is a nodal K3 surface Y with a half-even set $\{P_1\}$ of cardinality $t = 1$, then $\nu \le 13$, $k' \le 3$, equality holding if and only if $\nu = 13$.*

Moreover, its code $\mathcal{K}''$ is the orthogonal direct sum

$$\mathcal{K} \oplus \mathbb{F}_2 E_1 \subset \mathbb{F}_2^{\mathcal{S} \setminus \{P_1\}} \oplus \mathbb{F}_2 E_1.$$

These codes do in fact occur, as double projections from a nodal K3 of degree $d \equiv 6 \, mod(8)$, and with a codeword in $\mathcal{K}'$ of weight 4.

They are shortenings of the one with $\nu = 13$, $k' = 3$.

Proof $\mathcal{S} \setminus \{P_1\}$ has at most 14 elements, and contains the support of $\mathcal{K}$, hence $k \le 3$, hence $k' \le 4 \Rightarrow \nu \le 14$.

But if $\nu = 14$, then $\mathcal{S} \setminus \{P_1\}$ has 13 elements, hence $k \le 2$, a contradiction.

Hence $\nu \le 13$, and if equality holds, then $k = 2$, $k' = 3$. Otherwise, if $\nu \le 12$, $k \le 1$.

The rest is obvious by our previous results. □

Proposition 1.72 *If Y is a nodal K3 surface Y with a half-even set $\mathcal{B} := \{P_1, P_2\}$ of cardinality $t = 2$, then $\nu \le 14$, $k' \le 4$, equality holding if and only if $\nu = 14$.*

These codes do in fact occur, as projections from a nodal K3 of degree $d \equiv 6 \, mod(8)$, *and with a codeword in* $\mathcal{K}'$ *of weight* 4. *They are shortenings of the one with* $\nu = 14, k' = 4$.

Proof Every even set $\mathcal{N}$ of cardinality 8 either contains $\mathcal{B}$ or is disjoint from it. Since these sets correspond to complements of affine subspaces containing an affine subspace $\mathcal{A}$, it follows that $k \leq 3$, hence $k' \leq 4$ and $\nu \leq 14$.

If $\nu = 14, k' = 4$, then the union of $\mathcal{B}$ and $\mathcal{A}$ is an affine plane.

The rest follows easily as before. □

We have already seen that the possible codes of nodal K3 surfaces of degree d are not exactly those that we have for $d = 2 = 4 = 6 = 8$, and to classify them all it is now possible knowing that $t = 15$ occurs precisely for $d \equiv 14 \ (mod\ 16)$, $t = 14$ occurs precisely for $d \equiv 12 \ (mod\ 16)$, and $t = 13$ occurs precisely for $d \equiv 10 \ (mod\ 16)$.

1.7.4 Main Theorem on Varieties of Nodal K3 Surfaces

Theorem 1.73 *The connected components of the Nodal Severi variety* $\mathcal{F}_{K3}(d, \nu)$ *of nodal K3' s of degree d (d is an even number) and with* ν *nodes are in bijection with the isomorphism classes of their extended codes* $\mathcal{K}'$ *(equivalently, of the pair of codes* $\mathcal{K} \subset \mathcal{K}'' \subset \mathbb{F}_2^{\nu}$*).*

The weights of the code $\mathcal{K}$ *are either* 8 *or* 16, *and, if d is not divisible by* 4, *the code* $\mathcal{K}$ *admits only the weight* 8.

The weights t of vectors in $\mathcal{K}'' \setminus \mathcal{K}$ *satisfy the following relation:*

$$2t - d \equiv 0 \ (mod\ 8).$$

If $d \equiv 4 \ (mod\ 8)$ *and* $t \in \{6, 10\}$, *we get for* $\mathcal{K}'$ *all the shortenings of the extended Kummer code.*

The case where $d \equiv 4 \ (mod\ 8)$ *and there is some* $t = 14$ *occurs exactly when* $d \equiv 12 \ (mod\ 16)$, *and then we have* $\nu = 13$, $k = 2$ *and a unique code* $\mathcal{K}'$.

The case where $d \equiv 4 \ (mod\ 8)$ *and there is some* $t = 2$ *occurs only for* $\nu \leq 14$, *and we get all the shortenings of the code* $\mathcal{K}'$ *which occurs for* $\nu = 14$, $k' = 4$; *the latter is obtained via the projection from a node of a nodal K3 with* 15 *nodes, with degree* $d \equiv 6 \ (mod\ 8)$, *and with a codeword in* $\mathcal{K}'$ *of weight* 4.

If $d \equiv 6 \ (mod\ 8)$, *then* $t + 1$ *is divisible by* 4, *and we get all the shortenings* $\mathcal{K}'$ *of the two codes that we get for* $\nu = 15$, *namely, the case where* $\mathcal{K}'$ *is the strict Kummer code* $\mathcal{K}_{Kum}$, *and the case where there is a weight* 4 *vector in* $\mathcal{K}'$: *in the latter case* $\mathcal{K}$ *is generated by the characteristic functions of complements of hyperplanes* $H \subset \mathbb{P}^3_{\mathbb{F}_2}$, *and* $\mathcal{K}''$ *is generated by* $\mathcal{K}$ *and by the characteristic function of a line (a subset* $\mathcal{A}$ *consisting of 3 collinear points).*

The special case where $d \equiv 6 \ (mod\ 8)$ *and there is a* $t = 15$ *occurs exactly for* $d \equiv 14 \ (mod\ 16)$, *and then* $\nu = 15$, $k' = 5$, *and* $\mathcal{K}'$ *is the strict Kummer code.*

If $d \equiv 0$ (mod 8) then $t \in \{4, 8, 12\}$ and we get all the shortenings $\mathcal{K}'$ of the K8 code, the code for which $\mathcal{K}''$ is generated by the strict Kummer code and by the characteristic function f_π of an affine plane π.

If $d \equiv 2$ (mod 8) and $t \in \{5, 9\}$, we get all the shortenings $\mathcal{K}'$ of the code generated by the linear functions on $\mathbb{F}_2^4$ and by the quadratic function $x_1 y_1 + x_2 y_2 + 1$.

The case where $d \equiv 2$ (mod 8) and there is some $t = 13$ occurs exactly when $d \equiv 10$ (mod 16), and then we have $\nu = 13$, $k = 2$ and a unique code $\mathcal{K}'$.

The case where $d \equiv 2$ (mod 8) and there is some $t = 1$ occurs only for $\nu \leq 13$, and we get all the shortenings $\mathcal{K}'$ of the one for $\nu = 13$, $k' = 3$, which occurs via the projection from a node of a nodal K3 with 14 nodes, of degree $d \equiv 4$ (mod 8) and which has some $t = 2$.

Proof We have just classified all the codes $\mathcal{K}'$ which can occur, and proven the existence of a primitive embedding of the associated lattices $\mathcal{L}'$ inside the K3 lattice Λ: either establishing geometrically the existence of a nodal K3 surface with the given $\mathcal{K}'$, or as an application of Nikulin's Theorem 1.12.2 guaranteeing the existence of such a primitive embedding.

The proof follows now the identical pattern of the proof of Theorem 6, first using Theorem 1.43 to show the unicity of the primitive embedding of $\mathcal{L}'$, and then applying verbatim as in Theorem 6 the Torelli Theorem for K3 surfaces, especially using that for each $\mathcal{K}'$ we may find a nodal K3 surface defined over $\mathbb{R}$ and with all the nodes real points (see Theorem 1.51 and observe that the class of codes $\mathcal{K}'$ of nodal K3 surfaces defined over $\mathbb{R}$ and with all the nodes real points is closed under the operation of smoothing and of projection from one node). □

An important topological consequence is the following result

Theorem 1.74 *Let Y be a nodal K3 surface with ν nodes: then the fundamental group of the smooth part satisfies:*

$$\pi_1(Y^*) \cong \mathbb{Z}^4 \rtimes \mathbb{Z}/2, \ \nu = 16,$$

$$\pi_1(Y^*) \cong \mathcal{K} \cong (\mathbb{Z}/2)^k, \ \nu \leq 15,$$

in particular $\pi_1(Y^*) = 0, \ \nu \leq 7$.

Proof In the case $\nu = 16$ we have a Kummer surface, hence the stated result is shown as in Corollary 1.36.

The proof of the previous Theorem 1.73 shows that every nodal K3 surface is obtained from a Kummer surface via a sequence of operations:

(1) smoothing of one node
(2) projection from one node
(3) equisingular deformation.
(4) does not change the fundamental group $\pi_1(Y^*)$, whereas (1) and (2) have the same effect on $\pi_1(Y^*)$, which we now describe.

Namely, projection from one node P_i replaces a neighbourhood of the node P_i by a neighbourhood of the exceptional (-2)-curve E_i lying over it; whereas smoothing of P_i replaces a neighbourhood of the node P_i by the Milnor fibre of the singularity. By the cited theorem of Brieskorn both have the effect of dividing by the subgroup normally generated by the generator γ_i of the local fundamental group at P_i, which has order at most 2.

If we start from a Kummer surface, as shown in Corollary 1.36, we divide $\mathbb{Z}^4 \rtimes (\mathbb{Z}/2)$ by an element of order 2, hence we get as quotient a finite group $(\mathbb{Z}/2)^k$, hence the fundamental group equals the first homology group.

Proceeding further, since a quotient of an Abelian group is Abelian, we get always $\pi_1(Y^*) \cong \mathcal{K} \cong H_1(Y^*, \mathbb{Z})$.

Since every code vector has weight 8 for $\nu \leq 15$, we conclude that $\pi_1(Y^*)$ is trivial for $\nu \leq 7$. □

Remark 1.75 For a more general normal K3 surface Y it was shown in [19] that $\pi_1(Y^*)$ is either finite or a finite extension of $\mathbb{Z}^4$, extending previous results (as [30]).

1.8 Quadratic Functions, the Extended Kummer Code and its Shortenings

Assume that $\mathbb{F}$ is a perfect field of characteristic 2, for instance $\mathbb{F}_2$.

Then every quadratic function $q(x)$ on a vector space $\mathbb{F}^n$ can be written as the sum

$$q(x) = T(x) + L(x)^2,$$

where $L(x)$ is a linear function, and $T(x)$ is a triangular quadratic form

$$T(x) = \sum_{i<j} a_{i,j} x_i x_j.$$

If $T(x)$ is non identically zero then there is a coefficient $a_{i,j} \neq 0$; without loss of generality we may assume that $a := a_{1,2} \neq 0$.

Then we may write

$$\frac{1}{a} T(x) = x_1 (x_2 + \sum_{j \geq 3} b_{1,j} x_j) + T'(x_2, \ldots, x_n),$$

and setting $z_2 := (x_2 + \sum_{j \geq 3} b_{1,j} x_j)$, we have:

$$\frac{1}{a} T(x) = x_1 z_2 + q'(z_2, \ldots, x_n) = x_1 z_2 + T'(z_2, \ldots, x_n) + L'(z_2, \ldots, x_n)^2.$$

Taking as new variable $w_2 := az_2$, we reach the form:

$$q(x) = x_1 w_2 + T''(w_2, x_3, \dots, x_n) + L''(x)^2.$$

We can take as new variable y_1 the sum of x_1 with $\sum_j a''_{2,j} x_j$, so that we reach the normal form

$$q(x) = x_1 y_1 + T''(x_3, \dots, x_n) + L''(x)^2.$$

By induction we find new coordinates such that the quadratic form $q(x)$ takes the quasi-normal form

$$q(x) = x_1 y_1 + x_2 y_2 + \cdots + x_k y_k + L^2,$$

where L is a linear form.

If the linear form L is linearly independent of $x_1, y_1, x_2, y_2, \dots, x_k, y_k$, then we reach the normal form

$$(0)\ q(x) = x_1 y_1 + x_2 y_2 + \cdots + x_k y_k + x_{k+1}^2.$$

Otherwise we have the semi-normal form

$$q(x) = x_1 y_1 + x_2 y_2 + \cdots + x_k y_k + L_1(x_1, \dots, x_k)^2 + L_2(y_1, \dots, y_k)^2.$$

Since we may view the spaces with coordinates $x := (x_1, \dots, x_k)$ and $y := (y_1, \dots, y_k)$ as dual spaces, we can change basis in the x space and then take the dual basis in the y-space, so that either

(i)

$$q(x) = x_1 y_1 + x_2 y_2 + \cdots + x_k y_k,$$

or

(ii)

$$q(x) = x_1 y_1 + x_2 y_2 + \cdots + x_k y_k + x_1^2 + y_1^2,$$

or

(iii)

$$q(x) = x_1 y_1 + x_2 y_2 + \cdots + x_k y_k + x_1^2,$$

or

(iv)

$$q(x) = x_1 y_1 + x_2 y_2 + \cdots + x_k y_k + x_1^2 + y_2^2.$$

One sees however that cases (iii) and (iv) can be brought to the normal form (i).

Hence we have only the normal forms (0), (i), (ii), which are respectively called **parabolic, hyperbolic, elliptic**.

1.8.1 *The Case* $\mathbb{F} = \mathbb{F}_2$

Then every polynomial function of degree 2 is the sum of a constant and of a quadratic function (a linear function is also equal to its square).

Then each polynomial function of degree 2 has either the affine normal form

$$(0)\ q(x) = x_1 y_1 + x_2 y_2 + \cdots + x_k y_k + x_{k+1}^2,$$

or

$$(i)\ q(x) = x_1 y_1 + x_2 y_2 + \cdots + x_k y_k,$$

or

$$(ii)\ q(x) = x_1 y_1 + x_2 y_2 + \cdots + x_k y_k + 1,$$

since $x_1 y_1 + x_1^2 + y_1^2 = (x_1 + 1) + (y_1 + 1) + 1$.

If $n = 2k$, the respective functions $q(x)$ in cases (i), (ii) differ since in the even case (i) the number of zeros equals $2^{k-1}(2^k + 1)$, in the second case it equals $2^{k-1}(2^k - 1)$.

If we consider the vector space generated by the space $\mathcal{K}$ of affine functions and by a quadratic function $q \notin \mathcal{K}$, we see therefore that we may take as normal form for q the form i), here k is the rank of q, this is the minimal number of products of linear forms whose sum is q.

For example, if $n = 4$ the support of quadratic form $x_1 y_1$ has cardinality 4, whereas the support of the quadratic form $x_1 y_1 + x_2 y_2$ has cardinality 6.

Definition 1.76 The first order Reed-Muller code $\mathcal{K}_1$ over $\mathbb{F}_2$ is the vector space of affine functions on an affine space $\mathbb{F}_2^n$.

The second order Reed-Muller code $\mathcal{K}_2$ over $\mathbb{F}_2$ is the vector space generated by the space of affine functions on an affine space $\mathbb{F}_2^n$, and by the quadratic forms. The extended Kummer code is the code generated by the first order Reed-Muller code and by a nondegenerate quadratic form $q(x)$ (that is, of type i) for $n = 2k$, and of type 0) for $n = 2k + 1$).

We see immediately from the above discussion that, for $n = 2k$, the weights of $\mathcal{K}_1$ are just 2^{2k}, attained once, and the others are just 2^{2k-1}. Whereas the weights of vectors in $\mathcal{K}_2 \setminus \mathcal{K}_1$ are $2^{k-1}(2^k + 1), 2^{k-1}(2^k - 1)$.

1.8.2 Shortenings of the Extended Kummer Code

We want now to list the equivalence classes of shortenings of the (projection of the) extended Kummer code. In a subsequent subsection we shall determine their incidence hierarchy, obtained via the procedure of shortening.

Before we do this, it is good to consider that, for $n = 4$, we have two 2-dimensional vector spaces W_1, W_2 and $\mathcal{K}''_{Kum}$ is generated by $\mathcal{K}$, the vector space of affine functions on $W_1 \times W_2$, and by a non-degenerate bilinear function $\beta : W_1 \times W_2 \to \mathbb{F}_2$.

In particular, the group $Aff(W_1) \times Aff(W_2)$ acts on $W = W_1 \times W_2$ sending $\mathcal{K}''_{Kum}$ to an equivalent code. If we want the code $\mathcal{K}''_{Kum}$ to be left invariant, the linear parts must be of the form (A^t, A^{-1}), and we shall speak of special orthogonal transformations: this implies for instance that if we exchange e_1, e_2 in W_1, the same must be done in W_2, while if we cyclically permute $e_1 \mapsto e_2 \mapsto e_3$ in W_1, in W_2 we must have the inverse permutation.

Moreover, we can also exchange the roles of W_1, W_2, and we can also use the symmetries

$$(x, y) \mapsto (x + y, y), (x, y) \mapsto (x, x + y),$$

$$(x_1, x_2, y_1, y_2) \mapsto (x_1 + y_1, x_2, y_1, y_2), (x, y) \mapsto (x_1, x_2, x_1 + y_1, y_2),$$

which leave the subspace $\mathcal{K}''$ invariant.

- **One point shortening**: by the use of translations we may assume that we are taking $W \setminus \{0\}$ as subset of W: then we get the space generated by β and the linear functions, of dimension 5, and with support the whole $W \setminus \{0\}$. This case will be labelled (00).
- **Two points shortening**: the first case occurs if they have one coordinate equal, via translations, we may assume (possibly exchanging the roles of W_1, W_2) that this coordinate is the second, and it is equal to 0.

Using then a translation on W_1 and a special orthogonal transformations, we achieve that the two points are $0 = (0, 0)$ and $(e_1, 0)$.

Otherwise, we may assume that the first point is $0 = (0, 0)$ and that the first coordinate of the second equals e_1, while its second coordinate is non zero. Applying the symmetry $(x, y) \mapsto (x, y) \mapsto (x, x + y)$, we exclude the case of the point (e_1, e_1), and reduce the case of (e_1, e_3) to (e_1, e_2).

In turn, we can reduce the second case where the two points are $0 = (0, 0)$ and (e_1, e_2) to the first one, via a transformation

$$\Phi : (x_1, x_2, y_1, y_2) \mapsto (x_1, x_2, y_1 + x_2, y_2 + x_1).$$

In fact,

$$\Phi(0) = 0, \Phi(e_1, e_2) = (e_1, 0).$$

This transformation leaves $\beta = x_1 y_1 + x_2 y_2$ invariant, because

$$x_1(y_1 + x_2) + x_2(y_2 + x_1) = x_1 y_1 + x_2 y_2.$$

The condition of vanishing on the points $0 = (0, 0)$ and $(e_1, 0)$ yields the 4-dimensional subspace:

$$\mathcal{K}' = \langle x_2, y_1, y_2, x_1 y_1 + x_2 y_2 \rangle.$$

This case will be labelled (i).

- **Three points shortenings**: first of all we may then assume that we delete both points $0 = (0, 0)$ and $(e_1, 0)$, and a third point, which, using the allowed symmetries, can be one of the following points:

  (1) $(e_2, 0)$ in the case where all the three points have one coordinate equal: this case yields the subspace generated by $y_1, y_2, x_1 y_1 + x_2 y_2$.
  (2) $(0, e_1)$ in the case where the coordinates of the three points are just two (in the set $\{0, e_1\}$): this case yields the subspace generated by $x_2, y_2, x_1 y_1 + x_2 y_2$.

  While in the case where there are three coordinates for the three points, we may assume that these are in the set $\{0, e_1, e_2\}$, and we have the further options:

  (3) $(0, e_2)$, yielding the subspace generated by $x_2, y_1, x_1 y_1 + x_2 y_2$,
  (4) (e_1, e_2), yielding the subspace generated by $x_2, y_1, x_1 y_1 + x_2 y_2$,
  (5) (e_2, e_1), yielding the subspace generated by $x_1, y_2, x_1 y_1 + x_2 y_2$.

Indeed the case where there are four coordinates leads to (e_2, e_3) which is equivalent to (e_2, e_1), and (e_3, e_2) which is equivalent to (e_1, e_2).

Now, cases (1), (3), (4), (5) are obviously equivalent, since in all these cases we take β and two elements of the standard basis, occurring in different monomials of β.

Cases (1) and (2) are distinct, since in case (1) the vector subspace U of W, of dimension 2, where the linear forms in $\mathcal{K}$ vanish, is contained in the zero set of $\mathcal{K}''$, hence the support of $\mathcal{K}''$ is the complement of the subspace U and the effective length of $\mathcal{K}''$ is $n = 12$; while in case (2) the support of $\mathcal{K}''$ is the complement of the set of three points, i.e. $U \setminus \{(e_1, e_1)\}$ and the effective length of $\mathcal{K}''$ is $n = 13$.

Hence we have two non equivalent shortenings for three points, cases (1) and (2).

- **Four points shortenings**: we can assume that we are deleting either three points of case (1) and a fourth point, or three points of case (2) and a fourth point.

  (I) We delete the subspace $U := \{y = 0\}$ of case (1): then we get the same code $\mathcal{K}''$, of dimension 3, appearing in (1).

  In all other cases we get a subspace $\mathcal{K}''$ of dimension 2.

  (II) We delete the subspace $U := \{x_2 = y_2 = 0\}$ of case (2): then $\mathcal{K}'' = \mathcal{K}$, since both subspaces have dimension 2.

  (III) We are left with the following restrictions: $dim(\mathcal{K}) = 1$, $dim(\mathcal{K}'') = 2$, which we have shown to give two combinatorial possibilities $a = 2, b = 4$, or $a = 4, b = 2$. The corresponding codes are not isomorphic, since for $a = 2$ the weight 10 occurs, but it does not occur for $a = 4$.

The two codes correspond to the subspaces respectively generated by

(III-1) $x_2 + y_2, x_1 y_1 + x_2 y_2$,
(III-2) $y_2, x_1 y_1 + x_2 y_2$.

In the first case (III-1) the support of $\mathcal{K}''$ is the complement of $\{x_2 + y_2 = x_1 y_1 + x_2 y_2 = 0\}$ and has cardinality 12, in the second case (III-2) the support of $\mathcal{K}''$ is the complement of $\{y_2 = x_1 y_1 = 0\}$ and has cardinality 10.

- **Shortenings by at Least Five Points**
  Clearly case (III-2) leads to five and six point shortenings which do not change $\mathcal{K}''$.

  We need therefore only to consider shortenings which reduce the dimension of $\mathcal{K}''$ to 1.

  There are some obvious cases,

(a) $\mathcal{K} = \mathcal{K}''$, of dimension 1, effective length 8, hence $\nu = 8, 9, 10, 11$;
(b) $\mathcal{K} = 0$, $dim(\mathcal{K}'') = 1$, generated by $x_1 y_1 + x_2 y_2$, effective length 6, hence $\nu = 6, 7, 8, 9, 10, 11$;
(c) $\mathcal{K} = 0$, $dim(\mathcal{K}'') = 1$, generated by $x_1 y_1 + x_2 y_2 + 1$, effective length 10, hence $\nu = 10, 11$.

Altogether we have 23 nontrivial cases, for 11 distinct nontrivial codes (again here $k' := dim(\mathcal{K}'), k := dim(\mathcal{K})$, while n' is the effective length of $\mathcal{K}''$), plus 11 trivial cases. Observe that the B-inequality $n \leq \nu \leq 10 + k' (\leq 11 + k)$ is always verified, since we have shortenings of the extended Kummer code, which satisfies the inequality.

1.8.3 List of the Shortenings of the Extended Kummer Code

	$\mathcal{K}''$	k'	k	ν	n
(0)	$\langle 1, x_1, x_2, y_1, y_2, x_1y_1 + x_2y_2\rangle$	6	5	16	16
(00)	$\langle x_1, x_2, y_1, y_2, x_1y_1 + x_2y_2\rangle$	5	4	15	15
(i)	$\langle x_2, y_1, y_2, x_1y_1 + x_2y_2\rangle$	4	3	14	14
(1)	$\langle y_1, y_2, x_1y_1 + x_2y_2\rangle$	3	2	12, 13	12
(2)	$\langle x_2, y_2, x_1y_1 + x_2y_2\rangle$	3	2	13	13
(II)	$\langle x_2, y_2\rangle$	2	2	12	12
$(III-1)$	$\langle x_2 + y_2, x_1y_1 + x_2y_2\rangle$	2	1	12	12
$(III-2)$	$\langle y_2, x_1y_1 + x_2y_2\rangle$	2	1	10, 11, 12	10
(a)	$\langle y_2\rangle$	1	1	$8, 9, 10, 11$	8
(b)	$\langle x_1y_1 + x_2y_2\rangle$	1	0	$6, 7, 8, 9, 10, 11$	6
(c)	$\langle x_1y_1 + x_2y_2 + 1\rangle$	1	0	10, 11	10
(d)	$\langle 0\rangle$	0	0	$0, 1, \ldots, 10$	0

Corollary 1.77 *The Nodal Severi variety $\mathcal{F}(4, \nu)$ has exactly one irreducible component for $\nu = 0, 1, 2, 3, 4, 5, 14, 15, 16$, exactly two irreducible components for $\nu = 6, 7, 13$, exactly three irreducible components for $\nu = 8, 9$, exactly four irreducible components for $\nu = 11, 12$, exactly five irreducible components for $\nu = 10$. Hence its stratification is made of 34 strata, where each stratum is in the boundary of another if and only if the corresponding extended code admits the other as a shortening.*

Remark 1.78 Following up the work of Rohn [79, 80], see also [57], Endraß determined all the nontrivial extended codes of nodal quartic surfaces in $\mathbb{P}^3$ with ν nodes in [45, Theorem 3.3] by applying elementary coding theory. He also showed that these codes were attained by some nodal quartic surfaces.

Our classification above confirms the classification by Endraß.

Theorems 6 and Corollary 1.77 are a sharpening of the results of Rohn and Endraß, because they show that $\mathcal{K}'$ determines an irreducible component of $\mathcal{F}(4, \nu)$; moreover, our statement is more conceptual since it shows that these codes are exactly all the shortenings of the extended Kummer code, and puts an incidence hierarchy among them.

Observe that the list also contemplates the 5 cases where $\mathcal{K} = \mathcal{K}'$.

(*) Removing those cases, there are exactly 18 cases where

$$0 \neq \mathcal{K} \neq \mathcal{K}'.$$

1.8.4 The Incidence Relation for the Shortenings of the Extended Kummer Code

We shall now determine when the codes in the above list are a shortening of each other: for this purpose it will be sufficient to see when one is obtained form the other by a one point shortening, which need not diminish the effective length n of $\mathcal{K}''$, but certainly lowers ν by 1.

In the course of producing the list we have shown that certain one point shortenings occur, now we have also to determine when they cannot occur.

For a one point shortening to occur, we need ν to drop to $\nu - 1$; moreover, this certainly happens in the trivial cases where $n < \nu$ and the code remains the same. Moreover, a one point shortening occurs when we take a subspace of codimension 1 in $\mathcal{K}''$.

Since $(0) \mapsto (00)$, $(00) \mapsto (i)$, $(i) \mapsto (1), (2)$ we are left with inspecting the cases where $\nu \leq 13$.

Let us start w ith $\nu = 13$.

Lemma 1.79 *The code (1) admits only a one point shortening to (1) (with* $\nu = 12$*) or to (III-2).*

Proof The first assertion is clear, since (1) has $\nu = 13$, $n = 12$, and this one point shortening corresponds to take the point of $U \setminus \mathrm{supp}(\mathcal{K}'')$.

Else, we take a point (ξ, η) outside of U, hence with $\eta \neq 0$.

Using a special orthogonal transformation we may assume that $\eta = e_1$.

We may further assume that $\xi = 0$ since translations in the ξ-plane leave the space U invariant.

Now, the shortening corresponding to $(0, 0, 1, 0)$ is exactly (III-2). □

Lemma 1.80 *The code (2) does not admit a one point shortening to (1), but it does for (III-2), (II) (III-1).*

Proof The first assertion is clear, since both codes have dimension 3, and $n = \nu = 13$ for (2), which implies that any shortening has strictly smaller dimension.

The other three assertions follow since, by inspection of the list, we find that in all three cases we are taking a codimension 1 subspace of $\mathcal{K}''$.

More precisely, for (II) we take the point of $U \setminus \mathrm{supp}(\mathcal{K}'')$, for $(0, 1, 0, 0)$ we get (III-2), for $(1, 1, 1, 1)$ we get (III-1). □

Let us pass now to $\nu = 12$.

We make the obvious remark that (1) can only be 1-shortened to (III-2), since it has $k' = 3$, and (a), (b), (c) have $k' = 1$ (and inspection of the respective bases shows that this 1-shortening).

Whereas (II) has $n = \nu$, hence it can only be 1-shortened to $k = k' = 1$, that is, only to case (a), and one sees immediately that this occurs.

(III-2) has $\delta := \nu - n = 2$, hence it can only be 1-shortened lowering k' by 1 and keeping δ not smaller, hence it cannot be 1-shortened to (c); we see that it is 1-shortened to (a), (b).

(III-1) cannot be 1-shortened to (III-2) because k' cannot increase. We see now that it is 1-shortened to (a), (b), (c), taking the respective points $(1, 0, 1, 0)$, $(0, 1, 0, 0)$, $(1, 1, 1, 0)$: for the last we are left with the quadratic form

$$x_1 y_1 + x_2 y_2 + x_2 + y_2 = x_1 y_1 + (x_2 + 1)(y_2 + 1) + 1.$$

After this, there are only trivial 1-shortenings (that is, preserving the code $\mathcal{K}''$), or shortenings to the zero code.

1.9 The K8 Code, Kummer Code of Degree 8, its Shortenings and Codes of K3 Surfaces of Degree Divisible by 8

We shall first describe here the possible codes of K3 surfaces of degree divisible by 8.

In this situation we have a pair of codes $\mathcal{K} \subset \mathcal{K}'' \subset \mathbb{F}_2^{\nu}$, with $\nu \leq 16$, such that

(i) the weights of vectors $v \in \mathcal{K}$ are divisible by 8
(ii) the weights of vectors $u \in \mathcal{K}'' \setminus \mathcal{K}$ belong to the set $\{4, 8, 12\}$
(iii) setting $k' := dim(\mathcal{K}''),\ k := dim(\mathcal{K})$, we have $k' \leq k + 1$, and
(iv) $\nu \leq 10 + k'$.

Let us recall (Propositions 1.31 and 1.35) that property (i) implies that $k \leq 5 \Rightarrow k' \leq 6$, and $\mathcal{K}$ is a shortening of the Kummer code $\mathcal{K}_{Kum}$.

We have seen (Corollary 1.36) that the code of a Kummer surface with a polarization of degree divisible by 8 leads to a well determined code, that we shall now call the K8 code $\mathcal{K}_8''$.

Definition 1.81 Let V be an affine space of dimension 4 over $\mathbb{F}_2$, and let π be a 2-dimensional affine subspace of V.

Then we define the **K8 code** as the code $\mathcal{K}_8''$ generated by the Kummer code $\mathcal{K}_{Kum}$, consisting of the affine functions $\phi : V \to \mathbb{F}_2$, and by the characteristic function f_π of the plane π.

Hence a basis of $\mathcal{K}_8''$ is $1, x_1, x_2, x_3, x_4, f_\pi$.

Remark 1.82 Consider a function $u \in \mathcal{K}_8'' \setminus \mathcal{K}$, where $\mathcal{K}_8''$ is the K8-code.

Then, setting $u := f_\pi + \phi$, the weight of u is 4 for $\phi = 0$, 12 for $\phi = 1$, else it is equal to

$$w(u) := 4 + 8 - 2 \cdot |\pi \cap \{x | \phi(x) = 1\}| = 4 + 2(|\{x \in \pi | \phi(x) = 0\}|).$$

Hence, defining $W := \{x | \phi(x) = 0\}$, we have:

- $w(u) = 12$ if $\pi \subset W$,
- $w(u) = 4$ if $\pi \cap W = \emptyset$, else
- $w(u) = 8$ if $\emptyset \neq (\pi \cap W) \neq \pi$.

according to the cardinalities 4, 0, 2 for $\pi \cap W$.
Hence all the shortenings of the K8 code satisfy properties (i)–(iv).

We show now the converse, that the codes satisfying properties (i)–(iv) are precisely the shortenings of the K8 code.

To simplify our discussion, let us split the result into three parts.

Theorem 1.83 *Let $\mathcal{K} \subset \mathcal{K}''$ be a pair of codes satisfying properties (i)–(iv) above. Then*

(I) for $k' = 6$, we have that $\mathcal{K}''$ is the K8 code,
(II) for $k = k'$ we have $k' \leq 2$ and we have a shortening of the K8 code,
(III) also for $k' = k + 1 \leq 5$ we have a shortening of the K8 code.

Proof

(I) we claim that there must be $u \in \mathcal{K}'' \setminus \mathcal{K}$ with weight 4.
In fact, $k = 5$, $k' = 6$, hence there $\exists u \in \mathcal{K}'' \setminus \mathcal{K}$. If its weight is 12, replace u by $u + 1$. If instead the weight of all vectors in $\mathcal{K}''$ is divisible by 8, then by Proposition 1.31 $\mathcal{K}'' = \mathcal{K}$, contradicting $k' = 6$.
Such a function u is the characteristic function of a set $\mathcal{A}$ of cardinality 4 such that each affine subspace intersects $\mathcal{A}$ in 0, 2, 4 points. If the four points of $\mathcal{A}$ were affinely independent, then there would be an affine function vanishing on three of the points, and not on the fourth, a contradiction. Hence the points of $\mathcal{A}$ are affinely dependent, hence $\mathcal{A}$ is a 2-dimensional affine subspace π.

(II) We have $k \leq 5$ and $k' = k$, hence $k' \leq 5$ and, by iv), we have $\nu \leq 15$. We take then the set $\mathcal{S}$ of cardinality ν as a subset of V minus one point, which we take as the origin 0 of a vector space structure. Then $\mathcal{K}''$ is a subspace of the space of linear forms, hence $k' \leq 4$, and then $\nu \leq 10+k' \leq 14$. Hence the complement of $\mathcal{S}$ contains at least two points, say $0, v$; but then since the functions in $\mathcal{K}$ vanish in $0, v$ we have $k \leq 3$, hence $k' \leq 3$, and then $\nu \leq 10+k' \leq 13$. Hence the complement of $\mathcal{S}$ contains three points, hence $k \leq 2$, so that $k' \leq 2$, and then $\nu \leq 10 + k' \leq 12$.

For $k' = 2$, the complement of $\mathcal{S}$ contains 4 points, so that these points form a 2-dimensional affine subspace U. Choose an affine plane π such that π and U intersect precisely at one point, say 0; hence U is the vector subspace spanned by e_1, e_2, π is the vector subspace spanned by e_3, e_4. $\mathcal{K}$ is generated by the linear forms x_3, x_4, and, if we define $\mathcal{K}''$ to be the shortening of the K8 code associated to $V \setminus U$, then the elements of $\mathcal{K}'' \setminus \mathcal{K}$ are functions of the form

$$u = f_\pi + \phi = f_\pi + a + \sum_i b_i x_i,$$

where the condition of vanishing on U implies $a = 1$ (vanishing on 0), and moreover we must have $1 + b_1 = 1 + b_2 = 1 + b_1 + b_2 = 0$, a contradiction. Hence $\mathcal{K}'' = \mathcal{K}$, and we have the shortening of the K8 code associated to the complement of the set U.

If we take the further shortening associated to the complement of the set $U \cup \{e_3\}$ we get $k = k' = 1$, and $\mathcal{K}''$ is generated by the linear form x_4.

The respective codes belonging to this class shall be respectively denoted $II_8 - 2$ **and** $II_8 - 1$.

(III) Consider now the case $\mathcal{K} \neq \mathcal{K}''$, with $k' \leq 5$, hence $\nu \leq 10 + k' \leq 15$.

Hence $\mathcal{K}''$ is generated by $\mathcal{K}$ and by a function f, the characteristic function of a set $\mathcal{A}$.

We have $k \in \{0, \dots, 4\}$, and $\mathcal{K}$ is isomorphic to the space of linear forms vanishing on a linear subspace U of dimension equal to $4 - k$. In particular the support of $\mathcal{K}$ consists of $V \setminus U$, of cardinality $16 - 2^{4-k}$.

Subcase III-i We assume here that $\mathcal{K}''$ does not admit the weights 4, 12, that is, $a_4 = a_{12} = 0$.

Then $\mathcal{K}''$ is a shortening of the strict Kummer code, and if $k' = 5$ we get the whole strict Kummer code, hence $\nu = 16$, a contradiction.

Hence $k' \leq 4$, and $\mathcal{K}''$ is a space of linear forms vanishing on a subspace U' of dimension $4 - k'$. We have on one hand $\nu \leq 10 + k'$, on the other $\nu \geq 16 - 2^{4-k'}$: hence

$$6 \leq k' + 2^{4-k'} \Rightarrow k' = 1, 2.$$

We get exactly two codes;

for $k' = 1$, $\mathcal{K}''$ is generated by a function $f_\pi + x_1$, with x_1 attaining both values 0, 1 on π,

for $k' = 2$, $\mathcal{K}''$ is generated by a linear function x_4 and by $f_\pi + x_1$ as above, and such that also $x_1 + x_4$ attains both values 0, 1 on π.

The respective codes shall be denoted c_8) **and** $III_8 - 0$).

Subcase (III-ii-0) We assume here that $a_4 \neq 0$, $a_{12} = 0$ and the weight of elements of $\mathcal{K}'' \setminus \mathcal{K}$ is always 4.

If $k > 0$, take a nonzero vector $v \in \mathcal{K}$, and $u \in \mathcal{K}'' \setminus \mathcal{K}$: then the support of u is contained in the one of v; hence $k \leq 1$, and we have exactly two codes, for $k = 0$ we get the span of f_π, for $k = 1$ the span of f_π, x_4 where $\pi \cap U \neq \emptyset$.

The respective codes shall be denoted b_8) **and** $III_8 - 2$).

Observe now that we cannot have $a_{12} \geq 2$; because, if u, u' have support of cardinality 12, then the intersection of the supports has cardinality 8, hence $\nu \geq 16$, a contradiction.

Subcase (III-iii) we assume here that $a_{12} \neq 0 \Leftrightarrow a_{12} = 1$.

Since $\nu \geq 12$, we have $k' \geq 2$, hence we may choose vectors v_1, v_2 with respective weights $w(v_1) = 8$, $w(v_2) = 12$. Letting $a = |\operatorname{supp}(v_1) \cap \operatorname{supp}(v_2)|$, looking at the weight of $v_1 + v_2$,

$$w(v_1 + v_2) = (8 - a) + (12 - a) = 2(10 - a) \in \{4, 8\} \Leftrightarrow a \in \{8, 6\},$$

we see that it equals 4 if and only if the support of v_1 is contained in the one of v_2, and that it equals 8 if $a = 6$.

The case $a_{12} = 1, a_4 = 0$ is not possible.

In fact, since $a = 6, n \geq 14$, hence $k' \geq 4, k \geq 3$.

We have an injective projection of $\mathcal{K}$ into $\mathbb{F}_2^{\mathcal{S} \setminus \mathcal{A}} = \mathbb{F}_2^{\nu - 12}$, hence $\nu - 12 \geq k \Rightarrow \nu \geq 15 \Rightarrow k + 1 \geq 5 \Rightarrow \nu \geq 16$ and we have reached a contradiction.

Subcase (III-iii-4) we assume here that $a_{12} = 1, a_4 \neq 0$.

Since $a_4 \neq 0$, then we may assume $w(v_1 + v_2) = 4$, hence the support of $v_1 + v_2$ and those of all other weight 4 vectors are contained in $\mathcal{A}$, the support of v_2.

Hence the subcode $\mathcal{K}'''$ generated by v_2 and by the weight 4 vectors is supported in $\mathcal{A}$, hence it has dimension at most 3 ($\mathcal{K}''' \cap \mathcal{K}$ has dimension at most 2).

If $\mathcal{K}''' = \mathcal{K}''$ we have exactly two codes, according to $k = 1, 2$: in fact, for $k = 2$ we have a partition of the set $\mathcal{A}$ into three disjoint sets of 4 elements, each one being the support of a weight 4 vector.

These correspond to the cases (4-3-i) and $III_8 - 1$) in the list (which contains the description showing that they are shortenings of the K8 code).

There remains to be treated the case where $a_{12} = 1, a_4 \neq 0$, and there is a vector in $\mathcal{K}'' \setminus \mathcal{K}$ with weight 8, in particular $k' \geq 3$.

For $k = 2, 3, 4$, we have $|\mathcal{S}| =: \nu \leq 11 + k$.

Take our weight 12 vector, its support is a subset $\mathcal{A}$ with 12 elements of $\mathcal{S} \supset V \setminus U$.

For $k = 4$, then $U = \{0\}$ and $\mathcal{S} = V^* := V \setminus \{0\}$. The subset $\mathcal{A}$ has the property that, for all linear forms x_W with zero set a hyperplane W, $f_{\mathcal{A}} + x_W$ has weight $\in \{4, 8\}$.

Setting $\mathcal{B} := \mathcal{S} \setminus \mathcal{A}$, $\mathcal{B}$ consists of three non zero vectors, and the weight of $f_{\mathcal{A}} + x_W$ equals $10 - 2b$, if we set $b := |\mathcal{B} \cap W|$. Hence it is only possible $b = 1, 3$, but not $b = 2$.

Hence every linear form vanishing on two points of $\mathcal{B}$ vanishes on the third, hence $\mathcal{B}$ equals $\pi \setminus \{0\}$, where π is a plane containing the origin 0.

Similarly, for $k = 3$, then U has 2 elements, and $\mathcal{S}$ at most 14, hence $\mathcal{S} = V \setminus U$.

Again the complement $\mathcal{B}$ of $\mathcal{A}$ has 2 elements, and, for all linear forms vanishing on U, $f_{\mathcal{A}} + x_W$ has weight $8 - 2b$, hence we can only have $b = 0, b = 2$. Again $\mathcal{B}$ equals $\pi \setminus U$, where π is a plane containing U.

For $k = 2$, the two sets $\mathcal{S} \supset (V \setminus U)$ have respectively $\leq 13, 12$ elements, hence $\mathcal{B}$ has cardinality $h \leq 1$, and, for all linear forms vanishing on U, $f_{\mathcal{A}} + x_W$ has weight $h + 4 - 2b \geq 4$, so the only possibility is $h = 0$, which means that $\mathcal{S} = V \setminus U$ and $\nu = 12$.

For the three cases above, we have that $\mathcal{K}''$ is spanned by $1 + f_\pi$ and by the linear forms vanishing on U, a linear subspace of dimension $4 - k$, such that $U \subset \pi$. In all these cases, we have that $\mathcal{K}''$ is the shortening associated to the complement of the set $\{0, e_1, \dots, e_{4-k}\}$ formed by the origin and a basis of U.

These cases are respectively labelled (12-1), (12-2), (12-3) in our list.

Subcase (III-ii-1) we assume here that $a_4 \neq 0, a_{12} = 0$ and the weights of elements of $\mathcal{K}'' \setminus \mathcal{K}$ are not always 4, in particular $k' \geq 2$.

Observe that if the weight of an element $v' \in \mathcal{K}'' \setminus \mathcal{K}$ equals 8, then for each nonzero vector $v \in \mathcal{K}$ the weight of $v' + v$ equals 4, respectively 8 if the supports of v, v' have intersections of respective cardinalities 6, 4.

In other words, we consider the case where $\mathcal{A}$ has 4 elements but there is no vector of weight 12, and we let W be the hyperplanes which are zero sets of nonzero vectors in $\mathcal{K}$: these are the hyperplanes containing the zero set U of $\mathcal{K}$.

$\mathcal{A}$ is a set of cardinality 4, $\mathcal{A} \subset \mathcal{S}$, and, for all linear forms x_W vanishing on U, $f_{\mathcal{A}} + x_W$ has weight (setting $a := |\mathcal{A} \cap W|$) $a + (8 - (4 - a)) = 4 + 2a$, hence a can only equal 0, 2.

$k = 4$: then there is certainly a linear form vanishing on at least 3 of the points of $\mathcal{A}$, a contradiction.

$k = 3$: then projection with centre U maps the points of $\mathcal{A}$ to a projective basis $\mathcal{A}'$ in the plane $\mathbb{P}^2_{\mathbb{F}_2}$ with coordinates x_2, x_3, x_4. We observe that, even if for each point $P' \in \mathcal{A}'$ there are two different choices for a point in $U * P' \setminus U$, these choices lead to isomorphic pairs of codes.

We can therefore assume that $\mathcal{A} = \{e_2, e_3, e_4, e_2 + e_3 + e_4\}$, which is an affine plane π. Hence we have the code spanned by f_π and the three linear forms x_2, x_3, x_4, which is the shortening relative to the complement of the set $\{0, e_1\}$.

For $k = 2$, there are two cases, according to $\mathcal{S} = (V \setminus U)$ or $\mathcal{S} \neq (V \setminus U)$, where we may assume wlog that $\mathcal{S} = (V \setminus U) \cup \{0\}$.

The hyperplanes $W \supset U$ are a pencil, so they are three ($\{x_3 = 0\}$, $\{x_4 = 0\}$, $\{x_3 + x_4 = 0\}$) and if $\mathcal{S} = (V \setminus U)$ each of them contains none or two of the points of $\mathcal{A}$: hence in this case we may assume that $\mathcal{A} \subset \{x_3 = 0\} \cup \{x_3 + x_4 = 0\}$. Take π equal to the plane $\pi := \{0, e_1, e_3, e_1 + e_3\}$ and consider the function $f_\pi + 1 + x_2$, vanishing on U and on π: its support is then the set $\mathcal{A} := \{x_2 = 0\} \setminus \pi$ which has cardinality 4 since $x_2 = 0$ on π. We get the shortening associated to the complement of the set $\{0, e_1, e_2\}$.

While, if $\mathcal{S} = (V \setminus U) \cup \{0\}$ there is also the possibility that $0 \in \mathcal{A}$ and each hyperplane in the pencil contains exactly one point of $\mathcal{A}$. This case is realized by setting $\mathcal{A}$ equal to the plane $\pi := \{0, e_3, e_4, e_3 + e_4\}$, and we have the shortening associated to the complement of $U^* = \{e_1, e_2, e_1 + e_2\}$.

For $k = 1$, since we shall have $a_4 = 1, a_8 = 2$, it follows that $a = 2, n = 10$ (but $\nu \leq 12$).

These cases are just respectively labelled (4-2), (4-3-ii), $II I_8 - 3$ **in the list.**

□

1.9.1 *The List of the Shortenings of the K8 code*

We shall now give the list of all the shortenings of the K8 code, following the proof of the previous theorem. Altogether we have 15 codes with $\nu = n > 0$, and if we also allow $\nu \geq n$, we have 36 nontrivial codes and 47 strata.

To ease the notation, we define as usual U to be the zero set of the functions in $\mathcal{K}$.

Recall moreover that π is an affine plane in $\mathbb{F}_2^4$, while a_m denotes the number of codewords of weight m (and observe that this number can only decrease if we take a shortening).

	$\mathcal{K}''$	k' k	ν	n		
$(K8)$	$\langle 1, x_1, x_2, x_3, x_4, f_\pi \rangle$	6 5	16	16		
$(12-1)$	$\langle 1+f_\pi, x_1, x_2, x_3, x_4 \rangle$	5 4	15	15	$0 \in \pi$	$a_{12} = 1$
$(12-2)$	$\langle 1+f_\pi, x_2, x_3, x_4 \rangle$	4 3	14	14	$\pi \supset U$	$a_{12} = 1$
$(12-3)$	$\langle 1+f_\pi, x_3, x_4 \rangle$	3 2	12, 13	12	$\pi = U$	$a_{12} = 1$
$(4-2)$	$\langle f_\pi, x_2, x_3, x_4 \rangle$	4 3	14	14	$\pi \cap U = \emptyset$	$a_{12} = 0$
$(4-3-i)$	$\langle f_\pi + 1 + x_2, x_3, x_4 \rangle$	3 2	12, 13	12	$\pi \cap U = \{0, e_1\}$	$a_{12} = 0$
$(4-3-ii)$	$\langle f_\pi + 1 + x_2, x_3, x_4 \rangle$	3 2	13	13	$\pi \cap U = \{0\}$	$a_{12} = 0$
(II_8-2)	$\langle x_3, x_4 \rangle$	2 2	12	12	$\pi \cap \langle e_1, e_2 \rangle = \{0\}$	
(II_8-1)	$\langle x_4 \rangle$	1 1	8, 9, 10, 11	8	$\pi \cap \langle e_1, e_2 \rangle = \{0\}$	
(III_8-0)	$\langle f_\pi + x_1, x_4 \rangle$	2 1	12	12	$\pi = \{x_2 = x_3 = 0\}$	$a_4 = a_{12} = 0$
(III_8-1)	$\langle 1+f_\pi, x_4 \rangle$	2 1	12	12	$\pi \subset U$	$a_{12} = 1$
(III_8-2)	$\langle f_\pi, x_4 \rangle$	2 1	$8, \ldots, 12$	8	$\pi \cap U = \emptyset$	$a_4 = 2, a_{12} = 0$
(III_8-3)	$\langle f_\pi, x_4 \rangle$	2 1	10, 11, 12	10	$\|\pi \cap U\| = 2$	$a_4 = 1, a_{12} = 0$
(b_8)	$\langle f_\pi \rangle$	1 0	$4, \ldots, 11$	4		
(c_8)	$\langle f_\pi + 1 + x_4 \rangle$	1 0	8, 9, 10, 11	8		
(d_8)	$\langle 0 \rangle$	0 0	$0, 1, \ldots, 10$	0		

Appendix: Binary Codes of Quartic Type Without the B-Inequality

Sascha Kurz and Michael Kiermaier

This appendix uses the computer to classify all extended codes with the same weights as for quartic surfaces, but without assuming the B-inequalities.

This appendix complements the results of Sect. 1.8, and contains in particular a computer verification of the partial assertion (*) made in Remark 1.78, with exception of the cases where the weight 16 occurs.

It classifies all extended codes with the same weights as for quartic surfaces, but without assuming the B-inequality.

Codes with exactly one nonzero weight have been classified in [12]. For two weights w, $2w$ a classification can be found in [58].

Lemma 1.84 *Let* $\mathcal{K} = \mathcal{K}_k$ *be the* k*-dimensional binary code whose nonzero codewords have weight* 8*, i.e.,* $\mathcal{K}_k = 2^{4-k} S_k$ *for* $0 \leq k \leq 4$*, where* $\mathcal{K}_0$ *is the empty* *code and S_k is the simplex code. For an extended code $\mathcal{K}''$ with $k' := \dim(\mathcal{K}'') = \dim(\mathcal{K}) + 1$, $\mathcal{K} \subset \mathcal{K}''$, $\nu(\mathcal{K}'') \leq 16$, and the codewords in $\mathcal{K}'' \setminus \mathcal{K}$ have weights in $\{6, 10\}$ we have exactly the following possibilities (here $\left[\underline{n}, k'\right]_2$ denotes a binary code of effective length n and of dimension k'):*

- *(b) : for $k = 0$ a $\left[\underline{6}, 1\right]_2$ code with weight enumerator $W(z) = 1z^0 + 1z^6$:*

$$\begin{pmatrix} 111111 \end{pmatrix}$$

- *(c) : for $k = 0$ a $\left[\underline{10}, 1\right]_2$ code with weight enumerator $W(z) = 1z^0 + 1z^{10}$:*

$$\begin{pmatrix} 1111111111 \end{pmatrix}$$

- *(III-2) : for $k = 1$ a $\left[\underline{10}, 2\right]_2$ code with weight enumerator $W(z) = 1z^0 + 2z^6 + 1z^8$:*

$$\begin{pmatrix} 1111001111 \\ 0000111111 \end{pmatrix}$$

- *(III-1) : for $k = 1$ a $\left[\underline{12}, 2\right]_2$ code with weight enumerator $W(z) = 1z^0 + 1z^6 + 1z^8 + 1z^{10}$:*

$$\begin{pmatrix} 111111000011 \\ 000000111111 \end{pmatrix}$$

- *for $k = 1$ a $\left[\underline{14}, 2\right]_2$ code with weight enumerator $W(z) = 1z^0 + 1z^8 + 2z^{10}$:*

$$\begin{pmatrix} 11110011110000 \\ 00001111111111 \end{pmatrix}$$

- *(1) : for $k = 2$ a $\left[\underline{12}, 3\right]_2$ code with weight enumerator $W(z) = 1z^0 + 4z^6 + 3z^8$:*

$$\begin{pmatrix} 110001011010 \\ 001101010101 \\ 000011111111 \end{pmatrix}$$

- *(2) : for $k = 2$ a $\left[\underline{13}, 3\right]_2$ code with weight enumerator $W(z) = 1z^0 + 3z^6 + 3z^8 + 1z^{10}$:*

$$\begin{pmatrix} 1100111001011 \\ 0011110100111 \\ 1111110011101 \end{pmatrix}$$

- *for* $k = 2$ *a* $\left[\underline{14}, 3\right]_2$ *code with weight enumerator* $W(z) = 1z^0{+}2z^6{+}3z^8{+}2z^{10}$*:*

$$\begin{pmatrix} 11100011111000 \\ 00011111110100 \\ 00001100110011 \end{pmatrix}$$

- *for* $k = 2$ *a* $\left[\underline{15}, 3\right]_2$ *code with weight enumerator* $W(z) = 1z^0{+}1z^6{+}3z^8{+}3z^{10}$*:*

$$\begin{pmatrix} 111000111110000 \\ 000111111101000 \\ 001111011100111 \end{pmatrix}$$

- *for* $k = 2$ *a* $\left[\underline{16}, 3\right]_2$ *code with weight enumerator* $W(z) = 1z^0 + 3z^8 + 4z^{10}$*:*

$$\begin{pmatrix} 1110001111100000 \\ 0001111111010000 \\ 0110110011001111 \end{pmatrix}$$

- *(i) : for* $k = 3$ *a* $\left[\underline{14}, 4\right]_2$ *code with weight enumerator* $W(z) = 1z^0 + 6z^6 + 7z^8 + 2z^{10}$*:*

$$\begin{pmatrix} 10001001000111 \\ 01000101110100 \\ 00111100001111 \\ 00000011111111 \end{pmatrix}$$

- *for* $k = 3$ *a* $\left[\underline{15}, 4\right]_2$ *code with weight enumerator* $W(z) = 1z^0{+}4z^6{+}7z^8{+}4z^{10}$*:*

$$\begin{pmatrix} 100110011111000 \\ 010001111110100 \\ 001111100110010 \\ 110001101000001 \end{pmatrix}$$

- *for* $k = 3$ *a* $\left[\underline{16}, 4\right]_2$ *code with weight enumerator* $W(z) = 1z^0{+}2z^6{+}7z^8{+}6z^{10}$*:*

$$\begin{pmatrix} 1001100111110000 \\ 0100011111101000 \\ 0011111001100100 \\ 1111111000100011 \end{pmatrix}$$

- *(00) : for* $k = 4$ *a* $[\underline{15}, 5]_2$ *code with weight enumerator* $W(z) = 1z^0 + 6z^{10} + 15z^8 + 10z^6$*:*

$$\begin{pmatrix} 100001000101011 \\ 010010001001101 \\ 001011101111110 \\ 000111100001111 \\ 000000011111111 \end{pmatrix}$$

- *for* $k = 4$ *a* $[\underline{16}, 5]_2$ *code with weight enumerator* $W(z) = 1z^0 + 6z^6 + 15z^8 + 10z^{10}$*:*

$$\begin{pmatrix} 1001100111110000 \\ 0100011111101000 \\ 0011111001100100 \\ 1110101010100010 \\ 0011100010100001 \end{pmatrix}$$

Proof We have used the software package `LinCode` [64]. □

Remark 1.85 If we only consider those $[\underline{\nu}, k']_2$ codes from Lemma 1.84 that satisfy the *B-inequality* $\nu \le k' + 10$, then only 8 codes remain.

If we add the Kummer code $\mathcal{K}''_{Kum}$, see Proposition 1.31, with its unique extension, see Proposition 1.33, then we end up with 9 codes.

Taking zero columns into account, i.e., adding $x \ge 0$ zero columns to every $[\underline{n}, k']_2$ code (of effective length n), such that $n+x \le k'+10$, gives 18 possibilities.

Chapter 2
Cubic Hypersurfaces, Associated Discriminants and Low Degree Nodal Surfaces

Fabrizio Catanese, Yonghwa Cho, Stephen Coughlan, and Davide Frapporti

Abstract The projection with centre an appropriate linear subspace contained in a cubic hypersurface has an associated discriminant hypersurface, which often has special properties such as many singularities. We study the geometry of linear subspaces contained in singular cubics and the Fano schemes parametrising them. This is done in the general setting and also for special cubics with many nodes, such as the Segre and Goryunov–Kalker cubics. From this we infer properties of the associated discriminants, especially in the two-dimensional setting, where we get surfaces with many nodes. For example, we construct a Togliatti quintic surface as a projection from a line on a 4-dimensional GK-cubic. We also show that the Togliatti quintic and the Barth sextic are unobstructed, and we apply representation theory to study the deformations of Segre and GK-cubic hypersurfaces in all dimensions, discussing some open questions.

2.1 Linear Subspaces L Contained in Cubic Hypersurfaces

In this section we consider a cubic hypersurface X in $\mathbb{P}^n$ containing a linear subspace L of dimension k.

F. Catanese (✉)
Mathematisches Institut der Universität Bayreuth, Bayreuth, Germany
e-mail: Fabrizio.Catanese@uni-bayreuth.de

Y. Cho
Department of Mathematics, Gyeongsang National University, Jinju-si, South Korea
e-mail: yhcho@gnu.ac.kr

S. Coughlan
Department of Mathematics and Computer Studies, Mary Immaculate College, Limerick, Ireland
e-mail: stephen.coughlan@mic.ul.ie

D. Frapporti
Politecnico di Milano, Dipartimento di Matematica, Milano, Italy
e-mail: davide.frapporti@polimi.it

F. Catanese (ed.), *Varieties of Nodal Surfaces, Coding Theory and Discriminants of Cubic Hypersurfaces*, Lecture Notes of the Unione Matematica Italiana 31,
https://doi.org/10.1007/978-3-032-05899-7_2

Lemma 2.1 *Let $X \subset \mathbb{P}^n$ be a cubic hypersurface, and let $L \subset X$ be a linear subspace of dimension k. If L is contained in the smooth locus of X, then we have $2k \leq n - 1$, while $2k \leq n$ if X has isolated singularities along L.*

In the case of equality $n = 2k$ and if X has isolated singularities along L, then the schematic intersection $L \cap \mathrm{Sing}(X)$ consists of a 0-dimensional subscheme of length 2^k.

Proof Set $b + 1 := n - k$, and choose coordinates in $\mathbb{P}^n$

$$(z_0, z_1, \dots, z_k, x_0, x_1, \dots, x_b)$$

such that $L = \{x_0 = x_1 = \cdots = x_b = 0\}$. Let F be an equation of X: since $L \subset X$, we may write

$$F = \sum_{i,j=0}^{k} a_{i,j}(x) z_i z_j + \sum_{i=0}^{k} q_i(x) z_i + G(x),$$

where the $a_{i,j}(x)$, $q_i(x)$, $G(x)$ are homogenous polynomials of respective degrees $1, 2, 3$.

We can now describe the schematic intersection:

$$\begin{aligned} L \cap \mathrm{Sing}(X) &= \{(z, 0) \mid \frac{\partial}{\partial x_h}(\sum_{i,j} a_{i,j}(x) z_i z_j) = 0, \ \forall h = 0, \dots, b\} \\ &= \{(z, 0) \mid \sum_{i,j} a_{i,j;h} z_i z_j = 0, \ \forall h = 0, \dots, b\}, \end{aligned}$$

where $a_{i,j}(x) = \sum_h a_{i,j;h} x_h$. In other words, if we write

$$F = \sum_h x_h Q_h(z) + \sum_i q_i(x) z_i + G(x),$$

$L \cap \mathrm{Sing}(X)$ is defined in L by the $b+1$ quadratic equations $Q_h(z) = 0$, in particular it is not empty if $k \geq b+1 \Leftrightarrow 2k \geq n$, and has dimension at least $k - b - 1 = 2k - n$.

In case of equality, and if $L \cap \mathrm{Sing}(X)$ has dimension zero, it consists of the complete intersection of the k quadrics $Q_h(z) = 0$, $h = 0, \dots, b = k - 1$, hence the scheme has length 2^k by Bezout's theorem. □

Remark 2.2 The last assertion shall be particularly useful in the special case $k = 3$, $n = 6$: it says that if X is nodal, a deformation of X preserving L must also preserve the 8 nodes of X along L.

Corollary 2.3 *For each pair (k, n) such that $2k + 1 \leq n$ there exists a smooth cubic hypersurface $X \subset \mathbb{P}^n$ which contains a linear subspace L of dimension k.*

Proof In the above notation, take $X = \{F = 0\}$, where $F = \sum_h x_h Q_h(z) + \sum_i q_i(x)z_i + G(x)$, and the $n-k$ quadrics $Q_h(z)$ have no common zeroes in $L \cong \mathbb{P}^k$.

By Bertini's theorem X is smooth outside the base locus L of this linear system, while X has no singular points in L by the previous Lemma 2.1. □

Consider now the Grassmann variety

$$\mathcal{G} := Gr(k,n) := \{L \subset \mathbb{P}^n \mid L \cong \mathbb{P}^k\},$$

the space $\mathcal{H} \cong \mathbb{P}^N$ of cubic hypersurfaces in $\mathbb{P}^n$, and the incidence correspondence of subspaces contained in a cubic hypersurface

$$\mathcal{I} \subset \mathcal{G} \times \mathcal{H},\ \mathcal{I} := \{(L, X) \mid L \subset X\}.$$

Here $N + 1 = \binom{n+3}{3}$ and the projection $p : \mathcal{I} \to \mathcal{G}$ is a projective bundle with fibre isomorphic to $\mathbb{P}^{N-c}$, where $c := \binom{k+3}{3}$. In particular, the projection on the space $\mathcal{H}$ cannot be surjective unless $g := \dim(\mathcal{G}) = (k+1)(n-k) \geq c$. This inequality amounts to $6(n-k) \geq (k+3)(k+2) \Leftrightarrow 6n \geq k^2 + 11k + 6$, yielding $n \geq 3$ for $k = 1, n \geq 6$ for $k = 2, n \geq 8$ for $k = 3, n \geq 11$ for $k = 4$.

Even if our main interest here is to concentrate on the special cases $k = 1, n = 5$ and $k = 2, n = 6$, we proceed with some results which hold more generally, and also recalling the cases with smaller dimensions, for instance $k = 1, n \leq 4$.

Definition 2.4 The Fano scheme $F_k(X)$ of X is the fibre of $\mathcal{I}$ over X, namely $F_k(X) := \{L \in \mathcal{G} \mid L \subset X\}$.

On the Grassmann variety we have the exact sequence

$$0 \to \mathcal{U} \to V \otimes \mathcal{O}_{\mathcal{G}} \to \mathcal{Q} \to 0,$$

where $\mathcal{U}$ is the universal subbundle (it has rank $= k + 1$), and $\mathcal{Q}$ is the universal quotient bundle (it has rank $n - k$), and its dual sequence.

Taking $X = \{F = 0\}$, $F \in \mathrm{Sym}^3(V^\vee)$ induces a section $s_F \in H^0(\mathcal{G}, \mathrm{Sym}^3(\mathcal{U}^\vee))$ such that $F_k(X) = \{L \in \mathcal{G} \mid s_F(L) = 0\}$.

In the special case $k = 1$, s_F is the section of a rank 4 bundle, while for $k = 2$, s_F is the section of a rank 10 bundle, hence the expected codimension of $F_k(X)$ is equal to 4 for $k = 1$, 10 for $k = 2$.

In general the local dimension of $F_k(X)$ at each point is always at least $\dim(\mathcal{G}) - \binom{k+3}{3}$.

To see whether equality is attained, we study the local structure of $F_k(X)$, extending an argument of [36] who proved:

Lemma 2.5 (Collino) *Assume that $n = 6$, $k = 2$, and that L is contained in the smooth locus of X: then L is a smooth point of $F_2(X)$ of dimension 2 if and only if the 4 conics $Q_0(z), \ldots, Q_3(z)$ are linearly independent.*

Moreover, for all such cubic hypersurface X, $F_2(X)$ is connected, and, for general (smooth) X, $F(X)$ is a smooth irreducible surface.

In the following lemma, statements (i) and (ii) are not new (cf. [1]), and statement (iv) is a special case of the more general Theorem 2.1 of [39].

Lemma 2.6 *Assume that $L \in F_k(X)$, and take coordinates in $\mathbb{P}^n$*

$$(z_0, z_1, \dots, z_k, x_0, x_1, \dots, x_b)$$

such that $L = \{x_0 = x_1 = \dots = x_b = 0\}$ $(b + 1 := n - k)$.

Write the equation F of X as

$$F = \sum_h x_h Q_h(z) + \sum_i q_i(x) z_i + G(x).$$

(i) Then in the local chart of the Grassmannian $\mathcal{G} := Gr(k, n)$, the affine space of matrices B such that $L_B = \{x = Bz\}$, the Zariski tangent space of $F_k(X)$ is the linear subspace of linear syzygies among the quadrics Q_h,

$$T(F_k(X))_L = \{B | \sum_h Q_h(z) (\sum_j B_{h,j} z_j) \equiv 0\}.$$

(ii) In particular, we get a smooth point of $F_k(X)$ of the expected codimension $\binom{k+3}{3}$ if and only if the quadrics Q_h generate the space of cubic forms in L.

(iii) If L is contained in the smooth locus of X, then we get a smooth point of $F_k(X)$ of the expected codimension $\binom{k+3}{3}$ for $k = 1$, and for $k = 2$ unless the quadrics Q_h generate a vector space of quadratic forms of vector dimension 3. In this case the local codimension is at least 9.

(iv) If the expected dimension $(k+1)(b+1) - \binom{k+3}{3}$ of $F_k(X)$ is non-negative, then $F_k(X)$ is non-empty; if it is strictly positive, then $F_k(X)$ is connected.

Proof The local equations of the Fano scheme are written (according to the Taylor development in B) as:

$$F = \sum_h Q_h(z) (\sum_j B_{h,j} z_j) + \sum_i q_i(Bz) z_i + G(Bz) = 0.$$

Recall that $L \cap \text{Sing}(X)$ is defined in L by the $b+1$ quadratic equations $Q_h(z) = 0$, and if L is contained in the smooth locus of X then $b \geq k$ and the zero set of the linear system of quadrics spanned by the Q_h's is empty.

By rearranging the order, we may assume that $Q_0, \dots, Q_k$ have no common zeroes, hence they form a regular sequence.

We set

$$\beta_h(z) := (\sum_j B_{h,j} z_j).$$

By exactness of the Koszul complex, it follows that the subspace generated by $\sum_0^k \beta_h Q_h$ has dimension equal to $(k+1)^2$.

For $k = 1$ we get $(k+1)^2 = 4 = \binom{k+3}{3}$ and we are done.

For $k = 2$ we get $(k+1)^2 = 9 < 10 = \binom{k+3}{3}$ and, for $b \geq 3 > 2$, we get the full space of cubic forms unless all the other quadrics are linear combinations of Q_0, Q_1, Q_2.

In fact, let R be the polynomial ring $R := \mathbb{C}[z_0, z_1, z_2]$, and let I be the ideal $I := (Q_0, Q_1, Q_2)$, so that $\dim(I_3) = 9$.

By a theorem of Macaulay [68], there is a perfect pairing

$$R/I \times R/I \to R_3/I_3.$$

If now $Q_3 \notin I_2$, then there is a degree 1 polynomial f such that $Q_3 f \notin I_3$, hence $(Q_0, Q_1, Q_2, Q_3) = R_3$ as claimed. □

In the special case $n = 6, k = 2$, $\mathcal{G}$ has dimension 12, and the expected dimension of $F_k(X)$ is equal to 2; for $n = 5, 4, 3,\ k = 1$, the expected dimension of $F_k(X)$ is respectively equal to 4, 2, 0.

As known classically, for the case of cubic surfaces, if X is smooth, then it contains 27 lines, and a finite number of lines if X is normal, and not a cone.

For $X_3^2 \subset \mathbb{P}^3$ irreducible, $F_1(X)$ has dimension equal to 1 if and only if X is a cone, or X has a singular line. And $F_1(X)$ has dimension equal to 2 exactly if X is reducible (then it contains a plane).

Going up, if $X_3^3 \subset \mathbb{P}^4$ has isolated singularities, then, as we saw, it cannot contain a linear subspace of dimension ≥ 3, but it can contain some plane, a linear subspace Λ of dimension 2.

Consider the incidence variety

$$\{(\Lambda, H) \mid X \supset \Lambda \cong \mathbb{P}^2,\ \Lambda \subset H \cong \mathbb{P}^3\},$$

which is a $\mathbb{P}^1$-bundle over $F_2(X)$, and which has a finite map to $(\mathbb{P}^4)^\vee$. Hence, if $F_2(X)$ has dimension at least 1, then there is a surface $\Sigma \subset (\mathbb{P}^4)^\vee$ such that $X \cap H$ is reducible for $H \in \Sigma$, hence H is tangent to X on a curve, which is then a fibre of the Gauss map. Hence the dual variety is degenerate, and X is a cone.

We have shown: if $X_3^3 \subset \mathbb{P}^4$ has isolated singularities and is not a cone, then it contains only a finite number of planes.

Consider now the incidence variety

$$\{(L, H) \mid X \supset L \cong \mathbb{P}^1,\ L \subset H \cong \mathbb{P}^3\},$$

which is a $\mathbb{P}^2$-bundle over $F_1(X)$, and therefore has dimension at least 4 at each point.

Continuing to assume that $X_3^3 \subset \mathbb{P}^4$ has isolated singularities and is not a cone, we see that each plane $\Lambda \subset X$ yields a 4-dimensional subvariety of the incidence variety above.

We show now: if $X_3^3 \subset \mathbb{P}^4$ has isolated singularities and is not a cone, then $F_1(X)$ has dimension 2.

In fact, the fibre of the incidence variety over $(\mathbb{P}^4)^\vee$ always has dimension ≥ 0, and the general fibre has dimension 0, while the fibre has dimension at most 2, and equal to 2 only if we have a plane $\Lambda \subset X$. Moreover, the fibre has dimension 1 if and only if H is a point of the dual variety where the Gauss map has fibre of dimension 1, hence the fibre has dimension 1 over at most a curve.

The preceding discussion proves:

Lemma 2.7 *If $X_3^3 \subset \mathbb{P}^4$ has isolated singularities and is not a cone, then it contains only a finite number of planes. Moreover, $F_1(X)$ has dimension 2.*

The next case is important for our applications:

Theorem 2.8 *Let $X_3^4 \subset \mathbb{P}^5$ be a nodal cubic hypersurface. Then the Fano scheme $F_1(X)$ has dimension exactly 4 at each point and the Fano scheme $F_2(X)$ contains only a finite number of planes Λ passing through some node of X.*

Proof The incidence variety

$$\{(L, H) \mid X \supset L \cong \mathbb{P}^1,\ L \subset H \cong \mathbb{P}^4\},$$

is a $\mathbb{P}^3$-bundle over $F_1(X)$, and therefore has dimension at least 7 at each point.

Projecting to $(\mathbb{P}^5)^\vee$, we see that the fibre dimension is at most 2 unless $H \cap X$ is a cone or it has nonisolated singularities. The first possibility is excluded since at the vertex of a cone X would not have a nondegenerate quadratic singularity. The second possibility leads again to a positive dimensional fibre of the Gauss map. However, for a nodal hypersurface the Gauss map is nondegenerate, hence we have at most a surface where the fibre has positive dimension.

Hence it suffices to show that the fibre can never have dimension ≥ 6, which is obvious since otherwise the fibre would be the whole Grassmannian, hence $H \cap X = H$, a contradiction since X contains no linear subspace of dimension 3.

For the second assertion, assume that a plane $\Lambda \subset X$ passes through some node P of X. Projection from P sends Λ to a line R contained in a K3 surface S, complete intersection of type (2, 3) in $\mathbb{P}^4$, $S = \{Q = G = 0\}$, where $F = zQ(x) + G(x)$ is the Taylor development of the equation of X at P.

Then $R^2 = -2$, and R does not move in S, hence the same holds for Λ. □

Theorem 2.9 *Let $X_3^5 \subset \mathbb{P}^6$ be a nodal cubic hypersurface. Then the Fano scheme $F_3(X)$ consists of at most a finite number of points and the Fano scheme $F_2(X)$ has dimension exactly 2 at each point L such that there is no 3-dimensional linear subspace $\Lambda \subset X$ with $L \subset \Lambda$.*

Proof For the first assertion, remark that each point $\Lambda \in F_3(X)$ contributes a 3-dimensional subvariety of $F_2(X)$, where the general point is a subspace L contained in the smooth locus.

By applying (iii) of Lemma 2.6 we see that the local dimension at such points is at most 3, hence we get only a finite number of such 3-dimensional subspaces Λ.

For the second assertion, assume that the dimension of $F_2(X)$ at L equals 3. Then, by Lemma 2.6, the linear system of quadrics on L generated by the Q_h's has dimension 3, and the Zariski tangent space has dimension 3. Hence $F_2(X)$ is smooth of dimension 3 at L.

We write then the equation of X along L as usual, where Q_0, Q_1, Q_2 have no common zeros on L, and we may assume that $Q_3 \equiv 0$:

$$F = \sum_0^2 x_h Q_h(z) + \sum_i q_i(x) z_i + G(x).$$

Then local coordinates for $F_2(X)$ are given by the coefficients of $\beta_3(z)$, and since the tangent space is given by the subspace $\beta_0 = \beta_1 = \beta_2 = 0$, we assume that locally the coefficients of these three forms are functions of the coefficients of $\beta_3(z)$, which have a Taylor development beginning with order ≥ 2. For shorthand notation we write, for $j = 0, 1, 2$, $\beta_j(z) = f_j(\beta_3, z)$.

The condition that these subspaces are contained in X amounts to:

$$\sum_0^2 f_h(\beta_3, z) Q_h(z) + \sum_i q_i(B(\beta_3)z)) z_i + G(B(\beta_3)z) \equiv 0,$$

where $B(\beta_3)z := (f_0(\beta_3, z), f_1(\beta_3, z), f_2(\beta_3, z), \beta_3(z))$.

The important point here is however the following. Consider the universal family $\Gamma_2(X)$ of planes over $F_2(X)$,

$$\Gamma_2(X) \subset F_2(X) \times \mathbb{P}^6$$

and let Z be the image inside $\mathbb{P}^6$ of the projection p_2 on the second factor.

Note that $\beta_3, (z_0, z_1, z_2)$ are (mixed, affine resp. homogeneous) coordinates around L.

We claim that Z is a variety of dimension 3. To show this, it suffices to observe that, along each such subspace L the derivative of p_2 is, as we have shown, of rank 3, since the image of the derivative is the subspace $x_0 = x_1 = x_2 = 0$.

Since $\Gamma := \Gamma_2(X)$ has dimension 5, take a smooth point $z \in Z$ such that the fibre $W := p_2^{-1}(z)$ has dimension 2; then, since W is transversal to the fibres of $p_1 : \Gamma \to F_2(X)$, we have a 2-dimensional family of planes L_w contained in Z and passing through z. But each L_w is contained in the (projective) tangent space

$T := TZ_z \cong \mathbb{P}^3$. These are then all the planes in T passing through z, hence their union is just T, and we conclude that $T \subset Z$. Hence by equality of dimension $T = Z$, and since $L \subset T$ we have a contradiction. □

2.2 Discriminants Y of Cubic Hypersurfaces X for the Projection with Centre $L \subset X$

In the above situation of a k-dimensional linear subspace contained in a cubic hypersurface, $L \subset X$, consider the projection

$$\pi_L : \mathbb{P}^n \setminus L \rightarrow \mathbb{P}^b = \mathbb{P}^{n-k-1},$$

such that

$$\pi_L(z, x) = x.$$

This rational map extends to a morphism on the blow up $\tilde{\mathbb{P}}_L \subset \mathbb{P}^n \times \mathbb{P}^b$ with centre L, which contains the proper transform $\mathcal{C}$ of X. The fibre over $x \in \mathbb{P}^b$ is the join $\mathbb{P}_x$ of L with $(0, x)$, i.e., the $\mathbb{P}^{k+1}$ with coordinates $(z_0, z_1, \dots, z_k, z_{k+1})$ parametrizing the points

$$(z_0, z_1, \dots, z_k, z_{k+1}x_0, \dots, z_{k+1}x_b).$$

The intersection of this subspace with X yields the divisor $L + \mathcal{C}_x$, where L is defined by $z_{k+1} = 0$ and $\mathcal{C}_x$ is the quadric

$$\mathcal{C}_x := \{(z_0, \dots, z_{k+1}) \mid F_x(z) = \sum_{i,j=0}^{k} a_{i,j}(x) z_i z_j + \sum_{i=0}^{k} 2q_i(x) z_i z_{k+1} + G(x) z_{k+1}^2 = 0\}.$$

Definition 2.10 The discriminant hypersurface of the pair $L \subset X$ is

$$Y := \{x \in \mathbb{P}^b \mid \mathrm{Sing}(\mathcal{C}_x) \neq \emptyset\}.$$

This is the determinantal hypersurface

$$Y = \{x \mid \det A(x) = 0\}$$

where we define $A(x)$, respectively $A'(x)$, as the symmetrical $(k+2) \times (k+2)$, respectively $(k+1) \times (k+1)$ matrices:

$$A(x) := \begin{pmatrix} a_{00}(x) \dots a_{0k}(x) \; q_0(x) \\ \dots \quad \dots \quad \dots \quad \dots \\ a_{0k}(x) \dots a_{kk}(x) \; q_k(x) \\ q_0(x) \; \dots \; q_k(x) \; G(x) \end{pmatrix}, \quad A'(x) := \begin{pmatrix} a_{00}(x) \dots a_{0k}(x) \\ \dots \quad \dots \quad \dots \\ a_{0k}(x) \dots a_{kk}(x) \end{pmatrix}. \tag{2.1}$$

The matrix A' corresponds to the linear system of quadrics in $\mathbb{P}^k$, of projective dimension $\leq b$, generated by $Q_0, \dots, Q_b$, and whose base locus is $\mathrm{Sing}(X) \cap L$.

Remark 2.11

(1) Since $\mathcal{C}_x$ is the fibre of $\mathcal{C} \to \mathbb{P}^b$ over x, it follows that Y is the set of critical values of $\pi_L|_{\mathcal{C}}$.
(2) The intersection of X with the subspace $\mathbb{P}^b := \{(z, x) \mid z = 0\}$ equals $\{(0, x) \mid G(x) = 0\}$.
(3) Given a point $w := (z, x) \in X \setminus L$, there is a unique x such that $(w, x) \in \mathcal{C}_x$.
(4) A point $w = (z, 0) := (z_0, \dots, z_k, 0, \dots, 0)$ in $X \cap L$ yields a point (w, x) in $\mathcal{C}_x$ if and only if $z \in Q_x := L \cap \mathcal{C}_x = \{z \mid \sum_h x_h Q_h(z) = 0\}$. We have $\mathcal{C}_x \subset \mathbb{P}_x$ and, identifying $\mathbb{P}_x$ with $\mathbb{P}^{k+1}$, the point $(z_0, \dots, z_k, 0)$ corresponds to (w, x).

The point (w, x) is singular in $\mathcal{C}_x$ if and only if

$$A'(x)z = 0, \; \sum_i q_i(x) z_i = 0 \iff (z_0, \dots, z_k, 0) \in \ker A(x),$$

and, since $G(x) z_{k+1}^2$ vanishes to order 2 at our point (w, x), it is singular in $\mathcal{C}$ if and only if

$$(w, x) \in \mathrm{Sing}\, \mathcal{C}_x \text{ and } w \in \mathrm{Sing}\, X \text{ (that is, } Q_h(z) = 0, \; \forall h).$$

Proposition 2.12 *Given a point $w = (z, x) \in X \setminus L$, choose the unique $x \in \mathbb{P}^b$ such that $(w, x) \in \mathcal{C}_x$. Then $w \in \mathrm{Sing}(X)$ if and only if $(w, x) \in \mathrm{Sing}(\mathcal{C})$. Moreover, if $\mathcal{C}_x$ is smooth at (w, x) and $\mathcal{C}_x \neq \mathbb{P}_x$, then $w \notin \mathrm{Sing}(X)$.*

Proof Since $w \notin L$, X and $\mathcal{C}$ are locally isomorphic, hence the first assertion. If $\mathcal{C}$ were singular at (w, x), then the codimension b complete intersection $\mathcal{C}_x$ would also be singular at (w, x). □

It follows that if $w \in \mathrm{Sing}(X) \setminus L$, then its projection point x belongs to the discriminant hypersurface Y, i.e., $\mathrm{rank}(A(x)) \leq k + 1$.

Proposition 2.13 *Assume that $x' \in Y$ satisfies $\mathrm{rank}(A(x')) = k + 1$, that is, the matrix $A(x')$ has corank $c = 1$, and let $f(x) = 0$ be a local holomorphic equation for Y at x'.*

(i) There is at most one possible singular point w of $\mathcal{C}$ lying in $\mathcal{C}_{x'}$, and it has local equation

$$\zeta_1^2 + \dots \zeta_{k+1}^2 + f(x) = 0,$$

a so-called suspension of the singularity of Y *at* x'.

Moreover, w *corresponds to the point* $z := \ker(A(x'))$, *and it is a singular point (resp. a node) of* $\mathcal{C}$ *if and only if* x' *is a singular point (resp. a node) of* Y.

(ii) *Moreover,* w *comes from a point* $z \in L$ *if and only if* $\ker A(x') \in L \Leftrightarrow$ *there is a solution* $z \in L$ *to*

$$A'(x')z = 0, \sum_0^k q_i(x')z_i = 0.$$

(iii) *Assume that the point* w *comes from a point* $z \in L$. *Then:*

(1) *If* $w \in \operatorname{Sing}(\mathcal{C})$, *then* $z \in \operatorname{Sing}(X)$.
(2) *If* $k+1 \leq b-1$ *and* Y *has isolated singularities then* w *is not a singular point of* $\mathcal{C}$ *(and* x' *is not in* $\operatorname{Sing}(Y)$*).*
(3) *If instead* $k+1 = b$ *and* $x' \in \operatorname{Sing}(Y)$, *then* $z \in \operatorname{Sing}(X)$. *If the singularities of* Y *are nodes, then* $z \in \ker A(x')$ *for exactly two nodes of* Y.
(4) *Finally, if* $k \geq b$, *and the singularities of* Y *are nodes, then* $z \in \ker A(x')$ *for at most two nodes of* Y.

Proof Under our hypothesis we have a quadratic form in $(k+2)$-variables over the local ring $\mathcal{O}$ of the point $x' \in \mathbb{P}^b$ and its image in the residue field $\mathbb{C}$ has rank equal to $k+1$. Therefore it can be brought in diagonal form

$$\zeta_0^2 f(x) + \sum_1^{k+1} \zeta_i^2,$$

where $f(x)$ belongs to the maximal ideal.

It follows that $f(x) = 0$ is the local equation of $Y = \{\det(A(x)) = 0\}$, and that we have at most a unique singular point w of $\mathcal{C}$ lying in $\mathcal{C}_{x'}$, corresponding to $\zeta_1 = \cdots = \zeta_{k+1} = 0$. There we may set $\zeta_0 = 1$, and obtain the asserted local equation for $\mathcal{C}$.

Moreover, this singular point corresponds to the point $\ker(A(x'))$; the other assertions of (ii) follow right away from the preceding Remark 2.11.

Let us prove (iii).

Without loss of generality we may assume that z is the vector e_0. Hence $e_0 \in \ker A(x')$ is equivalent to requiring $a_{0,j}(x') = 0$ for all $j = 0, \dots, k$, and $q_0(x') = 0$. Hence the first row and column of $A(x')$ vanish. Write $B(x')$ for the submatrix of $A(x')$ where the first row and first column are erased. Since $A(x')$ has corank $c = 1$, it follows that the determinant of $B(x')$ is invertible.

Taking the Taylor development of $f(x) = a_{0,0}(x)\det(B(x')) + \phi(x)$ at the point x', we see that $\phi(x)$ belongs to the square of the ideal generated by $a_{0,1}(x), \dots, a_{0,k}(x), q_0(x)$, hence it vanishes to order at least 2 at x', while

$\det(B(x')) \neq 0$. We conclude that Y is smooth at x' if and only if $a_{0,0}(x)$ is not identically zero.

If $a_{0,0}(x)$ is identically zero, we see that $z \in \mathrm{Sing}(X)$, since, for e_0, $z_0 = 1$ and $z_1, \dots, z_k, x_0, \dots, x_b$ are local coordinates at the point e_0.

Moreover, in this situation, then $e_0 \in \ker A(x')$ for all the solutions of $a_{0,j}(x') = 0, 1 \leq j \leq k, q_0(x') = 0$. These are k linear and one quadratic equations, hence for $b \geq k+2$ we get that Y does not have isolated singularities.

Assume now that $b = k+1$ and that the singularities of Y are isolated. Then the locus Σ of points x' such that $e_0 \in \ker A(x')$ is a complete intersection of the above $k+1$ equations, hence either two points, or one with multiplicity 2. If the singularities of Y are nodes, since $\det A(x)$ is in the square of the ideal generated by these equations, these must be local parameters, hence we get two singular points, nodes by our assumption.

Finally, if $k \geq b$, we get that the finite set $\mathrm{Sing}(Y)$ contains the locus Σ, which is therefore finite; hence, being defined by one quadratic equation and several linear equations, it has cardinality at most 2. □

Remark 2.14

(I) In the case $b = 3$, we see that the second assertion of (iii) above applies for $d = \deg(Y) \leq 5$, while for $d = 6$ there is the possibility that there are some nodes of X in the plane L.

(II) For our purposes we may restrict ourselves to the case where $r := \mathrm{rank}(A(x))$ is always $\geq k$. Indeed, since the matrix $A(x)$ over the local ring $\mathcal{O}$ can be brought to block diagonal form with upper entry the $r \times r$ identity matrix I_r, $r := \mathrm{rank}(A(x'))$, and with lower entry a $c \times c := (k+2-r) \times (k+2-r)$ matrix with entries in the maximal ideal of $\mathcal{O}$. If the corank $c \geq 3$, then x' is a point of multiplicity ≥ 3 for Y.

Proposition 2.15 *Assume that $x' \in Y$ satisfies $\mathrm{rank}(A(x')) = k$, that is, the matrix $A(x')$ has corank $c = 2$.*

Then there are functions ξ_1, ξ_2, ξ_3 in the maximal ideal of $\mathcal{O}$ such that $\xi_1\xi_2 - \xi_3^2$ is a local holomorphic equation for Y at x'.

Over a neighbourhood of x', C is defined by an equation of the form

$$\xi_1\zeta_0^2 + 2\xi_3\zeta_0\zeta_{k+1} + \xi_2\zeta_{k+1}^2 + \sum_1^k \zeta_i^2 \in \mathcal{O}[\zeta_0, \dots, \zeta_{k+1}]$$

as a projective variety over $\mathcal{O}$, where $\mathcal{O}$ is the local ring of $x' \in \mathbb{P}^b$.

The singular points w of C lying in $C_{x'}$, are contained in a projective line $\zeta_1 = \dots = \zeta_k = 0$, and they are defined by the further equation $\xi_1\zeta_0^2 + 2\xi_3\zeta_0\zeta_{k+1} + \xi_2\zeta_{k+1}^2 = 0$.

In particular, there are no such singular points $w \in C_{x'}$ (that is, $(w, x') \in C_{x'}$) if ξ_1, ξ_2, ξ_3 are part of a system of local coordinates, i.e., $\mathrm{Sing}(Y)$ at x' is a smooth subvariety of codimension 3 in $\mathbb{P}^b$.

There is at most one such singular point if the multiplicity of Y at x' is either three or it is two and the tangent cone of Y at x' is a quadric of rank ≥ 2.

Proof Under our hypothesis we have a quadratic form in $(k+2)$-variables over the local ring $\mathcal{O}$ of the point $x' \in \mathbb{P}^b$ such that its image in the residue field $\mathbb{C}$ has rank equal to k.

Therefore it can be brought in block diagonal form as explained in the preceding Remark

$$\xi_1\zeta_0^2 + 2\xi_3\zeta_0\zeta_{k+1} + \xi_2\zeta_{k+1}^2 + \sum_1^k \zeta_i^2,$$

where ξ_1, ξ_2, ξ_3 are in the maximal ideal of $\mathcal{O}$.

It follows that $\xi_1\xi_2 - \xi_3^2$ is the local equation of $Y = \{\det(A(x)) = 0\}$, and the singular points w of $\mathcal{C}$ lying in $\mathcal{C}_x$, are contained in the locus $\zeta_1 = \cdots = \zeta_k = 0$.

If such a point w has $\zeta_0 \neq 0$, then the hypothesis that ξ_1, ξ_2, ξ_3 are part of a system of local coordinates implies that we can write ξ_1 as a function of other coordinates, hence we have a smooth point. Similarly for the other case $\zeta_{k+1} \neq 0$.

More generally, if we have a singular point w, then its coordinates satisfy $\zeta_1 = \cdots = \zeta_k = 0$ and ζ_0, ζ_{k+1} must yield a nontrivial relation of linear dependence among the gradients $\nabla\xi_i$,

$$\zeta_0^2\nabla\xi_1 + 2\zeta_0\zeta_{k+1}\nabla\xi_3 + \zeta_{k+1}^2\nabla\xi_2 = 0.$$

We may assume that the singular point w corresponds to $\zeta_{k+1} = 0$, hence we have $\nabla\xi_1 = 0$. If there is another singular point, we may assume that it corresponds to $\zeta_0 = 0$, hence we have $\nabla\xi_2 = 0$.

Therefore ξ_1, ξ_2 belong to the square of the maximal ideal, and we have either a singular point of multiplicity at least 4, or a uniplanar double point (the tangent cone is induced by ξ_3^2, and is a quadric of rank 1).

We have a third singular point if and only if $\nabla\xi_i = 0$, $\forall i = 1, 2, 3$, hence each ξ_i belongs to the square of the maximal ideal, and Y has a singular point of multiplicity at least 4. □

Therefore, under the assumptions that

$$\operatorname{corank}(A(x')) \leq 2, \ \forall x',$$

(in particular this holds if Y has no triple points), and that the singular points of Y have a quadratic part of rank equal to 3, we have found that the set of singular points of $X \setminus L$ maps injectively into the set of singular points of Y with $\operatorname{corank}(A(x')) = 1$.

Moreover, we have singular points of $\mathcal{C}$ only over singular points x' of Y where $\operatorname{corank}(A(x')) = 1$.

And by Proposition 2.13(i) there is exactly one such point in $\mathcal{C}_{x'}$. This point comes from a point of $X \setminus L$ if $k+2 \leq b$, while if $b = k+1$ this point can also be of the form (w, x') where $w \in L \cap \operatorname{Sing}(X)$. Then by Proposition 2.13(iii)-(3) there are exactly two such points (w, x'), (w, x'') with $x', x'' \in \operatorname{Sing}(Y)$ if we further assume $b = 3$ and Y with isolated singularities.

The easier consequence of the previous discussion is assertion (3) of the following:

Theorem 2.16 *Let $Y \subset \mathbb{P}^3$ be the discriminant of the projection π_L of a cubic hypersurface $X \subset \mathbb{P}^{k+4}$ from a linear subspace $L \subset X$ of dimension k (hence $\deg(Y) = k + 4$).*

Assume that the singular points of Y are nodes, that is, they have quadratic part of rank 3.

(1) If $L \cap \operatorname{Sing} X = \emptyset$, then there is a bijection between the sets $\operatorname{Sing}(X)$ and

$$\operatorname{Sing}(Y) \cap \{x \mid \operatorname{corank}(A(x)) = 1\},$$

induced by the projection π_L.

(2) If $k \leq 1$, then $L \cap \operatorname{Sing} X = \emptyset$, so (1) holds automatically.

(3) If $k = 2$, then there is a surjection

$$\operatorname{Sing}(Y) \cap \{x \mid \operatorname{corank}(A(x)) = 1\} \rightarrow \operatorname{Sing}(X)$$

such the inverse image of each point in $L \cap \operatorname{Sing}(X)$ consists of two such nodes. In this case, the singularities of X are all nodes.

Proof By Proposition 2.15, there are no singularities of $\mathcal{C}$ over the singular points $x \in Y$ with $\operatorname{corank}(A(x)) = 2$.

Over the singular points $x \in Y$ with $\operatorname{corank}(A(x)) = 1$ there is exactly one singular point of $\mathcal{C}$ by Proposition 2.13(i). Now by Proposition 2.13(iii)-(1), if this point comes from a point $z \in L$, then $z \in \operatorname{Sing}(X)$, and $k + 1 > b - 1 = 2$ by Proposition 2.13(iii)-(2). This proves (1) and (2).

We are now going to see that a point $w \in L \cap \operatorname{Sing}(X)$ gives rise to singular points of $\mathcal{C}$ lying in some fibre $\mathcal{C}_x$.

Recall that $w = (z, 0) = (z_0, \ldots, z_k, 0, \ldots, 0) \in L \cap \operatorname{Sing}(X)$ means that z is in the base locus of the system of quadrics in $\mathbb{P}^k$, $Q_0(z) = \cdots = Q_b(z) = 0$.

We look for $x \in \mathbb{P}^b$ such that $(z, 0) = (z_0, \ldots, z_k, 0) \in \ker A(x)$. The condition $A'(x)z = 0$ then determines k linear equations, whose solution set is a subspace of dimension $\geq b - k$ of $\mathbb{P}^b$. The equation $\sum_i q_i(x) z_i = 0$ determines a quadric in this subspace, so that we have certainly a solution x as soon as $b \geq k + 1$.

If such x exists, $\mathcal{C}$ is singular at (w, x) by Remark 2.11(4). We know that $\operatorname{corank} A(x) \neq 2$ by Proposition 2.15, while the corank of A is never ≥ 3 by our assumption.

If $\operatorname{corank} A(x) = 1$, then by Proposition 2.13(iii)-(3), we obtain two singular points (w, x'), (w, x'') of $\mathcal{C}$ associated to $w \in L \cap \operatorname{Sing}(X)$, where x', x'' are nodes with $\operatorname{corank} A = 1$.

Finally, for $k = 2$, we want to show that X has only nodes as singularities. This is true for the singularities in $\mathrm{Sing}(X) \setminus L$, and the singularities of $\mathcal{C}$.

Let us show that the singularities in $X \cap L$ (which produce two nodes after the blow up yielding $\mathcal{C}$) are also nodes.

We use the notation of Proposition 2.13(iii), hence $w = e_0$, and we may assume that $a_{00} = 0$, $a_{01} = x_1$, $a_{02} = x_2$, $q_0 = x_0x_3$; then at w, the quadratic part of the equation of X equals

$$x_1z_1 + x_2z_2 + x_0x_3.$$

Since $x_1, z_1, x_2, z_2, x_0, x_3$ are local coordinates at w, we get a node of X. □

Remark 2.17 Therefore, if $Y \subset \mathbb{P}^3$ is nodal of degree d and is the discriminant of the projection of X with centre $L \subset X$, then for $d \leq 6$, X has only nodes as singularities.

2.3 Nodal Maximizing Surfaces in Low Degree

If Y is a nodal surface of degree d in $\mathbb{P}^3$, then it is known that the number $\nu(Y)$ of nodes is smaller than the maximum possible $\mu(d)$, and that $\mu(1) = 0, \mu(2) = 1, \mu(3) = 4, \mu(4) = 16, \mu(5) = 31, \mu(6) = 65$.

A major concern in this paper is to show that if $3 \leq d \leq 6$ and the number of nodes is the maximum possible, then Y is a cubic discriminant.

Example 2.18 It is worthwhile to observe that, already for degree $d = 3$, $\mu(d)$ can be smaller, for surfaces with isolated singularities, than the total sum of the Milnor numbers of the singularities. In fact, if a cubic surface has exactly 4 singular points, it is projectively equivalent to the Cayley cubic with equation $\sigma_3(x) = 0$, where σ_3 is the third elementary symmetric function, and its singularities are just four nodes.

But there is the cubic surface $xyz = w^3$ which possesses 3 singular points of type A_2, hence realizing a total Milnor number (sum of the Milnor numbers of the singularities) equal to 6.

This leads to the following question: for a normal surface of degree d in $\mathbb{P}^3$, what is the maximum that the total Milnor number can achieve?

Generalizing the above example, we can take the surface

$$X := \{(x) | x_0^d = \Pi_1^d L_j(x_1, x_2, x_3)\}, L_1, \dots, L_d \text{ general linear forms}\},$$

which has $\frac{d(d-1)}{2}$ A_{d-1} singularities, hence $\mathrm{Milnor}(X) = \frac{1}{2}d(d-1)^2$.

Bruce [17] gave general upper bounds for $s_n(d)$, in particular he showed that $s_2(d) =: s(d) \leq \frac{1}{2}(d-1)^3$ if the degree d is odd, else $s(d) \leq \frac{1}{2d}((d-1)^3(d+1)+1)$ for d even.

Hence the above examples show that

$$s(d) < \text{Milnor}(d), \quad \forall d.$$

For $d = 4$, this gives a quartic surface with six A_3 singularities and total Milnor number 18, and it is natural to expect an example with Milnor number 19, since the Picard number of a K3 surface is at most 20 (see [18, 97]). We are not aware of general results on the function $\text{Milnor}_n(d)$.

If X has Rational Double Points, then an easy bound is given by $h^{1,1}(S) - 1$, where S is the minimal resolution of X. This yields the inequality

$$\text{Milnor}(X) \leq h^{1,1}(S) - 1 = b_2(S) - 2p_g(S) - 1 = e(S) - 2p_g(S) - 3 =$$

$$= 10\chi(\mathcal{O}_S) - K_S^2 - 1 = 9 + 10p_g(S) - K_S^2 =$$

$$= 9 + \frac{5}{3}(d-1)(d-2)(d-3) - d(d-4)^2 = \frac{2}{3}d^3 - 6d^2 + \frac{7}{3}d - 1 < \frac{2}{3}d^3.$$

In degree $d = 4$ the surfaces with 16 singularities have nodes as singularities and are the Kummer surfaces; they can be obtained projecting a 10-nodal cubic hypersurface $X \subset \mathbb{P}^4$ from a smooth point.

Going back to $d = 3$, (see [49]) the maximal number of nodes that a cubic hypersurface $X \subset \mathbb{P}^m$ can have is $\gamma(m)$, where $\gamma(3) = 4, \gamma(4) = 10, \gamma(5) = 15, \gamma(6) = 35$ (we have in general $\gamma(m) = \binom{m+1}{[m/2]}$).

For $m = 2h$, equality is attained by the Segre cubic

$$\Sigma(m-1) := \{x \in \mathbb{P}^{m+1} \mid \sigma_1(x) = \sigma_3(x) = 0\}$$
$$= \{x \in \mathbb{P}^{m+1} \mid s_1(x) = s_3(x) = 0\},$$

where σ_i is the i-th elementary symmetric function, and $s_i = \sum_j x_j^i$ is the i-th Newton function.

The singular locus of $\Sigma(m-1)$ consists of $\frac{1}{2}\binom{m+2}{h+1} = \binom{m+1}{h}$ nodes which form an $\mathfrak{S}_{2h+2}$-orbit, generated by $(1^{h+1} : -1^{h+1})$ where α^s means α is repeated s times.

Whereas, for $m = 2h+1$ odd, Goryunov produces

$$T(m-1) :=$$
$$= \{(x,z) \mid \sigma_1(x) = \sigma_3(x) + z\sigma_2(x) + \frac{1}{12}h(h+1)(h+2)z^3 = 0\}$$
$$= \{(x,z) \mid s_1(x) = 2s_3(x) - 3z\sigma_2(x) + \frac{1}{2}h(h+1)(h+2)z^3 = 0\}$$

where (x, z) denotes $(x_0, \ldots, x_m, z)$. This is projectively equivalent to Kalker's cubic hypersurface defined by

$$T(m-1) := (x_0 + \cdots + x_m + 2x_{m+1} = x_0^3 + \cdots + x_m^3 + 2x_{m+1}^3 = 0) \subset \mathbb{P}^{m+1}.$$

The symmetric group $\mathfrak{S}_{m+1}$ acts on $T(m-1)$ by permuting the first $m+1$ coordinates and fixing z, respectively x_{m+1}. The singular locus of $T(m-1)$ consists of $\binom{m+1}{h}$ nodes, forming the $\mathfrak{S}_{m+1}$-orbit of the point $(1^{h+1} : -1^{h+1} : 1)$.

We have the following result, the first part of which is due to Segre [83], another proof is to be found in [59], and we give an argument here based on a result which shall be established in a later section, where we shall also show that cubic 4-folds with 15 nodes form an irreducible family.

Theorem 2.19 *Any nodal maximizing cubic hypersurface X in $\mathbb{P}^4$ is projectively equivalent to the Segre cubic.*

Any equisingular deformation of the Segre cubic $\Sigma(m-1)$ in $\mathbb{P}^{2h}$ is projectively equivalent to the Segre cubic.

For $h \geq 3$, any equisingular deformation of the Goryunov cubic $T(m-1)$ in $\mathbb{P}^{2h+1}$ is projectively equivalent to the Goryunov cubic.

The Segre cubics $\Sigma(3)$, $\Sigma(5)$ and the Goryunov cubic $T(4)$ are unobstructed, while all other Segre and Goryunov cubics are obstructed.

Proof Here we prove the first assertion. The other assertions are proven in Sect. 2.6.

Let $P = (0, 0, 0, 0, 1)$ be a node of $X \subset \mathbb{P}^4$, and consider the Taylor development of its equation at the point P: $F(x, z) = zQ(x) + G(x)$.

Then by Corollary 1.46 the curve in $\mathbb{P}^3$ given by $Q(x) = G(x) = 0$ is a nodal curve contained in a smooth quadric and has $\gamma - 1$ nodes, where γ is the number of nodes of X. But a curve of bidegree (3, 3) on $Q = \mathbb{P}^1 \times \mathbb{P}^1$ has at most 9 nodes, equality holding if and only if it consists of three vertical and three horizontal lines. □

Remark 2.20 The unobstructedness of the Segre cubic in $\mathbb{P}^6$ (see Theorem 1.25) is equivalent to the second assertion, since cubics in $\mathbb{P}^6$ are a projective space of dimension 83, the group of projectivities of $\mathbb{P}^6$ has dimension 48, and the group of projective automorphisms of the 35-nodal Segre cubic is finite.

Remark 2.21 It is known [59] and [49] that, for $m = 5$, nodal maximizing cubics are not projectively unique.

The previous result leads to the following natural questions: is every nodal maximizing cubic hypersurface in $\mathbb{P}^m$ projectively equivalent to the Segre cubic, for m even? And for $m \geq 7$ odd, to the Goruynov cubic?

Given a point in a cubic hypersurface $X \subset \mathbb{P}^4$, we write its equation as usual as

$$\{(z, x_0, x_1, x_2, x_3) = (z, x) \mid a(x)z^2 + 2q(x)z + G(x) = 0\},$$

so that its discriminant is the quartic surface $\{x \mid a(x)G(x) - q(x)^2 = 0\}$.

The point $x = 0$ of X is smooth if $a(x)$ is not identically zero, equivalently if Y is not a quadric counted with multiplicity two.

If we make this assumption, the points where the corank of the matrix $A(x)$ drops by 2 are the points $a(x) = q(x) = G(x) = 0$, in general a set of 6 distinct points (nodes of Y) if hypothesis (1) of Theorem 2.16 is verified.

Hence under assumption (1) we obtain that $\nu(Y) = 6 + \nu(X)$, and, since a Kummer surface contains a half-even set of nodes of cardinality 6, we get the whole three dimensional family of Kummer surfaces by projecting the Segre cubic from a smooth point (see [93] and [27, Proposition 26]).

Remark 2.22 We obtain an explicit example of a Kummer surface as discriminant of a cubic hypersurface by choosing a point on the Segre cubic which does not lie in any one of the linear spaces $L' \cong \mathbb{P}^2$, $L' \subset \Sigma$ which are in the $\mathfrak{S}_6$-orbit of the subspace $x_i + x_{i+3} = 0, i = 0, 1, 2$.

Setting $x := (x_0, x_1, x_2, x_3)$, we take a point

$$(x_0, x_1, x_2, x_3, y, -y - \sum_0^3 x_i)$$

with

$$y^2 s_1(x) + y s_1(x)^2 = \frac{1}{3}(s_3(x) - s_1(x)^3),$$

for instance

$$(a, a, a, 5a, y, -y - 8a)$$

with $y^2 + 8ay + 16a^2 = 0$, so that we may take $a = 1,\ y = -4$.

In the case of Togliatti quintics Y (quintics with 31 nodes), it was proven by Beauville [9] that Y is the discriminant of a cubic hypersurface $X \subset \mathbb{P}^5$ for the projection from a line $L \subset X$, since Y possesses an even set of nodes of cardinality 16. More generally, we have:

Proposition 2.23 *Let Y be a nodal quintic surface with an even set of nodes of cardinality 16. Then Y is the discriminant of a cubic hypersurface $X \subset \mathbb{P}^5$ for the projection from a line $L \subset X$, L contained in the smooth locus of X, and X has exactly $\gamma(X) = \nu(Y) - 16$ nodes as singularities, where $\nu(Y)$ is the number of nodes of Y.*

Proof It follows from [21] that Y is a discriminant of a cubic hypersurface $X \subset \mathbb{P}^5$ for the projection from a line $L \subset X$.

And the even set of 16 nodes is exactly the set of points of Y where $\operatorname{corank}(A(x)) = 2$.

Theorem 2.16 implies that there is a bijection between the set $\operatorname{Sing}(X)$ and the set of nodes of Y which are not in the given even set.

Finally, the singularities of X are nodes by Proposition 2.13. □

Example 2.24 Consider a discriminant quintic surface, corresponding to a matrix:

$$A := \begin{pmatrix} a_{00}(x) & a_{01}(x) & q_0(x) \\ a_{01}(x) & a_{11}(x) & q_1(x) \\ q_0(x) & q_1(x) & G(x) \end{pmatrix}. \tag{2.2}$$

In this example, let us restrict to the special case where

$$a_{01}(x) \equiv 0,$$

in order to give explicit equations for the 16 nodes.

If $a_{01}(x) \equiv 0$, we may assume without loss of generality that

$$a_{00}(x) = x_0, a_{11}(x) = x_1,$$

$$q_0(x) = q_0(x_1, x_2, x_3), q_1(x) = q_1(x_0, x_2, x_3).$$

The equation of the quintic takes the simple form

$$a_{00}(x)a_{11}(x)G(x) - a_{00}(x)q_1(x)^2 - a_{11}(x)q_0(x)^2 =$$
$$x_0x_1G(x) - x_0q_1(x)^2 - x_1q_0(x)^2 = 0.$$

The even set of nodes corresponding to the matrix A (points where the rank drops by two) consists of the 16 points, union of the two sets of 8 points, of respective equations:

$$\{x_0 = q_0(x) = x_1G(x) - q_1(x)^2 = 0\} \cup \{x_1 = q_1(x) = x_0G(x) - q_0(x)^2 = 0\}.$$

We need to assume that there are no common solutions of the equations:

$$x_0 = x_1 = q_0(x) = q_1(x) = 0,$$

and that in each plane the conic and quartic intersect in 8 distinct points, for instance:

$$q_0(x_1, x_2, x_3) = x_1G(0, x_1, x_2, x_3) - q_1(0, 0, x_2, x_3)^2 = 0$$

yields eight points.

The following result, the converse to Proposition 2.23, is essentially the main result of Togliatti [91], for which we give a slightly different proof (some details were not verified in [91]).

Proposition 2.25 *Let X be a cubic hypersurface $X \subset \mathbb{P}^5$ which has exactly $\gamma(X)$ nodes as singularities. Then there exist lines L contained in the smooth locus of X, and such that the different nodes P_i of X determine distinct planes $P_i * L$ containing L.*

Let Y be the discriminant of the cubic hypersurface $X \subset \mathbb{P}^5$ for the projection from a general line $L \subset X$. Then Y is a quintic surface with singular scheme consisting of $\gamma(X)$ nodes plus a local complete intersection scheme of length 16*. For general X, Y is a nodal quintic surface with exactly* $16 + \gamma(X)$ *nodes as singularities.*

Proof Let P_i be a singular point of X, and let $F(x) = q(x) + g(x)$ be the affine Taylor development at P_i.

The lines $L \subset X$ passing through P_i form a cone with vertex P_i over a complete intersection in $\mathbb{P}^4$ of dimension 2, given by the vanishing of the two terms $q(x), g(x)$ of the Taylor development at P_i, which have no common factor. Hence these lines form a 2-dimensional family.

On the other hand, by Theorem 2.8, the Fano scheme of lines $L \subset X$ has dimension 4 at each point and the lines L meeting one of the lines $P_i * P_j$ form a family of dimension at most 3: hence the first assertion is proven.

The nodes of X project to singular points of Y, and these are nodes of Y by Proposition 2.13 unless the plane $P_i * L$ intersects X in L plus a double line, or is contained in X; the second possibility is excluded since by Theorem 2.8 there are only finitely many planes in X containing a node, so L is not general.

For the first possibility, we use a dimension count. If $\pi \cong \mathbb{P}^2$, and $P_i \in \pi$, the condition that $\pi \cap X$ contains a line L' through P_i with multiplicity 2 means that this line L' gives a point $x \in \mathbb{P}^4$ where (again $F = q + g$ is the Taylor development of F at P_i) both q and g vanish, and π gives a line tangent to both $\{q(x) = 0\}$ and $\{g(x) = 0\}$. Since $\{q(x) = g(x) = 0\}$ is a surface, we get a 3-dimensional family of lines in $\mathbb{P}^4$ unless the hypersurfaces $\{q(x) = 0\}$ and $\{g(x) = 0\}$ are everywhere tangent. Since the singularities of X are nodes, the quadric $\{q(x) = 0\}$ is smooth, and by the theorem of Lefschetz all divisors on it are complete intersections. Hence the hypersurfaces $\{q(x) = 0\}$ and $\{g(x) = 0\}$ cannot be everywhere tangent.

Since the family of such lines is 3-dimensional, the family of corresponding planes π through P_i is also 3-dimensional, hence the family of the residual lines L ($\pi \cap X = 2L' + L$) is also 3-dimensional and then L is not general.

Hence the nodes P_i project to points where the matrix $A(x)$ has corank $= 1$.

Observe now that, if there is a point x' where the matrix $A(x)$ has corank 3, i.e., the matrix is zero at this point, then there is a 2-dimensional subspace Λ in X and containing L. But we can observe that then the point $z = 0, x = x'$ is a triple point of X.

Now the other assertions will follow from Theorem 2.16 provided we show that the points where the rank of the matrix $A(x)$ drops by 2 are 16 determinantal nodes; equivalently, the quadratic part has rank 3 at each of them.

Now, the Fitting ideal of the 2×2-minors of $A(x)$ defines a scheme Σ which, since the rank at the points of Σ equals 1, is locally defined by three equations (see Proposition 2.15). Hence, if Σ has dimension 0, it is a local complete intersection scheme of length 16 (s Example 2.24).

To show that generally the scheme Σ consists of 16 different points, we observe first of all that this holds for a general X with $\gamma = 15$ nodes using the irreducibility of the Togliatti quintics, and since we are exhibiting one such surface with 31 distinct nodes, see Theorem 2.29 and Proposition 2.32.

Now, each X with $\gamma \leq 15$ nodes is a small deformation of a cubic with 15 nodes, since the corresponding K3 surfaces are partial smoothings of the one with 15 nodes, in view of Theorem 1.57. □

Remark 2.26
(i) The condition that the line L is in the smooth locus of the cubic hypersurface $X \subset \mathbb{P}^5$ is implied by (but, as we just saw in the example 2.24, does not imply) the condition that the three entries of the matrix $A'(x)$ are linearly independent, in which case we get a normal form

$$X := \{(z,x) | x_0 z_0^2 + x_1 z_1^2 + 2x_2 z_0 z_1 + 2q_0(x) z_0 + 2q_1(x) z_1 + G(x) = 0\}$$

where we can assume $q_0(x) = q_0(x_1, x_2, x_3)$, $q_1(x) = q_1(x_0, x_2, x_3)$.

Then the rank of A never drops by 3 provided the monomial x_3^2 appears in q_0, or in q_1, or at least the monomial x_3^3 appears in G. If all the three conditions are satisfied, then the points where the rank drops by 2 are points of the quadric $x_0 x_1 - x_2^2$ different from the vertex $x_0 = x_1 = x_2 = 0$, hence for them we can assume that either $x_0 = 1$, or $x_1 = 1$.

In general all solutions have $x_0, x_1, x_2 \neq 0$. Because, for instance, if $x_0 = 0$ then $x_2 = 0$ and then we must have $q_0(1, 0, x_3) = 0, q_1(0, 0, x_3)^2 - G(0, 1, 0, x_3) = 0$, which does not happen for general choice of q_0, G.

Bordering the minor x_0 we get the two equations

$$x_0 q_1 - q_0 x_2 = x_0 G - q_0^2$$

and setting $x_2 = 1$ and multiplying by x_1 the first equation, and by x_1^2 the second we obtain the equivalent equations

$$x_2 q_1 - q_0 x_1 = x_1 G - q_1^2,$$

obtained bordering the minor x_1.

Hence all the solutions are gotten, setting $x_0 = 1$, hence $x_1 = x_2^2$, as solutions of the two equations

$$q_1(1, x_2, x_3) - x_2 q_0(x_2^2, x_2, x_3) = 0, \;\; G(1, x_2^2, x_2, x_3) - q_0(x_2^2, x_2, x_3)^2 = 0.$$

In this way the points where the rank drops by 2 lie on the curve of contact C, obtained as the residual curve to $\Lambda := \{x_0 = x_2 = 0\}$ in the sextic curve

$$\Lambda + C := \{x_0 x_1 - x_2^2 = x_0 q_1(x) - x_2 q_0(x) = 0\}.$$

If C is a smooth curve (see [21], Lemma 2.3) we get 16 distinct points as soon as the solutions to the above equations occur with multiplicity 1.

This condition pertains to the Taylor development of the equation of X along the line L.

(ii) The quintics of Proposition 2.25 are unobstructed.

We use for this the smoothness result of Theorem 2.29 for general γ: $\mathcal{F}_4(3, \gamma)$ is smooth, due to the fact that K3 surfaces are unobstructed. Then the variety $\mathcal{P}_\gamma$ consisting of pairs (X, L), where L fulfills the properties of not passing through the nodes of X, not being contained in a 2-dimensional subspace $\Lambda \subset X$, not meeting lines connecting two nodes, is smooth.

Fix now coordinates (z, x) such that $L = \{x = 0\}$: then we get a variety $\mathcal{P}'_\gamma$ such that $\mathcal{P}_\gamma$ is a fibre bundle over $\mathcal{P}'_\gamma$ with fibre a linear group: hence $\mathcal{P}'_\gamma$ is also smooth.

The discriminant map, defined on $\mathcal{P}'_\gamma$, yields a family of quintic surfaces

$$Y \in \mathcal{F}'(5, 16 + \gamma)$$

which is smooth, because $\mathcal{P}_\gamma$ is a fibre bundle over it (since, for a fixed quintic surface Y, the discriminant matrix $A(x)$ is determined by a $\frac{1}{2}$-even set of nodes plus the choice of a minimal set of generators for the associated graded module, see [21]). Hence the singularities of Y are unobstructed.

Here is an alternative to the last argument of Proposition 2.25.

Corollary 2.27 *One obtains, for all* $\gamma = 0, \ldots, 15$ *nodal quintic surfaces with* $16+\gamma$ *nodes by taking discriminants of cubic hypersurfaces with* γ *nodes, projecting from a general line* $L \subset X$.

Proof It suffices to consider a particular cubic with 15 double points, for instance the Goryunov cubic X, described by the equations $\sigma_1(x) = 0, \sigma_3(x) + z\sigma_2(x) + 2z^3 = 0$, and show that the space of cubics in $\mathbb{P}^5$ surjects onto $\oplus_1^{15} \mathbb{C}_{P_i}$, where $P_1, \ldots, P_{15}$ are the nodes of X. This was done in statement iv) of Theorem 1.25.

Hence the Goryunov cubic X is unobstructed and one can find deformations of X having any number $\gamma \leq 15$ of nodes. Then we apply the previous proposition. □

The maximum number of nodes, $\nu(Y) = 31$, is obtained in the case $\gamma = 15$ (see [9]).

Observe that cubics with 15 nodes in $\mathbb{P}^5$ have been studied by Togliatti in [89] and [90], and by Kalker in his Leiden Thesis [59]. Togliatti showed that 15 is the maximum number of nodes, and that such hypersurfaces do in fact exist (that 15 is the maximum is also explained in the next section).

In the following, the first assertion was proven in Theorem 1.57:

Theorem 2.28 *The Nodal Severi variety* $\mathcal{F}_4(3, 15)$ *of cubic hypersurfaces in* $\mathbb{P}^5$ *with* 15 *nodes is irreducible. It has dimension* 40.

Proof The group of projectivities of $\mathbb{P}^5$ has dimension 35, while K3 surfaces with 14 nodes depend on 5 moduli. □

From the above we derive the following

Theorem 2.29 *The Nodal Severi variety $\mathcal{F}(5,31)$ of Togliatti quintics, i.e. quintic surfaces in $\mathbb{P}^3$ with* 31 *nodes, is smooth irreducible of dimension* 24.

Proof Every Togliatti quintic is (in many ways) the discriminant of a nodal cubic hypersurface $X \subset \mathbb{P}^5$ for the projection with centre a line $L \subset X$.

Hence $\mathcal{F}(5,31)$ is the image of the variety $\mathcal{P}$ of such pairs (L, X) which fibres onto $\mathcal{F}_4(3, 15)$, with fibre an open subset of $F_1(X)$ contained in the open subset of lines not intersecting the singular locus of X, not contained in one of the finite number of planes $\Lambda \subset X$, and such that L does not intersect any of the lines joining two nodes.

By Theorem 2.28, to show irreducibility of $\mathcal{F}(5,31)$, it suffices to show that the Fano scheme of lines L contained in the smooth locus of a nodal cubic hypersurface $X \subset \mathbb{P}^5$ with 15 nodes is irreducible of dimension 4.

We have seen in Theorem 2.8 that the open set consisting of lines not intersecting the singular locus of X is smooth and that $F_1(X)$ is of dimension exactly 4 at each point; its complement consists of the lines passing through the nodes, and these lines correspond to the K3 surfaces of degree 6 associated to the Taylor development of the equation of X at the nodes.

Therefore $F_1(X)$ is nonsingular in codimension 1. Moreover, since it is the zero scheme of the section of a rank 4 bundle, which has exactly codimension 4 in the Grassmannian, it is a local complete intersection. Hence it is normal and, being also connected, it is irreducible.

Concerning the dimension, $\mathcal{F}(5,31)$ has dimension at least $55 - 31 = 15 + 9$, hence at least 9 moduli; and, up to projectivities, it is dominated by a family of dimension 9 (5 moduli for the cubic hypersurfaces, plus 4 parameters for the dimension of $F_1(X)$). Hence $\mathcal{F}(5,31)$ has dimension exactly 24.

Moreover, since the morphism $\mathcal{P} \to \mathcal{F}(5,31)$ is a fibre bundle, we obtain that $\mathcal{F}(5,31)$ is smooth if and only if $\mathcal{P}$ is smooth, equivalently, if and only if $\mathcal{F}_4(3,15)$ is smooth, which follows from the fact that K3 surfaces are unobstructed. □

2.4 An Explicit Togliatti Quintic

In this section we enumerate the 2-planes on the Goryunov–Kalker cubic hypersurface $T(4) \subset \mathbb{P}^5$ and exhibit a discriminant quintic surface with 31 nodes by projecting from an appropriate line contained in $T(4)$.

2.4.1 Linear 2-Spaces on the GK-Cubic 4-Fold

Proposition 2.30 *The Goryunov–Kalker cubic hypersurface $T(4) \subset \mathbb{P}^5 \subset \mathbb{P}^6$ contains* 60 *linear subspaces of maximal dimension* 2. *These split into two $\mathfrak{S}_6$-orbits represented by*

$$(x_1 + x_2 = x_3 + x_4 = x_5 + x_7 = x_6 + x_7 = 0) \text{ and}$$

$$(x_1 + x_2 = x_3 + x_4 = x_5 + x_6 = x_7 = 0).$$

The first orbit consists of 45 planes. Each such plane passes through four ordinary double points and there are 12 such planes passing through each ordinary double point. The second orbit consists of 15 planes, which are inherited from the Segre cubic 3-fold $\Sigma(3) \cong T(4) \cap (x_7 = 0)$ and which are disjoint from $\operatorname{Sing} T(4)$.

Proof We consider the 2-planes in $\mathbb{P}^6$ as points in the Grassmannian $G := \mathrm{Gr}(3, 7)$.

It is immediate to verify that the above 60 planes are contained in $T(4)$. However, the top Chern class of the vector bundle $\mathcal{E}$ such that the Fano scheme $F_2(T(4)) \subset G$ is the zero set of a section of $\mathcal{E}$ is bigger than 60.

It turns out, from the computations below, that the Fano scheme $F_2(T(4)) \subset G$ of 2-planes contained in $T(4)$ is zero-dimensional and everywhere non-reduced.

Let us find the points of the set $F_2(T(4))$.

Let U_{ijk} denote the affine chart of G whose points are represented by 3×7 matrices of rank 3 in row echelon form where columns i, j, k form the 3×3 identity matrix. For example, a point in $U_{123} \cong \mathbb{A}^{12}$ is represented by a matrix

$$M := \begin{pmatrix} 1 & 0 & 0 & a_{14} & a_{15} & a_{16} & a_{17} \\ 0 & 1 & 0 & a_{24} & a_{25} & a_{26} & a_{27} \\ 0 & 0 & 1 & a_{34} & a_{35} & a_{36} & a_{37} \end{pmatrix}.$$

Given a point in U_{ijk} represented by the matrix M, the corresponding 2-plane is parametrized by the map (here the points of $\mathbb{P}^6$ are seen as row vectors)

$$\mathbb{P}^2 \to \mathbb{P}^6, \ (u, v, w) \mapsto (u, v, w)M.$$

Thus the plane M lies in $T(4)$ if and only if the equations of $T(4)$ vanish identically when evaluated at $(u, v, w)M$. This gives conditions on the coefficients a_{mn} defining the scheme $F_2(T(4)) \cap U_{ijk}$ in $\mathbb{A}^{12}$. We use the computer algebra system MAGMA [13] to find all points of this scheme.[1]

In principle, we should need to perform this process for each of the affine charts U_{ijk} and then exclude 2-planes that appear more than once, but since $\mathfrak{S}_6$ acts on $T(4)$ by permuting the coordinates $x_1, \ldots, x_6$, the affine charts of G split into two orbits. Thus it suffices to consider the charts U_{123} and U_{127} and then apply the $\mathfrak{S}_6$-action to obtain all 2-planes on $T(4)$. We obtain the two orbits of 2-planes described in the statement of the Proposition.

The remaining part of the Proposition may be checked directly from the equations. □

[1] See the script [33, MAGMA 24].

2.4.1.1 Lines on the GK-Cubic 4-fold

To construct a quintic surface in $\mathbb{P}^3$ with 31 nodes, we search for lines l contained in $T(4)$ satisfying the following two conditions:

(i) l is disjoint from every line L through a pair of nodes on $T(4)$;
(ii) l is not contained in a 2-plane of $F_2(T(4))$.

These are open conditions on the Fano scheme of lines on $T(4)$.

Lemma 2.31 *The line*

$$x_1 - 9x_5 - 2x_6 = x_2 - 7x_5 - 2x_6 = x_3 - 5x_5 = x_4 + 8x_5 + x_6 = 0$$

satisfies the above conditions.

Proof The proof of the statement, which we verified with a computer calculation, can also be obtained by hand, checking the various intersections and showing that the equation of $T(4)$ is contained in the above ideal. □

How did we find this line? Consider the Grassmannian of lines in $\mathbb{P}^6$ and parametrise affine charts $U_{ij} \cong \mathbb{A}^{10}$ in a similar way to the above Proposition 2.30. A typical point of U_{12} corresponds to a matrix

$$\begin{pmatrix} 1 & 0 & b_{13} & b_{14} & b_{15} & b_{16} & b_{17} \\ 0 & 1 & b_{23} & b_{24} & b_{25} & b_{26} & b_{27} \end{pmatrix}$$

and the equations defining $T(4)$ induce conditions on $b_{13}, \ldots, b_{26}$ defining the affine chart of the Fano scheme of lines. We use the computer to search for integral points of height ≤ 10 in this affine chart and check whether the required properties are fulfilled. The given line is the point corresponding to $\left(\begin{smallmatrix} 9 & 7 & 5 & -8 & 1 & 0 & -7 \\ 2 & 2 & 0 & -1 & 0 & 1 & -2 \end{smallmatrix}\right)$ in the Grassmannian.

Proposition 2.32 *The quintic hypersurface $V \subset \mathbb{P}^3$ defined by the vanishing of the determinant of the symmetric matrix*

$$\begin{pmatrix} 32y_1 - 24y_3 + 15y_4 & 2y_1 - 7y_3 - 3y_4 & m_{13} \\ & -4y_3 - 3y_4 & m_{23} \\ & sym & \frac{1}{3}(s_3 - \frac{1}{4}s_1^3) \end{pmatrix}$$

has 31 *nodes and no other singularities. Here the notation is $s_1 = \sum y_i$, $s_3 = \sum y_i^3$ and*

$$m_{13} = 9y_1^2 + 7y_2^2 + 5y_3^2 - 8y_4^2 - \frac{7}{2}s_1^2, \quad m_{23} = 2y_1^2 + 2y_2^2 - y_4^2 - s_1^2.$$

Proof We project $T(4)$ from the line obtained in Lemma 2.31. The coordinates on the projected $\mathbb{P}^3$ are

$$y_1 := x_1 - 9x_5 - 2x_6, \; y_2 := x_2 - 7x_5 - 2x_6, \; y_3 := x_3 - 5x_5, \; y_4 := x_4 + 8x_5 + x_6.$$

Thus we may eliminate $x_1, \ldots, x_4$ using $y_1, \ldots, y_4, x_5, x_6$ and write the equation of $T(4)$ as a quadratic form in x_5 and x_6. The matrix of this quadratic form is the one given above.

The nodes were calculated using a computer, but it should be possible to prove the assertion without a computer. The 16 'determinantal' nodes arising as the points where the corank of the symmetric matrix is 2 are defined over a complicated field extension of $\mathbb{Q}$. □

2.5 Unobstructed Surfaces: A Togliatti Quintic and Barth's Sextic

Let X be a nodal hypersurface of degree d in $\mathbb{P}^n$. The short exact sequence

$$0 \to \mathcal{J}(d) \to \mathcal{O}_{\mathbb{P}^n}(d) \to \mathcal{O}/\mathcal{J} =: \mathcal{T} \to 0$$

induces the evaluation map

$$\mathrm{ev}\colon H^0(\mathbb{P}^n, \mathcal{O}(d)) \longrightarrow H^0(\mathcal{T}).$$

Here $\mathcal{J}$ is the Jacobian ideal of X and $\mathcal{T}$ is supported on the singular locus of X. Indeed, since X has ordinary double points, $H^0(\mathcal{T}) = \bigoplus_{P \in \mathrm{Sing}(X)} \mathbb{C}_P$. According to Burns–Wahl the deformations are *unobstructed* if the evaluation map is surjective, i.e. the nodes impose linearly independent conditions.

2.5.1 *A Special Togliatti Quintic*

A Togliatti quintic $X_T \subset \mathbb{P}^3$ is defined by

$$2(x^5 - 5x^4w - 10x^3y^2 - 10x^2y^2w + 20x^2w^3 + 5xy^4 - 5y^4w + 20y^2w^3 - \\ 16w^5) + 5(x^2 + y^2 + bz^2 + zw + dw^2)^2 z = 0$$

where $b = -\frac{5-\sqrt{5}}{20}$ and $d = -(1+\sqrt{5})$. The following script shows that the 31 nodes are defined over $\mathbb{Q}(\sqrt{5+2\sqrt{5}})$ and that X_T is unobstructed.

```
Q:=Rationals();
RR<u1>:=PolynomialRing(Q);
K1<t>:=SplittingField(u1^2-5);
RR<u2>:=PolynomialRing(K1);
K<t2>:=SplittingField(u2^2 - (2*t+ 5));
P3<x,y,z,w>:=ProjectiveSpace(K,3);
```

```

b:=-(5-t)/20;
d:=-(1+t);

P:= (x^5 -5*x^4*w -10*x^3*y^2 -10*x^2*y^2*w +20*x^2*w^3 +
5*x*y^4-5*y^4*w +20*y^2*w^3 -16*w^5 );
Q:= (x^2+y^2+b*z^2+z*w+d*w^2);
F:= 2*P+5*z*Q^2;
TQ:=Scheme(P3,F);
Nodes:=Set(SingularPoints(TQ));

//We check that the singularities are the rights ones
#Nodes; HasSingularPointsOverExtension(TQ);
for pt in Nodes do if not IsNode(pt) then "wrong!"; end if; end for;

L:=LinearSystem(P3,5); // The space of quintic in P^3
M:=[];
for section in Sections(L) do // 56 sections
     line:=[];
     for point in Nodes do
          Append(~line, Evaluate(section,[point[j]: j in [1..4]]));
     end for;
     Append(~M, line);
end for;
Rank(Matrix(M)); // The Rank is 31, so the map is surjective
```

2.5.2 Barth's Sextic

The Barth sextic $X_B \subset \mathbb{P}^3$ is defined by

$$(\tau^2 x^2 - y^2)(\tau^2 y^2 - z^2)(\tau^2 z^2 - x^2) - \frac{1}{4}(2\tau + 1)w^2(x^2 + y^2 + z^2 - w^2)^2$$

where $\tau := \frac{1}{2}(1 + \sqrt{5})$ is the golden ratio. The following script verifies that X_B is unobstructed.

```
Q:=Rationals();
RR<z>:=PolynomialRing(Q);
K<t>:=SplittingField(z^2-z-1);
P3<x,y,z,w>:=ProjectiveSpace(K,3);

F:=(t^2*x^2-y^2)*(t^2*y^2-z^2)*(t^2*z^2-x^2)-
     ((2*t+1)/4)*w^2*(x^2+y^2+z^2-w^2)^2;
BS:=Scheme(P3,F);
Nodes:=Set(SingularPoints(BS));

//We check that the singularities are the rights ones
#Nodes; HasSingularPointsOverExtension(BS);
for pt in Nodes do if not IsNode(pt) then "wrong!";end if; end for;

L:=LinearSystem(P3,6); // The space of sextic in P^3
M:=[];
for section in Sections(L) do // 84 sections
     line:=[];
     for point in Nodes do
          Append(~line, Evaluate(section,[point[j]: j in [1..4]]));
     end for;
     Append(~M, line);
end for;
Rank(Matrix(M)); // The Rank is 65, so the map is surjective
```

2.6 Geometry and Deformations of Nodal Segre Cubic Hypersurfaces

Let $n = 2k$. Recall that the Segre cubic hypersurfaces are defined by

$$X_S\colon (x_1 + \cdots + x_n = x_1^3 + \cdots + x_n^3 = 0) \subset \mathbb{P}^{n-1}$$

and their nodes form the $\mathfrak{S}_n$-orbit of $(1^k : -1^k)$ where α^s means α is repeated s times.

The Goryunov–Kalker (GK) cubic hypersurfaces are defined by

$$X_{GK}\colon (x_1 + \cdots + x_n + 2x_{n+1} = x_1^3 + \cdots + x_n^3 + 2x_{n+1}^3 = 0) \subset \mathbb{P}^n$$

and their nodes are the $\mathfrak{S}_n$-orbit of the point $(1^{k-1} : -1^{k+1} : 1)$.

If $n = 4$, then the Segre hypersurface is a plane cubic curve with three nodes i.e. a triangle! Whereas, the GK surface is the Cayley cubic surface. So we will restrict ourselves to the case $n = 2k \geq 6$.

In this section we prove Theorem 2.19 namely we show

(i) The Segre cubic 3-fold ($n = 6$) is unique and is unobstructed.
(ii) The Goryunov–Kalker cubic 4-fold ($n = 6$) has non-trivial equisingular deformations and is unobstructed.
(iii) The Segre cubic 5-fold and the Goryunov–Kalker 6-fold are unobstructed and locally rigid.
(iv) All other Segre cubics and Goryunov–Kalker cubics are locally rigid and have obstructed deformations.

Let H be the hyperplane $(x_1 + \cdots + x_n = 0) \subset \mathbb{P}^{n-1}$ respectively $(x_1 + \cdots + x_n + 2x_{n+1} = 0) \subset \mathbb{P}^n$. The short exact sequence

$$0 \to \mathcal{J}(3) \to \mathcal{O}_H(3) \to \mathcal{O}_H/\mathcal{J} =: \mathcal{T} \to 0 \tag{2.3}$$

induces the evaluation map

$$\mathrm{ev}\colon H^0(H, \mathcal{O}_H(3)) \to H^0(\mathcal{T}).$$

Here $\mathcal{J}$ is the Jacobian ideal of $X \subset H$ and $\mathcal{T}$ is supported on the singular locus of X. Indeed, since X has ordinary double points, $H^0(\mathcal{T}) = \bigoplus_{P \in \operatorname{Sing} X} \mathbb{C}_P$. According to Burns–Wahl [18] the deformations are unobstructed if the evaluation map is surjective, i.e. the nodes impose linearly independent conditions on the space of cubic hypersurfaces.

We say that X is *locally rigid* if any equisingular deformation of X is projectively equivalent to X itself. This condition is fulfilled if the kernel of the evaluation map is spanned by the image of X under the action of the space of infinitesimal automorphisms. This occurs in every case except the GK-cubic 4-fold, where $H^0(\mathcal{J}(3))$ contains an extra 5-dimensional space of cubics.

The evaluation map is $\mathfrak{S}_n$-equivariant and we prove the theorem by decomposing the various vector spaces into irreducible $\mathfrak{S}_n$-representations and applying Schur's lemma.

The proof is organized as follows. In Sect. 2.6.1 we recall some well-known facts about the irreducible representations of the symmetric group $\mathfrak{S}_n$. In Sects. 2.6.2 and 2.6.3 we describe the evaluation map of the Segre cubics and the Goryunov–Kalker cubics, and we prove the theorem. In Sect. 2.6.4 we give the decomposition of $H^0(\mathcal{T})$ into irreducible representations.

2.6.1 Representation Theory of the Symmetric Group

We briefly recall some background on Young tableaux, tabloids and Specht modules. We refer to James' book [55] for proofs and further details.

There is a correspondence between representations of the symmetric group $\mathfrak{S}_n$ and partitions $\lambda := (\lambda_1, \ldots, \lambda_m)$ of n, where $\lambda_1 \geq \cdots \geq \lambda_m$. To a partition, one associates a Young diagram, consisting of m rows each containing λ_i boxes. A Young tableau is a Young diagram with shape λ where the boxes are labelled by $1, \ldots, n$. A Young tableau is called standard if it has strictly increasing rows and columns.

Definition 2.33 A *Young tabloid* of shape λ is an equivalence class of Young tableaux of shape λ, with respect to the relation where two tableaux are equivalent if they differ by a permutation of the labelling of the rows.

The symmetric group $\mathfrak{S}_n$ acts on the set of Young tableaux of shape λ by permuting the labelling. The *column stabiliser* $\mathfrak{C}_T$ of a Young tableau T is the subgroup of $\mathfrak{S}_n$ which permutes the labelling of T leaving the columns invariant.

Example 2.34 $T := \begin{smallmatrix}1&5&6\\2&4&\\3&&\end{smallmatrix}$ is a standard Young tableau of shape $(3, 2, 1)$. As a tabloid, T is equivalent to $\begin{smallmatrix}6&1&5\\4&2&\\3&&\end{smallmatrix}$. The column stabiliser of T is $\mathfrak{S}(\{1, 2, 3\}) \times \mathfrak{S}(\{4, 5\}) \cong \mathfrak{S}_3 \times \mathfrak{S}_2$.

Fix a partition λ and consider the $\mathbb{C}$-vector space W generated by Young tabloids of shape λ. That is, the elements are formal $\mathbb{C}$-linear combinations of Young tabloids.

Definition 2.35 For a fixed Young tableau T, we define the *Specht vector* e_T by

$$\mathsf{e}_T = \sum_{\sigma \in \mathfrak{C}_T} \operatorname{sign}(\sigma) \cdot [\sigma(T)]$$

where $[X]$ denotes the tabloid corresponding to the tableau X.

The *Specht module* S_λ is the linear subspace of W spanned by the $\mathfrak{S}_n$-orbit of e_T where T has shape λ. This is a cyclic $\mathfrak{S}_n$-module generated by all Specht vectors of shape λ.

Example 2.36 The column stabiliser of the tableau $\substack{1\,2\,3\,4\\5\,6}$ is isomorphic to $\mathbb{Z}/2 \times \mathbb{Z}/2$ generated by permutations (15) and (26). The corresponding Specht vector is

$$\mathsf{e}_{\substack{1\,2\,3\,4\\5\,6}} = \left[\substack{1\,2\,3\,4\\5\,6}\right] - \left[\substack{5\,2\,3\,4\\1\,6}\right] - \left[\substack{1\,6\,3\,4\\5\,2}\right] + \left[\substack{5\,6\,3\,4\\1\,2}\right]$$

As a sum of tabloids, the right-hand side is independent of row-ordering. The left-hand side depends on the tableau, and therefore the row-ordering too e.g.

$$\mathsf{e}_{\substack{2\,1\,3\,4\\5\,6}} = \left[\substack{2\,1\,3\,4\\5\,6}\right] - \left[\substack{5\,1\,3\,4\\2\,6}\right] - \left[\substack{2\,6\,3\,4\\5\,1}\right] + \left[\substack{5\,6\,3\,4\\2\,1}\right]$$

Theorem 2.37 ([55, §§4–7]) *In characteristic 0, the Specht modules S_λ give all the irreducible representations of $\mathfrak{S}_n$. A basis of S_λ is given by the Specht vectors corresponding to standard tableaux of shape λ.*

Remark 2.38 The partition $\lambda = (n)$ corresponds to the trivial representation $V_{(n)}$ of $\mathfrak{S}_n$. The symmetric group $\mathfrak{S}_n$ acts in a natural way on the subspace $(\sum_{i=1}^n z_i = 0) \lhd \mathbb{C}^n$ by permuting the coordinates. This subspace is an irreducible representation called the *standard representation* $V_{(n-1,1)}$ of $\mathfrak{S}_n$. The partition $n = (n-k) + 1 + \cdots + 1$ corresponds to the k-th exterior product of the standard representation (cf. [46, §4.1]).

2.6.2 *Representation Theory for the Segre Cubics*

Recall that H is the hyperplane $(\sum_{i=1}^n x_i = 0) \subset \mathbb{P}^{n-1}$. The symmetric group $\mathfrak{S}_n$ acts in a natural way on $\mathbb{P}^{n-1}$, so the vector space $V := H^0(\mathcal{O}_H(1))$ is isomorphic to the standard representation $V \cong V_{(n-1,1)}$ of $\mathfrak{S}_n$.

2.6.2.1 Representations of Cubic Polynomials

We decompose $\mathrm{Sym}^3 V$ into irreducible representations and present the generators in Table 2.1 below.

First we decompose $\mathrm{Sym}^3 V$ into symmetric and antisymmetric representations, with basis $f_{ijk} := x_i x_j x_k$ for $1 \le i < j < k \le n$ and $g_{ij} := x_i x_j (x_i - x_j)$ for $1 \le i < j \le n$ respectively. Indeed, since $\mathrm{Sym}^3 V$ is $\mathrm{Sym}^3 \langle x_1, \ldots, x_n \rangle$ modulo $(\sum x_i = 0)$, we have

$$2x_a^3 = x_a\Big(\sum_{j \neq a} x_j\Big)^2 - x_a^2\Big(\sum_{j \neq a} x_j\Big) = \sum_j (x_a x_j^2 - x_a^2 x_j) \mod \langle f_{klm} \rangle$$

Table 2.1 Cubics on the Segre hypersurface

Partition λ	Tableau T	Generator	Dimension
(n)	$c_1 \dots c_n$	$\mathfrak{f}_T = \sigma_3(x_C)$	1
$(n-1, 1)$	$a\ c_1 \dots c_{n-2}$ b	$\mathfrak{f}_T = (x_a - x_b)\sigma_2(x_C)$	$n-1$
$(n-2, 2)$	$a_1\ a_2\ c_1 \dots c_{n-4}$ $b_1\ b_2$	$\mathfrak{f}_T = (x_{a_1} - x_{b_1})(x_{a_2} - x_{b_2})\sigma_1(x_C)$	$\frac{n(n-3)}{2}$
$(n-3, 3)$	$a_1\ a_2\ a_3\ c_1 \dots c_{n-6}$ $b_1\ b_2\ b_3$	$\mathfrak{f}_T = \prod_{i=1}^{3}(x_{a_i} - x_{b_i})$	$\frac{n(n-1)(n-5)}{6}$
$(n-1, 1)$	$a\ c_1 \dots c_{n-2}$ b	$\mathfrak{g}_T =$ $(x_a - x_b)(2x_a x_b + (x_a + x_b)\sigma_1(x_C) - s_2(x_C))$	$n-1$
$(n-2, 1, 1)$	$a_1\ c_1 \dots c_{n-3}$ a_2 a_3	$\mathfrak{g}_T = (x_{a_1} - x_{a_2})(x_{a_2} - x_{a_3})(x_{a_3} - x_{a_1})$	$\binom{n-1}{2}$

and

$$x_a^2 x_b = -x_a x_b \sum_{j \neq a} x_j = -x_a x_b^2 \mod \langle f_{klm} \rangle .$$

Thus $\operatorname{Sym}^3 V = \langle f_{ijk} \rangle \oplus \langle g_{ij} \rangle$. We further decompose these two summands into irreducible representations.

Proposition 2.39 *Let $n \geq 6$. Then*

$$\operatorname{Sym}^3 V \cong V_{(n)} \oplus V_{(n-1,1)} \oplus V_{(n-2,2)} \oplus V_{(n-3,3)} \oplus V_{(n-1,1)} \oplus V_{(n-2,1,1)} .$$

The generators of each summand are presented in Table 2.1.

Each row of Table 2.1 corresponds to a summand V_λ of the decomposition: the first four rows are spanned by f_{ijk} and the bottom two by g_{ij}. The generator column refers to a cyclic generator of V_λ as an $\mathfrak{S}_n$-module, i.e. the $\mathfrak{S}_n$-orbit of such spans V_λ as a vector space. We use $\sigma_d(x_C)$ (resp. $s_d(x_C)$) to denote the elementary symmetric polynomial of degree d (resp. the d-th power sum) in the variables x_c indexed by c in C. Here C is the set $\{c_i\}$.

Proof In Lemma 2.41 we show that the set of generators corresponding to standard Young tableaux of shape λ form a basis for the irreducible representation V_λ. Since each row of Table 2.1 gives a distinct irreducible representation, and the sum of their dimensions is $\binom{n+1}{3} = \dim \operatorname{Sym}^3 V$, we are done. □

Remark 2.40 The generators for the first four rows of Table 2.1 were obtained by analogy with the construction of the Specht module $S_{(n-p,p)}$. Note that $\sum_{\tau \in \mathfrak{C}_T} \prod_{i=1}^{p} x_{\tau(a_i)}$ is equal to $\prod_{i=1}^{p}(x_{a_i} - x_{b_i})$. For $V_{(n-2,1,1)}$, the generator is the Vandermonde determinant

$$\mathfrak{g}_T = \det \begin{pmatrix} 1 & 1 & 1 \\ x_{a_1} & x_{a_2} & x_{a_3} \\ x_{a_1}^2 & x_{a_2}^2 & x_{a_3}^2 \end{pmatrix} = g_{a_1 a_2} + g_{a_2 a_3} + g_{a_3 a_1},$$

a well-known skew polynomial of degree 3. For the skew summand $V_{(n-1,1)}$ we computed the orthogonal complement to $V_{(n-2,1,1)}$ inside $\langle g_{ij}\rangle$:

$$\mathfrak{g}_T = g_{a_1a_2} - \sum_{i<j<k} ((g_{ij} + g_{jk} + g_{ki}) \cdot g_{a_1a_2})(g_{ij} + g_{jk} + g_{ki}) .$$

Lemma 2.41 *Let λ be a partition of n as in Proposition 2.39, let T be a Young tableau of shape λ, and let $\mathfrak{e}_T$ be the corresponding generator.*

(i) If $\tau \in \mathfrak{C}_T \subset \mathfrak{S}_n$ preserves the columns of T, then $\tau(\mathfrak{e}_T) = \mathfrak{e}_{\tau(T)} = \operatorname{sign}(\tau)\mathfrak{e}_T$.
(ii) If γ in $\mathfrak{S}_C \subset \mathfrak{S}_n$ permutes only the C-coordinates then $\gamma(\mathfrak{e}_T) = \mathfrak{e}_{\gamma(T)} = \mathfrak{e}_T$.
(iii) For any permutation ρ in $\mathfrak{S}_n$, we have $\rho(\mathfrak{e}_T) = \mathfrak{e}_{\rho(T)}$.
(iv) The set of all $\mathfrak{e}_T$'s, with T standard gives a basis of the representation V_λ.

Proof We give an elementary proof for the third row $\lambda = (n-2, 2)$ of Table 2.1. The proof in the other cases works in exactly the same way, but to write everything out is repetitive. Here we consider Young tableaux T of shape $(n-2, 2)$:

$$T = \begin{matrix} a_1 & a_2 & c_1 & \dots & c_{n-4} \\ b_1 & b_2 \end{matrix}$$

(i) if τ preserves the columns then $\tau(\mathfrak{e}_T) = (x_{\tau(a_1)} - x_{\tau(b_1)})(x_{\tau(a_2)} - x_{\tau(b_2)})\sigma_1(x_C) = \mathfrak{e}_{\tau(T)} = \operatorname{sign}(\tau)\mathfrak{e}_T$.
(ii)–(iii) Using i), it suffices to prove the statements for transpositions of the first row of T. There are three cases. If ρ is a permutation of the C-coordinates, then clearly $\rho(\mathfrak{e}_T) = \mathfrak{e}_{\rho(T)}$ because $\sigma_1(x_C)$ is invariant under permutations of C. If ρ swaps say a_1 with c_1, then $\rho(\mathfrak{e}_T) = (x_{c_1} - x_{b_1})(x_{a_2} - x_{b_2})\sigma_1(x_{\rho(C)}) = \mathfrak{e}_{\rho(T)}$. If ρ swaps two elements of A, then $\rho(\mathfrak{e}_T) = (x_{a_2} - x_{b_1})(x_{a_1} - x_{b_2})\sigma_1(x_C) = \mathfrak{e}_{\rho(T)}$.
(iv) Suppose now T is standard. Then the last two monomials of $\mathfrak{e}_T$ in lexicographic order are $x_{b_1}x_{b_2}x_{c_{n-3}} + x_{b_1}x_{b_2}x_{c_{n-4}}$. Since the entries of T increase along rows and columns, these trailing terms are pairwise distinct. Thus the standard $\mathfrak{e}_T$ are linearly independent.

There is a general algorithm applying the so-called Garnir relations recursively to write any nonstandard generator $\mathfrak{e}_T$ as a linear combination of standard generators (cf. [55, §7]). In our case the algorithm terminates particularly quickly because T has only two rows.

Step 1: Suppose $\mathfrak{e}_T$ is a nonstandard generator. Using i) we may assume the columns are ordered and using ii) we may assume that the C-part is ordered.

Step 2: If $a_2 > c_1$ then we use the Garnir relation $\mathfrak{e}_T - (a_2, c_1)\mathfrak{e}_T + (a_2, c_1, b_2)\mathfrak{e}_T = 0$ and possibly column permutations to write $\mathfrak{e}_T$ as a linear combination of two generators $\mathfrak{e}_{T_1}$ and $\mathfrak{e}_{T_2}$ with $a_2(T_i) < c_1(T_i)$ for $i = 1, 2$. We then apply Step 3 to each of these.

Step 3: If $a_1 > a_2$ then we use the Garnir relation $\mathfrak{e}_T - (a_1, a_2)\mathfrak{e}_T + (a_1, a_2, b_1)\mathfrak{e}_T = 0$ and possibly column/C-part permutations to write $\mathfrak{e}_T$ as a linear combination of generators having $a_1 < a_2$.

We then restart from Step 1 with each of the new generators.

We prove directly that the Garnir relations hold for the third row of Table 2.1. For step 2, we first write $C' := C \setminus \{c_1\}$ and observe that

$$\begin{aligned}(x_{a_2} - x_{b_2})\sigma_1(x_C) - (x_{c_1} - x_{b_2})\sigma_1(x_{C'\cup\{a_2\}}) + (x_{c_1} - x_{a_2})\sigma_1(x_{C'\cup\{b_2\}}) =\\ = \big((x_{a_2} - x_{b_2}) - (x_{c_1} - x_{b_2}) + (x_{c_1} - x_{a_2})\big)\,\sigma_1(x_{C'})\\ + (x_{a_2} - x_{b_2})x_{c_1} - (x_{c_1} - x_{b_2})x_{a_2} + (x_{c_1} - x_{a_2})x_{b_2} = 0 + 0\end{aligned}$$

Multiplying the first line by $x_{a_1} - x_{b_1}$, gives $\mathsf{e}_T - (a_2, c_1)\mathsf{e}_T + (a_2, c_1, b_2)\mathsf{e}_T = 0$. In Step 3, the C-part of T is unchanged so we get a common factor of $\sigma_1(x_C)$:

$$\begin{aligned}\mathsf{e}_T - (a_1, a_2)\mathsf{e}_T + (a_1, a_2, b_1)\mathsf{e}_T =\\ = ((x_{a_1} - x_{b_1})(x_{a_2} - x_{b_2}) - (x_{a_2} - x_{b_1})(x_{a_1} - x_{b_2}) +\\ (x_{a_2} - x_{a_1})(x_{b_1} - x_{b_2}))\sigma_1(x_C) = 0\end{aligned}$$

□

2.6.2.2 The Evaluation Map and its Kernel

Let $X_S \subset \mathbb{P}^{n-1}$ be the Segre cubic and let $H^0(\mathcal{T}) = \bigoplus_{P \in \mathrm{Sing}\, X_S} \mathbb{C}_P$ be the vector space of global sections of the skyscraper sheaf $\mathcal{T}$. A vector in $H^0(\mathcal{T})$ is a formal $\mathbb{C}$-linear combination of points. We denote the basis vector 1_P in $\mathbb{C}_P$ by a $2 \times k$ labelled array

$$\frac{p_1 \ \cdots \ p_k}{p_{k+1} \ \cdots \ p_n} \begin{array}{l}\leftarrow \text{positive}\\ \leftarrow \text{negative}\end{array}$$

where $\{p_1, \ldots, p_n\} = \{1, \ldots, n\}$. The first row corresponds to the coordinates equal to 1, and the second row to coordinates equal to -1. This implicitly involves a choice of positive and negative row of the array. Swapping the rows multiplies by -1 in $\mathbb{C}_P$. Clearly, any permutation of the x_i which preserves the rows corresponds to the same point in $\mathbb{P}^{n-1}$, and so does any permutation which swaps the two rows (after applying the $\mathbb{C}^*$-action on $\mathbb{P}^{n-1}$). For example, for $n = 6$ we have

$$\begin{aligned}\frac{1\ 3\ 4}{2\ 5\ 6} = \frac{3\ 4\ 1}{2\ 6\ 5} = (1 : -1 : 1 : 1 : -1 : -1) =\\ -1 \cdot (-1 : 1 : -1 : -1 : 1 : 1) = -1 \cdot \frac{2\ 5\ 6}{4\ 1\ 3}\end{aligned}$$

Remark 2.42 Each permutation $\sigma \in \mathfrak{S}_n$ is a bijection of the set $\operatorname{Sing} X_S$, but *not* of the set of vectors $\{1_P \mid P \in \operatorname{Sing} X_S\}$, e.g. let $\sigma = (16)(24)(35)$ and $P = (1 : -1 : 1 : 1 : -1 : -1)$, then $\sigma(1_P) = -1_P$.

Lemma 2.43 *The evaluation map* $\operatorname{ev}\colon \operatorname{Sym}^3 V \to H^0(\mathcal{T})$ *is* $\mathfrak{S}_n$*-equivariant.*

Proof Clearly it is enough to prove the statement for a generator e_T in $\operatorname{Sym}^3 V$. We note that if $\sigma(1_P) = -1_Q$ then $\mathsf{e}_T(P) \cdot \sigma(1_P) = \mathsf{e}_T(-\sigma^{-1}(Q)) \cdot (-1_Q) = \mathsf{e}_T(\sigma^{-1}(Q)) \cdot 1_Q$, since e_T has degree 3. Analogously, if $\sigma(1_P) = 1_Q$ then $\mathsf{e}_T(P) \cdot \sigma(1_P) = \mathsf{e}_T(\sigma^{-1}(Q)) \cdot 1_Q$. Therefore,

$$\sigma(\operatorname{ev}(\mathsf{e}_T)) = \sum_P \mathsf{e}_T(P) \cdot \sigma(1_P) = \sum_P \mathsf{e}_T(\sigma^{-1}P) \cdot 1_P =$$

$$\sum_P \mathsf{e}_{\sigma(T)}(P) \cdot 1_P = \operatorname{ev}(\sigma(\mathsf{e}_T)).$$

□

Proposition 2.44 *The evaluation map* $\operatorname{ev}\colon \operatorname{Sym}^3 V \to H^0(\mathcal{T})$ *has kernel isomorphic to* $V_{(n)} \oplus V_{(n-1,1)} \oplus V_{(n-2,2)} \oplus V_{(n-2,1,1)}$ *and image isomorphic to* $V_{(n-1,1)} \oplus V_{(n-3,3)}$.

Proof By Schur's Lemma and Lemma 2.43 it suffices to consider the evaluation map on generators of each irreducible representation. It is easy to see that the evaluation map is identically zero on $V_{(n)} \oplus V_{(n-2,1,1)}$ by considering the generators listed in Table 2.1.

Fix now a generator f_T of $V_{(n-2,2)}$. For each $i = 1, 2$ we may assume that the a_i- and b_i-coordinates of P have opposite sign, otherwise $x_{a_i} - x_{b_i}$ would vanish. Thus exactly one half of the remaining $c_1, \ldots, c_{n-4}$-coordinates of P are positive, and one half negative. Hence $\sigma_1(x_C)$ vanishes at P which implies $\mathsf{e}_T(P)$ also vanishes.

Clearly, a generator f_T of shape $(n-3, 3)$ is nonzero on P with a_i- and b_i-coordinates of opposite sign, for $i = 1, 2, 3$. Note that $n \geq 6$ and so f_T is not in the kernel.

Both copies of $V_{(n-1,1)}$ in $\operatorname{Sym}^3 V$ map isomorphically to the *same* subspace of $H^0(\mathcal{T})$. Indeed, let P be any point whose a- and b-coordinate have opposite sign, and let f_T, g_T generators for the symmetric and the antisymmetric summand, then $\mathsf{f}_T(P)$ and $\mathsf{g}_T(P)$ are both nonzero. On the other hand $\sigma_2(x_C) = 1 - k$: the coefficient of z^{2k-2} in the polynomial $(z^2 - 1)^{k-1}$. Whence

$$\begin{aligned}(n\mathsf{f}_T + (1-k)\mathsf{g}_T)(P) &= \pm 2[n\sigma_2(x_C) + (1-k)(-2 - s_2(x_C))] \\ &= \pm 2[n(1-k) + (1-k)(-2-(n-2))] = 0\end{aligned}$$

Thus the subspace generated by $n\mathsf{f}_T + (1-k)\mathsf{g}_T$ for T of shape $(n-1, 1)$ is in the kernel. That is a copy of the irreducible representation $V_{(n-1,1)}$. □

Corollary 2.45 *The evaluation map is not surjective for* $n \geq 10$. *Hence the Segre cubic in dimension* ≥ 7 *has obstructed equisingular deformations.*

Proof The image of the evaluation map has dimension $n-1+\frac{n(n-1)(n-5)}{6}=\binom{n-1}{3}$. For $n\geq 10$, this is smaller than the number of nodes $\binom{n-1}{k-1}$, which is the dimension of $H^0(\mathcal{T})$. □

2.6.2.3 Representation of Endomorphisms

If $X\colon (f(x)=0)$ is a hypersurface in $H\cong \mathbb{P}(V)$ then any linear endomorphism $\varphi\colon V\to V$ induces an infinitesimal deformation $\tilde{X}$ of X defined by $(f(x+\varepsilon\varphi(x))=0)$ in $H\times \operatorname{Spec}\mathbb{C}[\varepsilon]/(\varepsilon^2)$. Since $f(x+\varepsilon\varphi(x))=f(x)+\varepsilon\sum_i \varphi_i(x)f_{x_i}$, where f_{x_i} are the partial derivatives of f, we see that the generic fibre of $\tilde{X}$ also contains the singular locus of X. Thus we have a map from $\operatorname{End}V$ to the kernel of the evaluation map. Recall that X is locally rigid if the kernel of the evaluation map is equal to the image of $\operatorname{End}V$.

Lemma 2.46 *The* $\mathfrak{S}_n$*-representation* $\operatorname{End}V$ *decomposes as*

$$\operatorname{End}V\cong V_{(n)}\oplus V_{(n-1,1)}\oplus V_{(n-2,2)}\oplus V_{(n-2,1,1)}$$

Proof Every representation of $\mathfrak{S}_n$ has real character, thus we get $\operatorname{End}V\cong V^*\otimes V$ is isomorphic to $V\otimes V$, and so the decomposition of $\operatorname{End}V$ is

$$\operatorname{End}V\cong \operatorname{Sym}^2V\otimes \bigwedge^2 V\cong V_{(n)}\oplus V_{(n-1,1)}\oplus V_{(n-2,2)}\oplus V_{(n-2,1,1)}\,.$$

□

Theorem 2.47 *The Segre cubic* $X_S\subset\mathbb{P}^{n-1}$ *is locally rigid for all* $n\geq 6$.

Proof By Schur's Lemma, Proposition 2.44 and Lemma 2.46, we have an isomorphism between the image of $\operatorname{End}V$ and the kernel of the evaluation map for all $n\geq 6$, whence X is locally rigid. □

2.6.3 Representation Theory for the Goryunov–Kalker Cubics

Recall that now H is the hyperplane $(\sum_{i=1}^n x_i+2x_{n+1}=0)\subset\mathbb{P}^n$ and define $V:=H^0(\mathcal{O}_H(1))$. This time, $\mathfrak{S}_n$ acts by permuting the first n coordinates and fixing the last one, so V decomposes as the standard representation plus the trivial one: $V\cong V_{(n-1,1)}\oplus V_{(n)}$.

Proposition 2.48 *Let* $n\geq 6$. *Then* Sym^3V *decomposes into irreducible representations as:*

$$\operatorname{Sym}^3V\cong V_{(n)}^{\oplus 3}\oplus V_{(n-1,1)}^{\oplus 4}\oplus V_{(n-2,2)}^{\oplus 2}\oplus V_{(n-3,3)}\oplus V_{(n-2,1,1)}$$

Table 2.2 Extra cubics on the GK hypersurface

Partition λ	Tableau	Generator	Dimension
(n)	$c_1 \dots c_n$	$\mathsf{h}_T = \sigma_2(x_C)x_{n+1}$	1
$(n-1,1)$	$a\ c_1 \dots c_{n-2}$ b	$\mathsf{h}_T = (x_a - x_b)\sigma_1(x_C)x_{n+1}$	$n-1$
$(n-2,2)$	$a_1\ a_2\ c_1 \dots c_{n-4}$ $b_1\ b_2$	$\mathsf{h}_T = (x_{a_1} - x_{b_1})(x_{a_2} - x_{b_2})x_{n+1}$	$\frac{n(n-3)}{2}$
$(n-1,1)$	$a\ c_1 \dots c_{n-2}$ b	$\mathsf{k}_T = (x_a - x_b)x_{n+1}^2$	$n-1$
(n)	$c_1 \dots c_n$	$\mathsf{l}_T = x_{n+1}^3$	1

Generators for each summand are found in Tables 2.1 and 2.2.

Proof Using the identity $\operatorname{Sym}^m(A \oplus B) \cong \bigoplus_{i+j=m} \operatorname{Sym}^i A \otimes \operatorname{Sym}^j B$, which is well-known, together with $A \cong V_{(n-1,1)}$ and $B = \langle x_{n+1} \rangle$ and Proposition 2.39 we obtain:

$$\begin{aligned}\operatorname{Sym}^3 V &\cong \operatorname{Sym}^3 V_{(n-1,1)} \oplus \langle x_{n+1} \rangle \otimes \operatorname{Sym}^2 V_{(n-1,1)} \oplus \\ &\qquad \left\langle x_{n+1}^2 \right\rangle \otimes V_{(n-1,1)} \oplus \left\langle x_{n+1}^3 \right\rangle \\ &\cong \operatorname{Sym}^3 V_{(n-1,1)} \oplus (V_{(n)} \oplus V_{(n-1,1)} \oplus V_{(n-2,2)}) \oplus V_{(n-1,1)} \oplus V_{(n)} \\ &\cong V_{(n)}^{\oplus 3} \oplus V_{(n-1,1)}^{\oplus 4} \oplus V_{(n-2,2)}^{\oplus 2} \oplus V_{(n-3,3)} \oplus V_{(n-2,1,1)}\end{aligned}$$

Looking at the decomposition in the first line, we can easily find cyclic generators for the summands involving x_{n+1}. These are reported in Table 2.2. The summands not involving x_{n+1} appear in Table 2.1. □

The cubic X is locally rigid if the kernel of the evaluation map is equal to the orbit of the defining equation of X under projectivities. Thus, we are going to compare the kernel with $\operatorname{End} V$.

Lemma 2.49 *The* $\mathfrak{S}_n$*-representation* $\operatorname{End} V$ *decomposes as*

$$\operatorname{End} V \cong V_{(n)}^{\oplus 2} \oplus V_{(n-1,1)}^{\oplus 3} \oplus V_{(n-2,2)} \oplus V_{(n-2,1,1)}\,.$$

Proof As in Proposition 2.46, we have $\operatorname{End} V \cong V \otimes V$ and hence

$$\begin{aligned}\operatorname{End}(V_{(n-1,1)} \oplus V_{(n)}) &\cong (V_{(n-1,1)} \oplus V_{(n)}) \otimes (V_{(n-1,1)} \oplus V_{(n)}) \\ &\cong V_{(n)}^{\oplus 2} \oplus V_{(n-1,1)}^{\oplus 3} \oplus V_{(n-2,2)} \oplus V_{(n-2,1,1)}.\end{aligned}$$

□

Let $X_{GK} \subset \mathbb{P}^n$ be the Goryunov–Kalker cubic and let $H^0(\mathcal{T}) = \bigoplus_{P \in \operatorname{Sing} X_{GK}} \mathbb{C}_P$ be the vector space of global sections of the skyscraper sheaf $\mathcal{T}$.

Remark 2.50

(i) The symmetric group $\mathfrak{S}_n$ acts trivially on the last coordinate, we have a natural choice for a representative of the basis vector 1_P in $\mathbb{C}_P$, namely we denote the basis vector 1_P in $\mathbb{C}_P$ by the representative of P having $x_{n+1} = 1$. We remark that each P has $k+1$ negative coordinates.

(ii) Since the symmetric group $\mathfrak{S}_n$ acts trivially on the last coordinate, each permutation $\sigma \in \mathfrak{S}_n$ is a not only a bijection of the set $\operatorname{Sing} X_{GK}$, but *also* of the set of vectors $\{1_P \mid P \in \operatorname{Sing} X_{GK}\}$.

Lemma 2.51 *The evaluation map* $\operatorname{ev}\colon \operatorname{Sym}^3 V \to H^0(\mathcal{T})$ *is* $\mathfrak{S}_n$*-equivariant.*

Proof Since a permutation σ induces a bijection on the set $\{1_P\}$, the proof is immediate:

$$\begin{aligned}\sigma(\operatorname{ev}(\mathsf{e}_T)) &= \sum_P \mathsf{e}_T(P)\cdot\sigma(1_P) = \sum_P \mathsf{e}_T(\sigma^{-1}P)\cdot 1_P\\ &= \sum_P \mathsf{e}_{\sigma(T)}(P)\cdot 1_P = \operatorname{ev}(\sigma(\mathsf{e}_T)).\end{aligned}$$

□

Proposition 2.52 *Let* $n \geq 8$. *The evaluation map* $\operatorname{ev}\colon \operatorname{Sym}^3 V \to H^0(\mathcal{T})$ *has kernel isomorphic to* $V_{(n)}^{\oplus 2} \oplus V_{(n-1,1)}^{\oplus 3} \oplus V_{(n-2,2)} \oplus V_{(n-2,1,1)}$ *and image isomorphic to* $V_{(n)} \oplus V_{(n-1,1)} \oplus V_{(n-2,2)} \oplus V_{(n-3,3)}$.

Proof Using the determinantal description of a generator e_T of $V_{(n-2,1,1)}$ given in Remark 2.40 we see immediately that $\mathsf{e}_T(P)$ vanishes for any singular point P and thus e_T is in the kernel.

A generator f_T of shape $(n-3,3)$ is nonzero on a point P with a_i- and b_i-coordinates of opposite sign, for $i = 1, 2, 3$. Clearly, such a singular point exists on X if and only if $k-1 \geq 3$, i.e. $n \geq 8$. If $n = 6$ then $V_{(n-3,3)}$ is contained in the kernel of the evaluation map.

Both copies of $V_{(n-2,2)}$ in $\operatorname{Sym}^3 V$ map isomorphically to the *same* subspace of $H^0(\mathcal{T})$. For a given tableau T consider the generators f_T and h_T. For each $i = 1, 2$ we may assume that the a_i- and b_i-coordinates of P have opposite sign, otherwise $x_{a_i} - x_{b_i}$ would vanish. Thus exactly $k-3$ of the remaining $c_1, \ldots, c_{n-4}$-coordinates of P are positive, and $k-1$ are negative. Hence $\sigma_1(x_C) = -2$ and $x_{n+1} = 1$, which implies that f_T and h_T do not vanish at P, but $\mathsf{f}_T + 2\mathsf{h}_T$ does.

The remaining two cases are similar:

Two copies of $V_{(n)}$ are in the kernel and one copy is in the image, because l_T evaluates to 1 but $\mathsf{f}_T + 2(k-1)\mathsf{l}_T$ and $\mathsf{h}_T + (k-2)\mathsf{l}_T$ evaluate to zero for any singular point P. Three copies of $V_{(n-1,1)}$ are in the kernel and one copy is in the image, because $\mathsf{f}_T + (k-4)\mathsf{k}_T$, $\mathsf{g}_T + k\mathsf{k}_T$, $\mathsf{h}_T + 2\mathsf{k}_T$ evaluate to 0 for any P, although k_T evaluates to ± 2 for certain P. □

As consequence we get

Theorem 2.53

(a) *The Goryunov–Kalker cubic $X_{GK} \subset \mathbb{P}^n$ is locally rigid for all $n \geq 8$.*
(b) *The evaluation map is not surjective for $n \geq 10$. Hence the Goryunov–Kalker cubic in dimension ≥ 8 has obstructed equisingular deformations.*

Proof

(a) By Schur's Lemma and Propositions 2.48 and 2.52, we have an isomorphism between the image of End V and the kernel of the evaluation map.
(b) The image of the evaluation map has dimension $n + \frac{n(n-3)}{2} + \frac{n(n-1)(n-5)}{6} = \binom{n}{3}$. For $n \geq 10$, this is smaller than the number of nodes $\binom{n}{k-1}$, which is the dimension of $H^0(\mathcal{T})$.

□

Remark 2.54 By the proof of Proposition 2.52, when $n = 6$ the kernel of the evaluation map is bigger than End V:

$$\ker(\mathrm{ev}) \cong \mathrm{End}\, V \oplus V_{(3,3)}$$

whence X is not locally rigid. The extra summand $V_{(3,3)}$ has basis

$$(x_1 - x_4)(x_2 - x_5)(x_3 - x_6)\,,\ (x_1 - x_3)(x_2 - x_5)(x_4 - x_6)\,,$$
$$(x_1 - x_3)(x_2 - x_4)(x_5 - x_6)\,,\ (x_1 - x_2)(x_3 - x_5)(x_4 - x_6)\,,$$
$$(x_1 - x_2)(x_3 - x_4)(x_5 - x_6)\,.$$

We predicted the existence of this extra summand by computing Euler–Poincaré characteristics of the short exact sequence (2.3) in the case $n = 6$. Here, the saturation of the Jacobian ideal $\mathcal{J}$ requires five cubic generators. In general, the extra generators of the saturation have degree $n - 3$.

We have seen in Corollary 1.46 that the projection from an ordinary double point P on a cubic 4-fold X has an associated K3 surface S of degree 6, and vice versa. The K3 surface is the intersection of the cubic with the projectivised tangent cone at P. Alternatively, $S \subset \mathbb{P}^4$ is defined by equations $A_2(x_1, \ldots, x_5) = B_3(x_1, \ldots, x_5) = 0$ where the Taylor expansion of X at the node $P_{x_6} := (0, 0, 0, 0, 0, 1)$ is $Ax_6 + B$. If X has 15 ordinary double points, then S has 14 A_1 singularities. Such K3 surfaces with 14 A_1 singularities have $19 - 14 = 5$ moduli, which should account for the extra summand in the kernel.

2.6.4 *Representations of Points*

In this section we decompose the vector space $H^0(\mathcal{T}) = \bigoplus_{P \in \mathrm{Sing}\, X_S} \mathbb{C}_P$ into irreducible $\mathfrak{S}_n$-representations. Even though we were able to prove Theorem 2.19

without the results of this section, we think that it may be of independent interest. The proofs are in the same spirit as those for cubic polynomials, so we just sketch them here.

2.6.4.1 Representations of Segre Points

A vector in $H^0(\mathcal{T}) = \bigoplus_{P \in \mathrm{Sing}\, X_S} \mathbb{C}_P$ is a formal $\mathbb{C}$-linear combination of points. We denote the basis vector 1_P in $\mathbb{C}_P$ by a $2 \times k$ labelled array as in Sect. 2.6.2.2.

For the remainder of this section we consider Young diagrams of shape $\lambda = (n-p, p)$ with p odd. These always have two rows, and to fix notation, we label a general tableau of shape $(n-p, p)$ as follows:

$$\begin{array}{llllll} a_1 & \dots & a_p & c_1 & \dots & c_{2l} \\ b_1 & \dots & b_p & & & \end{array}$$

We write A for the tuple $(a_1, \dots, a_p)$ and similarly B, C. Since n is even, we may write $2l := n - 2p$ for the length of C.

To a fixed Young tableau T, we associate a decomposition of the Segre cubic X_S as follows:

$$T \rightsquigarrow X_S : \left(\sum_{i=1}^{p} (x_{a_i} + x_{b_i}) + \sum_{j=1}^{2l} x_{c_j} = \sum_{i=1}^{p} (x_{a_i} + x_{b_i}) q_{a_i, b_i} + \sum_{j=1}^{2l} x_{c_j}^3 = 0 \right)$$

where $q_{a_i,b_i} := x_{a_i}^2 - x_{a_i} x_{b_i} + x_{b_i}^2$. Note that $X_C := X_S \cap \mathbb{P}^{2l-1}_{c_1,\dots,c_{2l}}$ is the Segre cubic of dimension $2l - 3$. Thus X_C has an $\mathfrak{S}_{2l}$-orbit of nodes generated by $(1^l : -1^l)$.

Definition 2.55 We define the *Specht generator* $\mathsf{v}_T \in H^0(\mathcal{T})$ to be

$$\mathsf{v}_T := \sum_{\tau \in \mathfrak{C}_T} \sum_{P \in \mathrm{Sing}(X_C)} \mathrm{sign}(\tau) \frac{\tau(A) \cup P^+}{\tau(B) \cup P^-}$$

where P^+ denotes the set of positive coordinates of P, and P^- the negative. For this to be well-defined, we use the convention that P is in the *c_{2l}-distinguished affine chart* of X_C, this means that c_{2l} is in P^+ (i.e. $x_{c_{2l}} = 1$).

Lemma 2.57 implies that v_T is independent of this choice of distinguished affine chart.

Example 2.56 For example, let $T = \begin{smallmatrix} 1 & 2 & 3 & 4 & 5 & 6 & 7 \\ 8 & 9 & 10 \end{smallmatrix}$. Then X_C is the plane cubic with three nodes represented by $\frac{47}{56}, \frac{57}{46}, \frac{67}{45}$ in $\mathbb{P}^3_{x_4,x_5,x_6,x_7}$, so that

$$\mathsf{v}_{\substack{1\,2\,3\,4\,5\,6\,7\\8\,9\,10}} = \frac{1\ 2\ 3\ 4\ 7}{8\ 9\ 10\ 5\ 6} + \frac{1\ 2\ 3\ 5\ 7}{8\ 9\ 10\ 4\ 6} + \frac{1\ 2\ 3\ 6\ 7}{8\ 9\ 10\ 4\ 5} + \dots$$

$$- \frac{8\ 9\ 10\ 4\ 7}{1\ 2\ 3\ 5\ 6} - \frac{8\ 9\ 10\ 5\ 7}{1\ 2\ 3\ 4\ 6} - \frac{8\ 9\ 10\ 6\ 7}{1\ 2\ 3\ 4\ 5}$$

has $|\mathfrak{C}_T| \cdot |\operatorname{Sing}(X_C)| = 2^3 \cdot \frac{1}{2}\binom{4}{2} = 24$ summands and the last entry of C, $c_4 = 7$, always appears in the first row.

Lemma 2.57 *Let T be a Young tableau of shape $\lambda = (n-p, p)$ with p odd. The v_T satisfy the following properties:*

(i) Let $\tau \in \mathfrak{C}_T$, then $\tau(\mathsf{v}_T) = \mathsf{v}_{\tau(T)} = \operatorname{sign}(\tau)\mathsf{v}_T$.
(ii) If $\gamma \in \mathfrak{S}_C$ permutes the C-coordinates, then $\gamma(\mathsf{v}_T) = \mathsf{v}_{\gamma(T)} = \mathsf{v}_T$.
(iii) For any $\rho \in \mathfrak{S}_n$, we have $\rho(\mathsf{v}_T) = \mathsf{v}_{\rho(T)}$.
(iv) The set of v_T for all standard tableaux T forms a basis of the representation V_λ.

Proof The proof is rather similar to that for polynomials (cf. Lemma 2.41). Thus we only prove (ii), because there the distinguished affine chart is important.

(ii) Suppose γ is a transposition of elements in the C-part. Let $T' := \gamma(T)$. Since T and T' differ only in their C-parts, we have $\mathfrak{C}_T = \mathfrak{C}_{T'}$. Also, $X_C = X_{C'}$ so the only difficulty is the distinguished affine chart.

If γ does not involve the last entry c_{2l}, then the affine chart is the same and since the column stabilisers and set of nodes are equal, we see that $\gamma(\mathsf{v}_T) = \mathsf{v}_T = \mathsf{v}_{\gamma(T)}$.

Now consider the transposition $\gamma = (c_x, c_{2l})$. Let $P = \frac{P^+}{P^-}$ be a node of X_C. If c_x is in P^+ then $P = \gamma(P)$ and so P lies in both distinguished affine charts, hence any summands of v_T involving P also appear in $\mathsf{v}_{T'}$ and $\gamma(\mathsf{v}_T)$. Now suppose c_x is in P^-. Then $\frac{P^-}{P^+}$ lies in the c_x distinguished affine chart of $X_{C'}$. Thus any summand of v_T involving P uniquely determines a complementary summand of $\mathsf{v}_{T'}$ appearing with opposite sign:

$$\operatorname{sign}(\tau)\frac{\tau(A) \cup P^+}{\tau(B) \cup P^-} \longleftrightarrow -\operatorname{sign}(\tau)\frac{\tau(B) \cup P^-}{\tau(A) \cup P^+}$$

The coefficient is -1 because we have applied the order two permutation $\prod_{i=1}^{p}(a_i, b_i)$ in $\mathfrak{C}_{T'}$ which has sign -1 because p is odd. The two sides of the above displayed formula are then equal as elements in $H^0(\mathcal{T})$, using the $\mathbb{C}^*$-action on $\mathbb{P}^{n-1}$. Hence $\mathsf{v}_T = \mathsf{v}_{\gamma(T)}$ and an analogous argument shows that $\gamma(\mathsf{v}_T) = \mathsf{v}_T$. □

We are now able to prove the following:

Proposition 2.58 *The $\mathfrak{S}_n$-representation $H^0(\mathcal{T})$ decomposes as*

$$\bigoplus_{P \in \operatorname{Sing} X_S} \mathbb{C}_P \cong V_{(n-1,1)} \oplus V_{(n-3,3)} \oplus \cdots \oplus \begin{cases} V_{(k,k)} & \textit{if}\, n \equiv 2 \mod 4 \\ V_{(k+1,k-1)} & \textit{if}\, n \equiv 0 \mod 4. \end{cases}$$

Proof By Frobenius' Formula ([46, §4.1]) the dimension of $V_{(n-p,p)}$ is $\binom{n}{p} - \binom{n}{p-1}$: the coefficient of $x^{n-p+1} y^p$ in $(x-y) \cdot (x+y)^n$.

Let $\hat{k}$ be the odd number k or $k-1$ according to the residue class of $2k = n$ mod 4. Using the identity $\sum_{j=0}^{s} (-1)^j \binom{n}{j} = (-1)^s \binom{n-1}{s}$ we get

$$\sum_{\substack{p=1, \\ p \text{ odd}}}^{\hat{k}} \dim V_{(n-p,p)} = \sum_{j=0}^{\hat{k}} (-1)^{j+1} \binom{n}{j} = \binom{n-1}{k-1} = \dim H^0(\mathcal{T}).$$

□

2.6.4.2 Representations of GK Points

Let $X_{GK} \subset \mathbb{P}^n$ be the Goryunov–Kalker cubic and let $H^0(\mathcal{T}) = \bigoplus_{P \in \operatorname{Sing} X_{GK}} \mathbb{C}_P$ be the vector space of global sections of the skyscraper sheaf $\mathcal{T}$.

We denote the basis vector 1_P in $\mathbb{C}_P$ by the representative of P having $x_{n+1} = 1$ (see Remark 2.50). In analogy to the Segre-case, we associate to this a 2-rows labelled array:

$$\begin{array}{ll} \underline{p_1 \ \dots \ p_{k-1}} & \leftarrow \text{positive} \\ p_k \ \dots \ \ p_n & \leftarrow \text{negative} \end{array}$$

where $\{p_1, \dots, p_n\} = \{1, \dots, n\}$. Here we do not write the last coordinate, since it is fixed to be 1. The first row corresponds to the remaining $k-1$ coordinates equal to 1, and the second row to the $k+1$ coordinates equal to -1.

For the remainder of this section we consider Young diagrams of shape $\lambda = (n-p, p)$

$$\begin{array}{llllllll} a_1 & \dots & a_p & c_1 & \dots & c_{2l} \\ b_1 & \dots & b_p \end{array}$$

and we use the same notation as in the Segre-case. Note that now p can be even.

To a fixed Young tableau T, we associate a decomposition of the Goryunov–Kalker cubic X_{GK} as follows:

$$T \rightsquigarrow X_{GK}: \left(\sum_{i=1}^{p} (x_{a_i} + x_{b_i}) + \sum_{j=1}^{2l} x_{c_j} + 2x_{n+1} = \right.$$

$$\left. \sum_{i=1}^{p} (x_{a_i} + x_{b_i}) q_{a_i,b_i} + \sum_{j=1}^{2l} x_{c_j}^3 + 2x_{n+1}^3 = 0 \right)$$

where $q_{a_i,b_i} := x_{a_i}^2 - x_{a_i}x_{b_i} + x_{b_i}^2$. Note that $X_C := X_{GK} \cap \mathbb{P}^{2l}_{c_1,\dots,c_{2l},x_{n+1}}$ is the GK-cubic of dimension $2l-2$. Thus X_C has an $\mathfrak{S}_{2l}$-orbit of nodes generated by $(1^{l-1} : -1^{l+1} : 1)$.

Definition 2.59 In the GK-case, we define the *Specht generator* $\mathsf{w}_T \in H^0(\mathcal{T})$ to be

$$\mathsf{v}_T := \sum_{\tau \in \mathfrak{C}_T} \sum_{P \in \mathrm{Sing}(X_C)} \mathrm{sign}(\tau) \frac{\tau(A) \cup P^+}{\tau(B) \cup P^-}$$

where P^+ denotes the set of positive coordinates of P, and P^- the negative. Note that also here we are forgetting the coordinate x_{n+1} which is always 1.

Example 2.60 Let $T := \begin{smallmatrix} 1 & 2 & 3 & 4 & 5 \\ 6 \end{smallmatrix}$.

Then X_C is a cubic surface with four nodes represented by $\frac{2}{345}, \frac{3}{245}, \frac{4}{235}, \frac{5}{234}$ in $\mathbb{P}^4_{x_2,x_3,x_4,x_5,x_7}$, so that

$$\mathsf{w}_T = \frac{1\,2}{6\,3\,4\,5} + \frac{1\,3}{6\,2\,4\,5} + \frac{1\,4}{6\,2\,3\,5} + \frac{1\,5}{6\,2\,3\,4} - $$

$$- \frac{6\,2}{1\,3\,4\,5} - \frac{6\,3}{1\,2\,4\,5} - \frac{6\,4}{1\,2\,3\,5} - \frac{6\,5}{1\,2\,3\,4}$$

has $|\mathfrak{C}_T| \cdot |\mathrm{Sing}(X_C)| = 2 \cdot \binom{4}{1} = 8$ summands.

Lemma 2.61 *Let T be a Young tableau of shape $\lambda = (n-p, p)$. The w_T satisfy the following properties:*

(i) Let $\tau \in \mathfrak{C}_T$, then $\tau(\mathsf{w}_T) = \mathsf{w}_{\tau(T)} = \mathrm{sign}(\tau)\mathsf{w}_T$.
(ii) If $\gamma \in \mathfrak{S}_C$ permutes the C-coordinates, then $\gamma(\mathsf{w}_T) = \mathsf{w}_{\gamma(T)} = \mathsf{w}_T$.
(iii) For any $\rho \in \mathfrak{S}_n$, we have $\rho(\mathsf{w}_T) = \mathsf{w}_{\rho(T)}$.
(iv) The set of w_T for all standard tableaux T forms a basis of the representation V_λ.

Proof The proof is analogous to that for the Segre-case (cf. Lemma 2.57), or even simpler, since we do not have to consider a distinguished affine chart in view of the special role of x_{n+1}. □

Proposition 2.62 *The vector space* $H^0(\mathcal{T}) \cong \bigoplus_{P \in \operatorname{Sing} X} \mathbb{C}_P$ *decomposes as*

$$\bigoplus_{P \in \operatorname{Sing} X_{GK}} \mathbb{C}_P \cong V_{(n)} \oplus V_{(n-1,1)} \oplus V_{(n-2,2)} \oplus \cdots \oplus V_{(k+1,k-1)} .$$

Proof As in Proposition 2.58 we have $\dim V_{(n-p,p)} = \binom{n}{p} - \binom{n}{p-1}$, hence we get the telescopic sum

$$\sum_{p=0}^{k-1} \dim V_{(n-p,p)} = \sum_{p=0}^{k-1} \binom{n}{p} - \binom{n}{p-1} = \binom{n}{k-1} = \dim H^0(\mathcal{T}) .$$

□

Chapter 3
Nodal Quintic Surfaces

Fabrizio Catanese and Alessandro Verra

Abstract In this chapter, we study the codes of nodal quintic surfaces. The most interesting situation is when the quintic has an even set of 20 nodes. We prove that in such a case the quintic can have at most 28 nodes. It follows that its associated code is a shortening of one of exactly two codes. One has length 26 and the other has length 28. We show that for length 28 we obtain an irreducible component of the variety of quintics with 28 nodes, which can be described as the discriminant of a special web of 3-dimensional quadrics.

If there is no even set of 20 nodes, then the code is a shortening of the biduality code. We also discuss some open questions, as the existence of quintics corresponding to shortenings of the code of length 26, and the determination of the fundamental group of the smooth locus.

3.1 Codes of Nodal Quintic Surfaces

In this section we shall describe the codes of nodal quintics, in particular we shall give new proofs of the code theoretical part of Beauville's theorem $\mu(5) = 31$.

Recall that ([9], see also section 3.7 of [20]) on quintic surfaces even sets of nodes have cardinality 16 or 20 (these are the weights of $\mathcal{K}$), and they correspond to representations of the surface as the determinant of a symmetric matrix with (homogeneous) polynomial entries.

If all the weights are 16, then $\mathcal{K}$ is a shortening of the biduality code (code of a Togliatti quintic with 31 nodes), and these codes all occur because there is an unobstructed quintic with 31 nodes (see Theorems 1.25 and 3.6).

The major interest is then to investigate the codes where the weight 20 occurs.

F. Catanese (✉)
Mathematisches Institut der Universität Bayreuth, Bayreuth, Germany
e-mail: Fabrizio.Catanese@uni-bayreuth.de

A. Verra
Dipartimento di Matematica, Universita' degli Studi di Roma Tre, Roma, Italy

F. Catanese (ed.), *Varieties of Nodal Surfaces, Coding Theory and Discriminants of Cubic Hypersurfaces*, Lecture Notes of the Unione Matematica Italiana 31,
https://doi.org/10.1007/978-3-032-05899-7_3

This is the first basic result in this direction.

Theorem 3.1 *Assume that Y is a nodal quintic surface, and that $\mathcal{N}$ is an even set of nodes of cardinality* 20.

Then the number ν of nodes of Y is at most 28.

We refer the reader to Tables 3.1, 3.2 in the Appendix to this chapter for a classification of all the codes with weights 16 or 20, with a sublist of those satisfying the B-inequality.

From this list it appears that not all the above codes which satisfy the B-inequality do indeed occur, since we shall show that if the weight 20 occurs, then $\nu \leq 28$.

With the inequality $\nu \leq 28$ there are only two interesting codes where the weight 20 occurs, namely we have:

Lemma 3.2 *Assume that $\mathcal{K} \subset \mathbb{F}_2^\nu$ has only weights* 16 *or* 20, *that the weight* 20 *occurs and that $\nu \leq 28$.*

Then $k := dim(\mathcal{K}) \leq 2$, and if $k = 2$ there are two unique codes, one with weights $(20, 20, 16)$, *the other with weights* $(20, 16, 16)$. *In the first case the length $n = 28$, in the second the length $n = 26$.*

Proof Let $v \in \mathcal{K} \subset \mathbb{F}_2^{\mathcal{S}}$ be a vector with weight 20, let $A := \mathrm{supp}(v)$, $B := \mathcal{S} \setminus A$.

If there is another vector u different from v, 0, then set

$$a := |\,\mathrm{supp}(u) \cap A|,\, b := |\,\mathrm{supp}(u) \cap B|.$$

Then $b \leq 8$ and if $a + b = 20$, then $20 - a + b = 16, 20 \Rightarrow a = b + 4, b$.

The second case is excluded since then $a = b = 10$, a contradiction. In the first case $a = 12, b = 8$.

If instead $a + b = 16$, and $20 - a + b = 16$, then $a = 10, b = 6$.

To finish the proof it suffices to show that $k < 3$, i.e., that $k = 3$ is not possible.

Assume that there are two vectors v, u of respective weights 20, 20, hence $a = 12, b = 8$, or of respective weights 20, 16, hence $a = 0, b = 6$.

Assume that there is another vector $v' \in \mathcal{K}$, linearly independent of v, u.

Denote by $A_1 := A \cap \mathrm{supp}(u)$, $|A_1| = 12, 10$, set $A_2 := A \setminus A_1$, so $|A_2| = 8, 10$.

Then we define a_1, a_2, b' through the cardinalities of the respective intersections of $\mathrm{supp}(v')$ with A_1, A_2, B. In particular, $b' = 6$ or 8, correspondingly $a_1 + a_2 = 10$ or 12.

Since also the following numbers belong to the set $\{16, 20\}$:

$$20 - a_1 + a_2 - b',\;\; 16 + a_1 - a_2 - b',$$

and they equal in the first case

$$14 - 2a_1 + 10 = 24 - 2a_1, \text{or} \;\; 2a_1 \Rightarrow 24 = 24 - 2a_1 + 2a_1 \geq 32,$$

a contradiction. In the second case we get

$$12 - 2a_1 + 12 = 24 - 2a_1, \text{or } 2a_1 - 4 \Rightarrow 20 = 24 - 2a_1 + 2a_1 - 4 \geq 32,$$

again a contradiction. □

Proof of Theorem 3.1 There is a symmetric matrix of linear forms

$$A(x) = (a_{i,j}(x)), \ i, j = 1, \dots 5,$$

such that Y is equal to $\{(x)|det(A(x)) = 0\}$, while

$$\mathcal{N} = \{(x)| \operatorname{corank}(A(x)) = 2\}.$$

Y is the projection in $\mathbb{P}^3$ of a complete intersection $\hat{Y}$ inside $\mathbb{P}^3 \times \mathbb{P}^4$ of 5 hypersurfaces of bidegree (1, 1).

$$\hat{Y} := \{(x, y)| \sum_{j=1}^{5} a_{i,j}(x) y_j = 0\}.$$

In particular, $K_{\hat{Y}} \cong \mathcal{O}_{\hat{Y}}(1, 0)$ and $p : \hat{Y} \to Y$ is a partial resolution of Y, of the nodes in $\mathcal{N}$.

In fact, at the points x where $\operatorname{corank}(A(x)) = 2$ the fibre of p is a $\mathbb{P}^1$, and it is easy to verify that $\hat{Y}$ is smooth at the points of this exceptional curve.

At the other points P of Y, where $\operatorname{corank}(A(P)) = 1$, the matrix is equivalent (as a quadratic form with coefficients in the local ring $\mathcal{O}_{\mathbb{P}^3, P}$) to a diagonal matrix $Diag(1, 1, 1, 1, f(x))$ where then $f(x) = 0$ is the local equation of Y at P.

Then the equations of $\hat{Y}$ in the fibre over P are:

$$y_1 = y_2 = y_3 = y_4 = 0, \ f(x) = 0,$$

hence we have a local isomorphism $\hat{Y} \cong Y$.

In this spirit we analyse the singularities of $\hat{Y}$. To simplify our computations, assume that P is the point e_0, that $\operatorname{corank}(A(P)) = 1$, and that $A(e_0) = Diag(1, 1, 1, 1, 0)$. Hence

$$A(x) = x_0 I_4 + x_1 A(1) + x_2 A(2) + x_3 A(3),$$

and over e_0 lies the point e'_5.

Taking in $\mathbb{P}^3 \times \mathbb{P}^4$ affine local coordinates

$$(1, x_1, x_2, x_3, x_4), (y_1, y_2, y_3, y_4, 1),$$

the equations of $\hat{Y}$ write as:

$$y_1 + \sum_1^3 x_i(\sum_1^4 A(i)_{1,j}y_j + A(i)_{1,5}) = 0,$$

$$y_2 + \sum_1^3 x_i(\sum_1^4 A(i)_{2,j}y_j + A(i)_{2,5}) = 0,$$

$$y_3 + \sum_1^3 x_i(\sum_1^4 A(i)_{3,j}y_j + A(i)_{3,5}) = 0,$$

$$y_4 + \sum_1^3 x_i(\sum_1^4 A(i)_{4,j}y_j + A(i)_{4,5}) = 0,$$

$$\sum_1^3 x_i(\sum_1^4 A(i)_{5,j}y_j + A(i)_{5,5}) = 0,$$

so that we can eliminate y_1, y_2, y_3, y_4 and we have a singular point if and only if

$$A(1)_{5,5} = A(2)_{5,5} = A(3)_{5,5} = 0.$$

The above conditions amount to requiring that e'_5 is a base point for the system of quadrics.

Therefore (e_0, e'_5) is a singular point of $\hat{Y}$ if and only if e'_5 is a base point for the system of quadrics, and a singular point for the quadric corresponding to $A(e_0)$, which has rank 4.

This implies that e'_5 is a base point for the 4 quadrics, of multiplicity at least 2. Conversely, if e'_5 is a base point for the 4 quadrics, of multiplicity at least 2, then the four gradients are not linearly independent, hence there is a quadric which is singular at e'_5.

Let $\mathcal{B}$ be the base locus of the web of quadrics ($\mathbb{P}^3$ of quadrics, spanned by the 4 quadrics $A(e_i)$).

If there were a component Γ in $\mathcal{B}$ of dimension at least 1, then at the general point $e' \in \Gamma$ the gradients would span a space of dimension 3, hence there would be a singular quadric containing e' and $\hat{Y}$ (hence also Y) would have infinitely many singular points, a contradiction.

Therefore $\mathcal{B}$ is finite and, by Bezout's theorem, it consists of 16 points counted with multiplicity. Therefore there can be at most 8 points of multiplicity ≥ 2.

The conclusion is that $\hat{Y}$ has at most 8 singular points, hence $\nu(Y) \leq 20+8 = 28$. □

Remark 3.3 The same proof works for sextics having a half-even set of cardinality 35, corresponding to a symmetric matrix of linear forms: then $\nu \leq 35 + 16 = 51$.

Corollary 3.4 *If the number of nodes of a quintic is at least* 29, *then the codewords in* $\mathcal{K}$ *have all weight* 16.

In particular Beauville's theorem: $\mu(5) \leq 31$, *with equality if and only if* $\mathcal{K}$ *is the simplex (biduality code), follows immediately.*

Proof The first assertion was already shown.

The second follows from the first two MacWilliams identities, and the B-inequality.

As we mentioned already, the MacWilliams identity reads out as a series of identities, the first two are:

$$\sum_{i>0} a_i = 2^k - 1, \sum_{i>0} i a_i = 2^{k-1} n.$$

If we have a code $\mathcal{K}$ where all the weights are just 2^m, let $k = dim(\mathcal{K})$, n its effective length.

In this case we get

$$2^m(2^k - 1) = 2^{k-1} n \Rightarrow n = (2^k - 1)2^{(m-k+1)} \Rightarrow k \leq m + 1.$$

Here $m = 4$, moreover, by the B-inequality $k \geq \nu - 26$, hence $\nu \leq 31$, equality if and only if $k = 5$, $n = 31$, and the code is seen to be the simplex code. Because an easy counting argument shows that a basis of $\mathcal{K}$ yields a bijection of the nodal set $\mathcal{S}$ with $\mathbb{F}_2^5 \setminus \{0\}$. □

The next question is whether the two codes with $k = 2$ of Lemma 3.2 are realized.

For the code with length 28, the answer is positive by virtue of the following theorem, whose proof shall be given in the following section two (see especially Corollary 3.16).

Theorem 3.5 *The subset of the Severi variety* $\mathcal{F}(5, 28)$ *such that* $\mathcal{K}$ *admits the weight* 20 *is an irreducible component, and the length of* $\mathcal{K}$ *equals* 28.

We have then the following main result for the codes of nodal quintics:

Theorem 3.6 *The codes of nodal quintic surfaces are all the shortenings of the biduality code (simplex) of dimension* 5, *all the shortenings of the code* $\mathcal{K}$ *of dimension* 2, *effective length* 28, *and weights* (20, 20, 16), *and possibly also the code* $\mathcal{K}$ *of dimension* 2, *effective length* 26, *and weights* (20, 16, 16).

Proof That all the shortenings occur follows from the unobstructedness of the Goryunov cubic, by Theorem 3.6, and by the fact that a general symmetroid (determinant of a 5×5 matrix of linear forms) has 20 nodes forming an even set.

We conclude then by Lemma 3.2. □

Remark 3.7 It is not yet clear to us whether the code $\mathcal{K}$ of dimension 2 and weights (20, 16, 16) does actually occur.

Remark 3.8 A more refined and very interesting question is for instance the one of determining the fundamental groups $\pi_1(Y^*)$ of nodal maximizing surfaces of degree d ($\nu(Y) = \mu(d)$). We saw the answer for $d \leq 4$, but for $d = 5, 6$ this is an open problem.

The case $d = 5$ appears rather difficult, at least for $d = 6$ there is a strategy, by viewing the Barth sextic as a Galois covering of the plane with group $(\mathbb{Z}/2)^3$.

3.2 Prym Varieties and Nodal Quintics with 28 Nodes, and Code Admitting the Weight 20

The following construction is based on the results of [96]:

Theorem 3.9 *Let C be a general canonical curve of genus 5, the complete intersection of three quadrics Q_0, Q_1, Q_2 in $\mathbb{P}^4$,*

$$C = \{Q_0 = Q_1 = Q_2 = 0\},$$

and let $\eta \in \text{Pic}(C)_2 \setminus \{0\}$ be a non trivial 2-torsion divisor class.

Let D be a general divisor in the linear series $K_C + \eta$.

Let Q_3 be a quadric in $\mathbb{P}^4$ such that $div_C(Q_3) = 2D$ on C.

Then, if we let Y be the discriminant of the web of quadrics in $\mathbb{P}^4$ generated by Q_0, Q_1, Q_2, Q_3, Y is a nodal quintic surface with 28 nodes and such that the length of the code $\mathcal{K}(Y)$ (which admits the weight 20) equals 28.

Proof The linear system $|K_C + \eta|$ is base point free, since if P were a base point, we would have that $h^1(\mathcal{O}_C(K_C + \eta - P)) \neq 0$, hence the linear system $|P + \eta|$ contains an effective divisor, and there is a point P' such that $2P \equiv 2P'$; since $P \neq P'$, the curve C is hyperelliptic, contradicting that it is canonically embedded in $\mathbb{P}^5$.

Claim 3.10 The linear system $|K_C + \eta|$ is birational onto its image for a general choice of the curve C.

Proof of the Claim 3.10 Otherwise the morphism ϕ associated to $|K_C + \eta|$, $\phi : C \to \mathbb{P}^3$ has image of degree strictly smaller than 8, and greater or equal to 3, hence its image is a curve Γ of degree 4; and ϕ factors through a double covering $p : C \to C'$, where C' cannot have genus 0 since C is not hyperelliptic.

Hence C' has genus 1, and is isomorphic to Γ which is a complete intersection of two quadrics. Since the covering is branched in 8 points, such curves C depend on 8 moduli, while curves of genus 5 have 12 moduli.

Consequence of Claim 3.10 The monodromy group of the hyperplane sections of C (the divisors D) is a transitive subgroup.

Choose now a general divisor $D \in |K_C + \eta|$, and let Y be the discriminant of the web of quadrics in $\mathbb{P}^4$ generated by Q_0, Q_1, Q_2, Q_3, where $div_C(Q_3) = 2D$.

In particular, D consists of 8 distinct points $P_1, \dots, P_8$ and all the quadrics in the web intersect C in the divisor $2D$. And we have

Claim 3.11 For each of the points P_i, there is a unique quadric $Q(i)$ in the web such that $P_i \in Ker(Q(i))$.

Proof of Claim 3.11 There are local coordinates x, y, z, t at P such that $Q_0 = x$, $Q_1 = y$, $Q_2 = z$, $Q_3 = t^2 + ax + by + cz + g$, where $a, b, c, \in \mathbb{C}$ and g is in the ideal (x, y, z) and vanishes of order at least 2 at P.

Hence the desired quadric is $Q_3 - aQ_0 - bQ_1 - cQ_2$.

We observe now that the 8 quadrics $Q(i)$ are all distinct, provided that the following claim holds:

Claim 3.12 Each of the quadrics $Q(i)$ has rank at least 4, for D general.

To prove Claim 3.12, by the consequence of Claim 3.10, it suffices to show that these quadrics cannot all have rank ≥ 3.

Claim 3.13 For D general, there are exactly 20 distinct quadrics of rank 3 in the web.

Observe that Claims 3.12 and 3.13, by virtue of the argument given in the proof of Theorem 3.1, show the desired assertion: that Y possesses 20 nodes where $\text{corank}(A(x)) = 2$, plus 8 other nodes, corresponding to the base points of the web through which passes a quadric which is singular at the point.

We begin now the main argument by observing first that the locus Σ of rank ≤ 3 quadrics in the web is finite: because the net generated by Q_0, Q_1, Q_2 contains no such quadric, for general choice of the three quadrics (the quadrics of corank ≥ 2 form a subvariety of codimension 3 in the space of quadrics in $\mathbb{P}^4$).

We consider now the variety

$$\mathcal{S} := \{(D, Q) | D \in |K_C + \eta|, div_C(Q) = 2D, \text{rank}(Q) \leq 3\}.$$

$\mathcal{S}$ fibres onto the linear system $\in |K_C + \eta|$ with finite fibres Σ, and we want to show that $\mathcal{S}$ is reduced, so that the general fibre consists of 20 distinct points, thus proving claim 4.

Indeed, from the work of [96], it follows that $\mathcal{S}$ splits into two irreducible components, which map with respective degrees 8, 12.

Then we shall prove claim 3.12 simply by showing that the component mapping of degree 8 does not consist of pairs where the quadric Q has vertex $ker(Q)$ containing a point of C.

A first remark is that, for general D, no quadric Q with $div_C(Q) = 2D$ can have rank equal to 2. Because otherwise there are two hyperplanes H_1, H_2 such that $2D = div_C(Q) = div_C(H_1) + div_C(H_2)$.

Writing $div_C(H_i) = 2D_i' + B$, we obtain, assuming that for each D there were such a rank 2 quadric, by monodromy invariance of such decomposition, and by transitivity of the monodromy group, that $B = 0$; then $D = D_1' + D_2'$, and D_1', D_2'

would be effective thetacharacteristics. Since C is not hyperelliptic, by Clifford's theorem, follows that $h^0(C, \mathcal{O}(D_i')) \leq 2$, in particular the general divisor $D \in |K + \eta| \cong \mathbb{P}^3$ cannot be written as the sum of two such effective divisors D_1', D_2'.

We can therefore restrict ourselves to the consideration of rank 3 quadrics.

Prym Interpretation of Rank 3 Quadrics

Assume now that Q is a rank 3 quadric, such that $Q \cdot C = 2D$, $D \in |K_C + \eta|$ (so in particular, C is not contained in Q).

Projection from the vertex $\mathcal{V} := ker(Q)$ is a rational map to $\mathbb{P}^2$ with image a conic q of equation $x_1x_2 - x_3^2 = 0$, so that $\mathcal{V} = \{x_1 = x_2 = x_3 = 0\}$: we recall that the rational map is given by setting $\lambda := \frac{x_1}{x_3} = \frac{x_3}{x_2} \in \mathbb{P}^1$. We also denote by H_λ the tangent hyperplane to Q corresponding to the point λ of the conic q.

We consider now the double cover $\bar{C} \to C$ defined as follows:

$$\bar{C} \subset C \times \mathbb{P}^1, \ \bar{C} := \{(x, \lambda) | x \in H_\lambda\}.$$

For each point $x \in C \setminus Q$, $x \mapsto (x_1, x_2, x_3) \in \mathbb{P}^2 \setminus q$, and we have two distinct points y_1, y_2 of $\bar{C}$ lying over x.

Hence $\bar{C} \to C$ is certainly unramified outside Q. If instead $x \in Q$, observe that

$$\bar{C} = C \times_{\mathbb{P}^2} Z, \ Z := \{(x_1, x_2, x_3, z) \in \mathbb{P}^3 | z^2 = -(x_1x_2 - x_3^2)\}.$$

Since $div_C(Q) = 2D$, the normalization $p : \tilde{C} \to \bar{C}$ is an étale covering of C, corresponding to $\eta \in \text{Pic}(C)_2 \setminus \{0\}$: because $H \equiv K_C, D \equiv K_C + \eta$. And $\tilde{C}$ is connected since η is nontrivial.

Since $Z \cong \mathbb{P}^1 \times \mathbb{P}^1$ we get two maps $\psi_+, \psi_- : \tilde{C} \to \mathbb{P}^1$, exchanged by the covering involution $\iota : z \mapsto -z$, since $-z^2 + x_3^2 = x_1x_2$ yields

$$\psi_\pm(x, z) := \frac{x_3 \pm z}{x_2} = \frac{x_1}{x_3 - \pm z}.$$

Denote by $\tilde{H}_+, \tilde{H}_-$ the divisor classes such that ψ_+, ψ_- yield two pencils respectively inside $|\tilde{H}_+|, |\tilde{H}_-|$. We clearly have:

$$\tilde{H}_+ + \tilde{H}_- \equiv p^*(K_C) = K_{\tilde{C}}, \ \iota^*(\tilde{H}_+) = \tilde{H}_-,$$

and moreover

$$Norm(\tilde{H}_-) = Norm(\tilde{H}_+) = K_C,$$

since the images of the lines of a ruling of Z map to lines in $\mathbb{P}^2$.

Recall the definition of the Prym variety (see [72], [96]):

Definition 3.14 Let $p : \tilde{C} \to \bar{C}$ be an étale covering of C, corresponding to $\eta \in \mathrm{Pic}(C)_2 \setminus \{0\}$ (hence such that, if g is the genus of C, then $\tilde{C}$ has genus $2g - 1$).

In this situation $\{\tilde{L} | Norm(\tilde{L}) = K_C\} \subset Pic^{2g-2}(\tilde{C})$ splits into two connected components, distinguished by the parity of $h^0(\tilde{L}) = h^1(\tilde{L}) = h^0(K_{\tilde{C}} - \tilde{L}) = h^0(\iota^* \tilde{L})$.

$P(C, \eta) := Prym(\tilde{C}, C)$ is defined then as the connected component where the parity is even, and it is isomorphic to the Abelian subvariety of $Pic^0(\tilde{C})$ which is the image of $1 - \iota^*$.

And the Prym-theta divisor Ξ of $P(C, \eta)$ consists of the effective divisors inside $P(C, \eta)$.

Conversely, giving two pencils inside the respective linear systems $|\tilde{L}|$ and $\iota^*|\tilde{L}|$, exchanged by the covering involution ι, for $\tilde{L}$ such that $Norm(\tilde{L}) = K_C$, yields a subvariety $Z \cong \mathbb{P}^1 \times \mathbb{P}^1$ inside the linear system $p^*|K_C|$. It is explained in [96] p. 228 (2.23) that ι^* acts on Z, and on $\mathbb{P}^3$ with eigenvalues $(1, 1, 1, -1)$, so that intersecting with the $+1$ eigenspace we obtain a rank 3 conic $q \subset \mathbb{P}^2$, whose points parametrize divisors of C of the form $2D$, where $D \in |K_C + \eta|$.

Now, a point $y \in \tilde{C}$ lies in the union of the base loci of the respective pencils if and only if all divisors coming from Z vanish at $y + \iota(y)$, or, equivalently, the point $x := p(y)$ of the divisor D lies in the vertex of the quadric Q.

Recall now (see [96] theorem 2.5) that we have a finite covering $\mathcal{S} \to |K_C + \eta|$, which has two irreducible components, which we denote by W and U^{even}, corresponding to the respective cases $h^0(\tilde{L}) \equiv 1, 0 \; mod(2)$.

Here, W parametrizes the quotient of the variety of pairs $(\tilde{L}, V)$ where $h^0(\tilde{L}) \geq 3$ is odd, and $V \subset H^0(\tilde{L})$ is a two-dimensional subspace, by the action of ι. And similarly for U^{even}.

As a matter of fact, for C general, for any such $\tilde{L} \in |\Xi|$ we have $h^0(\tilde{L}) = 2$, hence it turns out that $U^{even}/\iota \cong \Xi / \pm 1$.

We establish now the needed property.

Claim 3.15 For a general C, for both cases $h^0(\tilde{L}) \geq 2$ even or odd, there exists a pair $(\tilde{L}, V)$ where $V \subset H^0(\tilde{L})$ yields a two-dimensional base point free subsystem.

Proof of the Claim Take a special C such that $C = Q_1 \cap Q_2 \cap Q_3$, where Q_1, Q_2, Q_3 are quadrics in $\mathbb{P}^4$, Q_3 has rank $= 3$, and $\mathrm{Sing}(Q_3) \cap C = \emptyset$. This is possible, just starting from C_3 and then taking two general quadrics Q_1, Q_2.

Now, the projection with centre the vertex $\mathrm{Sing}(Q_3)$ of Q_3 yields a morphism $\Psi : C \to \mathbb{P}^1$ corresponding to a thetacharacteristic $\mathcal{G}$ on C.

Since C is not hyperelliptic, and the degree of $\mathcal{G}$ is 4, it follows by the theorem of Clifford that $h^0(\mathcal{G}) \leq 2$, hence equality holds: $h^0(\mathcal{G}) = 2$.

Now, we consider

$$\{\mathcal{G} \otimes \eta | \eta \in \mathrm{Pic}(C)_2 \setminus \{0\}\}.$$

This set contains even and odd thetacharacteristics, in particular, since as we have shown $h^0(\mathcal{G} \otimes \eta) \leq 2$, we can choose an η such that $h^0(\mathcal{G} \otimes \eta) = 1$, and also we can choose an η such that $h^0(\mathcal{G} \otimes \eta) = 0$ (since $\mathrm{Sing}(\Theta_C)$ cannot contain $2^{g-1}(2^g + 1)$ points, see Lemma 4 , pag. 233 of [52], showing that at least one theta constant is non zero).

We set now $\tilde{L} := p^*(\mathcal{G})$, where $p : \tilde{C} \to C$ is the étale double covering associated to η.

By the projection formula

$$H^0(\tilde{L}) = H^0(\mathcal{G}) \oplus H^0(\mathcal{G} \otimes \eta).$$

So, the dimension $h^0(\tilde{L}) = 2 + h^0(\mathcal{G} \otimes \eta) = 3$, or 2, and in the second case we get a base point free pencil, while in the first we get a base point free pencil for general choice of V, hence for general choice of the divisor D.

By a monodromy argument, we get that for general choice of C and D there are no quadrics of rank 3 in the web satisfying the property $Sing(Q) \cap C \neq \emptyset$. □

Corollary 3.16 *In the Nodal Severi variety $\mathcal{F}(5, 28)$ the condition that $\mathcal{K}$ contains a codeword of weight 20 determines an irreducible component.*

Proof Y must be the discriminant of a web of quadrics with 8 distinct base points P_i, for which there is exactly one quadric in the web which is singular at P_i. Therefore, for general choice of Q_0, Q_1, Q_2 in the web, $C := Q_0 \cap Q_1 \cap Q_2$ is a smooth curve, and $div_C(Q_3) = 2D$, where D is not contained in a hyperplane, hence $D \equiv K_C + \eta$, with η a nontrivial 2-torsion divisor.

Therefore our family is dominated by the family of genus 5 canonical curves, given together with a Prym-canonical divisor D as above. □

Appendix: Binary Codes $\mathcal{K}$ with Only 16, 20 as Weights

Michael Kiermaier

In this appendix we attempt a description of the binary codes with weights either 16 or 20, and classify explicitly those codes among them which satisfy the B-inequality. It follows from Chap. 3 that some of these codes can not be realized by nodal quintic surfaces.

For quintics, the maximum number ν of nodes is 31. This has been shown by Beauville in [9], where he proved the following important algebro geometric results:

(i) B-inequality: the even sets of nodes of a quintic yield a linear binary $[\nu, k]$ code C with $k \geq \nu - 26$, and with
(ii) non-zero weights in $\{16, 20\}$.

The main purpose of this appendix is to classify the codes satisfying properties (i), (ii), respectively only property (ii).

This is interesting, because it shows, compared with the results of Part III, that not all of the codes satisfying (i) and (ii) do geometrically occur as codes of a nodal quintic surface (because of Theorem 3.1).

In passing we shall also give a third proof of the code theoretic part of this result (the second proof being the content of Corollary 3.4) as a consequence of the MacWilliams identities for the weight enumerator of linear codes, of course using (i) and (ii).

We shall call *admissible* all the codes which satisfy (i) and (ii), and we shall here classify them up to isomorphism.

The MacWilliams Identities

For a code C of length n, let a_i be the number of codewords of Hamming weight i. Then $(a_i)_i$ is the weight enumerator of C, which may also be written in polynomial form as $w_C(x) = \sum_i a_i x^i \in \mathbb{Z}[x]$ or as a homogeneous polynomial $w_C(x, y) = \sum_i a_i x^i y^{n-i} \in \mathbb{Z}[x, y]$. Clearly, $a_0 = 1$.

For a $\mathbb{F}_q$-linear code C of length n, the *dual code* is defined as the orthogonal subspace of $\mathbb{F}_q^n$ with respect to the standard bilinear form $\langle \mathbf{x}, \mathbf{y} \rangle = \sum_{i=1}^n x_i y_i$. We have $\dim(C) + \dim(C^\perp) = n$. We denote the weight distribution of $C^\perp$ by $(a_i^\perp)_i$. Then $a_0^\perp = 1$. For $n \geq 1$ we have $a_1^\perp = 0$ if and only if C is *full-length*, i.e., if C does not have an all-zero coordinate. For $n \geq 2$ we have $a_2^\perp = a_1^\perp = 0$ if and only if C is projective.

The MacWilliams identities connect the weight distribution of C and $C^\perp$. The usage of homogeneous weight enumerators allows the compact formula

$$w_{C^\perp}(x, y) = \frac{1}{\#C} w_C(x - y, x + (q - 1)y).$$

For the weight enumerators in univariate polynomial form, we get

$$w_{C^\perp}(x) = \frac{1}{\#C}(x + (q - 1))^n w_C\left(\frac{x - 1}{x + (q - 1)}\right).$$

We remind that the code size $\#C$ may also be written in terms of the (homogeneous) weight enumerator as $\#C = w_C(1) = w_C(1, 1)$.

Writing down the equations for the individual coefficients, after a few transformations one arrives at the *Krawtchouk form* of the $(i + 1)$-th MacWilliams identity

$$a_i^\perp = \frac{1}{\#C} \sum_{j=0}^n K_i(j; n, q) a_j,$$

where

$$K_m(x; n, q) = \sum_{j=0}^{m} (-1)^j \binom{x}{j} \binom{n-x}{m-j} (q-1)^{m-j}$$

is the m-th q-ary *Krawtchouk polynomial* of order n.

The Statement of Beauville

Let C be a linear binary $[n, k]$ code C with $k \geq n - 26$ and non-zero weights in $\{16, 20\}$. Let (a_i) be the weight enumerator of C and $(a_i^{\perp})$ the dual weight enumerator. A priori, we don't know if necessarily $a_1^{\perp} = 0$ (i.e. C is full-length). However by puncturing the zero-coordinates, all admissible codes can be reduced to an admissible full-length code of the same dimension and the same weight distribution. So in the following we will assume that $a_1^{\perp} = 0$, keeping in mind that for an admissible code, all extensions by zero coordinates up to the length $n = k+26$ are admissible, too. Also, we do not know for sure if $a_2^{\perp} = 0$ (i.e. C is.projective).

With $a_0 = a_0^{\perp} = 1$, the first two MacWilliams identities give

$$1 + a_{16} + a_{20} = 2^k$$
$$n + (n-32)a_{16} + (n-40)a_{20} = 0$$

Therefore

$$\begin{pmatrix} 1 & 1 \\ n-32 & n-40 \end{pmatrix} \begin{pmatrix} a_{16} \\ a_{20} \end{pmatrix} = \begin{pmatrix} 2^k - 1 \\ -n \end{pmatrix}$$

Left-multiplication with the inverse matrix

$$-\frac{1}{8} \begin{pmatrix} n-40 & -1 \\ -n+32 & 1 \end{pmatrix}$$

yields

$$a_{16} = -2^{k-3}n + 5(2^k - 1) \quad \text{and} \quad a_{20} = 2^{k-3}n - 4(2^k - 1).$$

From the integrality of these values we get $4 \mid n$ for $k = 1$ and $2 \mid n$ for $k = 2$. From $a_{16} \geq 0$ and $a_{20} \geq 0$ we get $4(2^k - 1) \leq 2^{k-3}n \leq 5(2^k - 1)$ or equivalently

$$32(1 - 2^{-k}) \leq n \leq 40(1 - 2^{-k}). \tag{3.1}$$

The third MacWilliams identity gives

$$2^k a_2^\perp = (\frac{1}{2}n^2 - \frac{1}{2}n) + (\frac{1}{2}n^2 - \frac{65}{2}n + 512)a_{16} + (\frac{1}{2}n^2 - \frac{81}{2}n + 800)a_{20}.$$

Plugging in the expressions for a_{16} and a_{20}, we get

$$\begin{aligned} a_2^\perp &= -\frac{1}{2}n^2 + \frac{71}{2}n - 640\left(1 - \frac{1}{2^k}\right) \\ &= -\frac{1}{2}\left(n - \frac{71}{2}\right)^2 - \frac{79}{8} + \frac{640}{2^k} \end{aligned}$$

The condition $a_2^\perp \geq 0$ implies $\frac{640}{2^k} - \frac{79}{8} \geq 0$ or equivalently $2^k \leq \frac{8 \cdot 640}{79} \approx 64.8$. So $k \leq 6$. In these cases, $a_2^\perp \geq 0$ is equivalent to restriction

$$|n - \frac{71}{2}| \leq \sqrt{\frac{1280}{2^k} - \frac{79}{4}}$$

on n. For $k \leq 5$, this is weaker than Inequality 3.1. For $k = 6$, Inequality 3.1 is sharpened to $n \in \{35, 36\}$. Therefore, the condition $k \geq n - 26$ is violated and we have $k \leq 5$, implying that $n \leq k + 26 \leq 31$, which is the result of Beauville.

Let $\mathcal{P}$ be the set of points of the projective space $\mathbb{P}^{k-1}_{\mathbb{F}_q}$. Up to isomorphism, a $\mathbb{F}_q$-linear code C of dimension k and full length n is determined by a multiset of n points in $\mathcal{P}$, given by the n rows of a generator matrix G (a multiset is a map $\mathcal{C} : \mathcal{P} \to \mathbb{Z}_{\geq 0}$). Two multisets yield the same code if and only if they are projectively equivalent (this amounts to a change of basis for C). Moreover, since C has dimension k, the multiset, that is, the support of the function $\mathcal{C}$, must span $\mathcal{P}$.

For $A \subseteq \mathcal{P}$ we consider the multiplicity function $\mathcal{C}(A) := \sum_{x \in A} \mathcal{C}(x)$, and observe that to a codeword $\mathbf{c} = G\mathbf{x}$ corresponds the hyperplane $H_{\mathbf{x}} := \mathbf{x}^\perp$, and the weight $w(\mathbf{c})$ equals $\mathcal{C}(\mathcal{P} \setminus H_{\mathbf{x}}) = n - \mathcal{C}(H)$. Two codewords lead to the same hyperplane if and only if they yield the same point in $\mathcal{P}$.

By Proposition 1.3 it follows that the weights of the code vectors determine the function $\mathcal{C}(x)$, hence the code C.

Another way to argue is that the incidence matrix of the points vs. hyperplanes of is invertible (see [63] section 1.2, page 35), hence, if the multiplicities of the hyperplanes (or their complements) are fixed, so are the multiplicities $\mathcal{C}(x)$ of the points.

Classification of the Admissible Codes

Table 3.1 shows all remaining tuples $(n, k, a_{16}, a_{20}, a_2^\perp)$ with $k \geq n - 26$. We see that, indeed, there always exists a code with the given parameters.

Table 3.1 Binary linear $[n, k]$-codes with weights in $\{16, 20\}$ and $n \leq k + 26$

№	n	k	a_{16}	a_{20}	$a_2^\perp$	Codes	Shorten.
1	31	5	31	0	0	$\mathrm{Sim}_2(5)$	2
2	30	4	15	0	15	$2\,\mathrm{Sim}_2(4)$	4
3	29	3	6	1	49	in pr. plane: line of multipl. 3, otherw. multipl. 5	6, 7
4	28	3	7	0	42	$4\,\mathrm{Sim}_2(3)$	7
5	28	2	1	2	122	points on pr. line of multipl. $(8^2 12^1)$	8, 9
6	26	2	2	1	105	points on pr. line of multipl. $(6^1 10^2)$	8, 9
7	24	2	3	0	84	$8\,\mathrm{Sim}_2(2) \equiv$ points on proj. line of multipl.(8^3)	9
8	20	1	0	1	190	$20\,\mathrm{Sim}_2(1) \equiv$ 20–fold repetition code	10
9	16	1	1	0	120	$16\,\mathrm{Sim}_2(1) \equiv$ 16–fold repetition code	10
10	0	0	0	0	0	Zero code	

All codes are unique up to isomorphism by the following argument. In all cases, it is obvious from the list that up to isomorphism there is only a single choice to assign the weights 16 or 20 to the hyperplanes.

Therefore the Radon transform of the multiset $\mathcal{C}(x)$ is unique, hence, by Radon duality, $\mathcal{C}(x)$ is unique, hence the code—if one exists—is unique.

So we have:

Theorem 3.17 *If there is a quintic with ν nodes, its even sets of nodes form a binary linear $[\nu, k]$ code with $k \geq \nu - 26$. After removing the zero coordinates, the code is isomorphic to one of the codes in Table 3.1, where each line represents a single isomorphism type of a linear code.*

Remark on the Classification Without the Dimension Restriction

Without the restriction $k \geq \nu - 26$, further hypothetical parameters of codes show up, see Table 3.2.

The two codes with $k = 6$ are projective two-weight codes, which have been classified in [92]. For $n = 35$ there are 7 isomorphism types and for $n = 36$ (via the complement with $n = 27$) there are 5 isomorphism types. The most symmetric ones have a nice geometric description: for n=36, take the complement in $\mathbb{P}G(5, 2)$ of the points on a full rank elliptic quadric. For $n = 35$, take the points on a full rank hyperbolic quadric, which is the Klein quadric.

For code №32, take the three points on a line with multiplicities 2, 6 and 6 and the remaining 4 points with multiplicity 4. The code is uniquely determined by the above argument.

For code №31, the three lines corresponding to the codewords of weight 20 are either a line pencil or a line triangle. In the first case, the code is uniquely determined by a point of multiplicity 1 and the other six points of multiplicity 5. In

Table 3.2 Binary linear $[n, k]$-codes with weights in $\{16, 20\}$ and $n > k - 26$

№	n	k	a_{16}	a_{20}	$a_2^\perp$	Codes	#types
11	36	6	27	36	0	cpl. of full rank elliptic quadric	5
12	35	6	35	28	0	Klein quadric	7
13	38	5	3	28	7	$\mathrm{Sim}_2(5) + \mathrm{Sim}_2(3)$	?
14	37	5	7	24	9		?
15	36	5	11	20	10		?
16	35	5	15	16	10		?
17	34	5	19	12	9		?
18	33	5	23	8	7		?
19	32	5	27	4	4		?
20	37	4	1	14	29	$2\,\mathrm{Sim}_2(4) + \mathrm{Sim}_2(3)$	1
21	36	4	3	12	30	$2\,\mathrm{Sim}_2(4) + 2\,\mathrm{Sim}_2(3)$	1
22	35	4	5	10	30		?
23	34	4	7	8	29		?
24	33	4	9	6	27		?
25	32	4	11	4	24		?
26	31	4	13	2	20	see text	1
27	35	3	0	7	70	$5\,\mathrm{Sim}_2(3)$	1
28	34	3	1	6	69	$4\,\mathrm{Sim}_2(3) + 2\,\mathrm{Sim}_2(2)$	1
29	33	3	2	5	67	See text	1
30	32	3	3	4	64	See text	2
31	31	3	4	3	60	See text	2
32	30	3	5	2	55	See text	1
33	30	2	0	3	135	$10\,\mathrm{Sim}_2(2)$	1

the second case, the code is uniquely determined by a point triangle of multiplicity 3, the other three points on the secants with multiplicity 5 and the remaining point with multiplicity 7.

For code №30, the three lines corresponding to the codewords of weight 16 are either a line pencil or a line triangle. In the first case, the code is uniquely determined as a point of multiplicity 8 and the other six points of multiplicity 4. In the second case, take a point triangle with multiplicity 6, the other three points on the secants with multiplicity 4 and the remaining point with multiplicity 2.

For code №29, take the three points on a line with multiplicities 3, 3 and 7 and the remaining 4 points with multiplicity 5.

For code №28, take three collinear points of multiplicity 6 and the remaining four points with multiplicity 4.

For code №26, select a line L and a plane E through L in $\mathbb{P}G(3, 2)$. Take the three points on L with multiplicity 1, the four points on $E \setminus L$ with multiplicity 3 and the remaining eight points with multiplicity 2.

Chapter 4
Nodal Sextic Surfaces

Fabrizio Catanese, Yonghwa Cho, and Michael Kiermaier

Abstract The Barth sextic is a famous example of a sextic surface with 65 nodes, the maximum number. We show that every sextic surface with 65 nodes is the discriminant of a projection from a 5-dimensional cubic hypersurface, with centre a plane contained in it. More precisely, each sextic is a discriminant in two different ways: the cubic can have 31 nodes and the plane passes through 3 nodes, or the cubic has 32 nodes and the plane passes through 2 nodes. Somewhat surprisingly, a sextic with 65 nodes cannot be the projection from a cubic with more than 32 nodes.

We show that the automorphism group of the Barth sextic is the icosahedral group. Exploiting the action of this group on the set of nodes leads us to rediscover the so-called Doro-Hall graph and allows us to determine the extended code of the Barth surface. We explore the intimate connections between these objects and the geometry of the Barth sextic.

Finally, we present determinantal equations for the Barth sextic, based on the results of Chap. 2, on the knowledge of the extended code, and on a classification of half-even sets of nodes of small cardinality on sextics.

4.1 Half-Even Sets on Nodal Sextic Surfaces

This section is important for our purposes, even if we cannot yet completely classify the half-even sets of nodes on nodal sextic surfaces. Indeed, using homological algebra and earlier results, we get some information about the possible numbers attained by the cardinality t of a half-even set of nodes on a sextic.

F. Catanese (✉) · M. Kiermaier
Mathematisches Institut der Universität Bayreuth, Bayreuth, Germany
e-mail: Fabrizio.Catanese@uni-bayreuth.de; kiermaier@uni-bayreuth.de

Y. Cho
Department of Mathematics, Gyeongsang National University, Jinju-si, South Korea
e-mail: yhcho@gnu.ac.kr

F. Catanese (ed.), *Varieties of Nodal Surfaces, Coding Theory and Discriminants of Cubic Hypersurfaces*, Lecture Notes of the Unione Matematica Italiana 31,
https://doi.org/10.1007/978-3-032-05899-7_4

It satisfies $t \equiv -1\ (mod\ 4)$ and $t \geq 15$. Moreover, it can be $15, 27, 31, 35$ but cannot be $19, 23$; using the description of the code of the Barth sextic we see that also the cases $t = 39, 43$ occur. We cannot determine in general whether the cases with $t \geq 47$ occur, but we see that this occurrence is not allowed if the number of nodes $\nu = 65$.

Building from this we achieve here our original aim, to give a sufficient condition in order that a nodal sextic surface Y be the discriminant of a cubic hypersurface. The condition is that Y contains a half-even set $\mathcal{N}$ of cardinality 31, and such that a certain linear system, $|2H - L|$, has dimension zero (observe that, if $t = 31$, then necessarily $dim|2H - L| \geq 0$: we do not yet know whether the first condition alone is sufficient).

We set up again the notation: Y is a nodal sextic surface, $S := \tilde{Y}$ is its minimal resolution, and $\mathcal{N}$ is an half-even set of nodes.

We denote by $E_1, \dots, E_\nu$ the exceptional (-2)-curves, therefore there exists a divisor L on S such that

$$(*) \sum_{i \in \mathcal{N}} E_i + H \equiv 2L.$$

We denote by $t := |\mathcal{N}|$, and without loss of generality we assume that $\Delta := \sum_{i \in \mathcal{N}} E_i = \sum_1^t E_i$.

In this situation we attach to $(*)$ a double covering $f : Z \to S$ ramified on $\Delta + H$, where H is here any smooth hyperplane section, and such that

$$f_*(\mathcal{O}_Z) = \mathcal{O}_S \oplus \mathcal{O}_S(-L).$$

The inverse images of the (-2)-curves $E_1, \dots, E_t$ are (-1) curves, and can be contracted to smooth points, yielding $p : Z \to Z'$ and a finite double covering $f' : Z' \to Y$.

We have

$$K_Z = f^*(K_{\tilde{Y}} + L) = f^*(2H + L)$$

and, since by Kodaira-Ramanujam vanishing $H^1(Z, \mathcal{O}_Z(K_Z + nf^*H)) = 0$ for $n \geq 1$, we get

$$0 = H^1(Z, \mathcal{O}_Z(f^*((n+2)H+L))) = H^1(S, \mathcal{O}_S((n+2)H+L) \oplus H^1(S, \mathcal{O}_S((n+2)H)).$$

We obtain $H^1(S, \mathcal{O}_S(mH + L)) = 0$ for $m \geq 3$, and using

(i) Serre duality on S:

$$h^1(S, \mathcal{O}_S(mH+L)) = h^1(S, \mathcal{O}_S((2-m)H-L)) = h^1(S, \mathcal{O}_S((1-m)H+L-\Delta)$$

(ii) the exact cohomology sequence associated to the exact sequence

$$0 \to \mathcal{O}_S(aH + L - \Delta) \to \mathcal{O}_S(aH + L) \to \mathcal{O}_\Delta(L) \to 0,$$

and the vanishing of all cohomology groups of $\mathcal{O}_\Delta(L)$ since $h^j(\mathcal{O}_{\mathbb{P}^1}(-1) = 0$, yielding

$$(**)\ h^i(\mathcal{O}_S((a+1)H - L)) = h^i(\mathcal{O}_S(aH + L - \Delta)) = h^i(\mathcal{O}_S(aH + L))\ \forall i,$$

we conclude that

$$h^1(S, \mathcal{O}_S(mH + L)) = h^1(S, \mathcal{O}_S((2-m)H - L)) = h^1(S, \mathcal{O}_S((1-m)H + L),$$

hence we have the following vanishing result.

Proposition 4.1 *The first cohomology groups $H^1(S, \mathcal{O}_S(mH + L))$ and $H^1(S, \mathcal{O}_S(mH - L))$ are zero for all m if and only if $H^1(S, \mathcal{O}_S(mH + L)) = 0$ for $m = 1, 2$ or for $m = 0, -1$.*

Recall that [20] on Y we have the so called quadratic sheaf

$$\mathcal{F} := \pi_* \mathcal{O}_S(-L), \text{ such that } f'_*(\mathcal{O}_{Z'}) = \mathcal{O}_Y \oplus \mathcal{F}.$$

Corollary 4.2 *$h^1(S, \mathcal{O}_S(mH + L)) = 0$ for $m = 1, 2$ if and only if the following equivalent conditions hold:*

(1) the sheaf $\mathcal{F}$ is arithmetically Cohen-Macaulay,
(2) $H^1(\mathcal{F}) = H^1(\mathcal{F}(1)) = 0$,
(3) $\mathcal{N}$ is a symmetric set of nodes, that is, there is a symmetric matrix of homogeneous polynomials $A_{i,j}(x)$ such that

$$Y = \{x \mid \det(A(x)) = 0\},\ \mathcal{N} = \{x \mid \operatorname{corank}(A(x)) = 2\}.$$

We use now the Riemann-Roch theorem to calculate Euler-Poincaré characteristics.

Proposition 4.3

$$\begin{aligned}\chi(aH + L) :&= \chi(\mathcal{O}_S((aH + L)) \\ &= \chi(\mathcal{O}_S(((a+1)H - L)) \\ &= 11 + \frac{1}{4}[3(2a+1)(2a-3) - t].\end{aligned}$$

In particular,

$$\chi(\mathcal{F}) = \chi(\mathcal{F}(3)) = 11 + \frac{1}{4}(15 - t),$$

and accordingly we have

$$\chi(\mathcal{F}(3)) \geq 0 \Leftrightarrow t \leq 59,$$
$$\chi(\mathcal{F}) > 0 \Leftrightarrow t \leq 55.$$

Proof It was already shown that $\chi(\mathcal{O}_S((aH+L)) = \chi(\mathcal{O}_S(((a+1)H-L))$. Then

$$\chi(aH+L) := \chi(\mathcal{O}_S((aH+L)) = \chi(\mathcal{O}_S) + \frac{1}{2}(aH+L)(aH+L-K_S) =$$

$$= 11 + \frac{1}{8}((2a+1)H+\Delta)((2a-3)H+\Delta) = 11 + \frac{1}{4}[3(2a+1)(2a-3)-t].$$

We finish observing that, because of rational singularities,
$\chi(\mathcal{F}(m)) = \chi(mH-L)) = \chi((m-1)H+L)$. □

Lemma 4.4 *The cardinalities of the symmetric sets of nodes depend on the degrees $d_1 \leq d_2 \leq \cdots \leq d_p$ of the diagonal entries of the matrix $A(x)$, which must be odd numbers.*

- *degrees (1,5) : $t = 15$, in this case L is effective ($|L|$ has dimension 0);*
- *degrees (3,3) : $t = 27$, in this case L is not effective, but $L+H$ is effective ($|L+H|$ has dimension 1);*
- *degrees (1,1,1, 3) : $t = 31$, in this case L is not effective, but $L+H$ is effective ($|L+H|$ has dimension 0);*
- *degrees (1,1,1,1,1,1) : $t = 35$, in this case $L+H$ is not effective, but $L+2H$ is effective.*

Proof We use the inequalities of [21], and of [20], page 249, especially: $d_i + d_{p-1-i} > 0$.

This inequality implies, since $\sum_1^6 d_i = 6$, and since d_i is odd, that p is even, hence

$$\sum_{2i \leq p} (d_i + d_{p-i}) = 6.$$

Since $d_i + d_{p-i} > 0 \geq d_i + d_{p-1-i} > 0$, $d_i + d_{p-i} \geq 2$, therefore p equals one of the three numbers 2, 4, 6.

From $d_i \leq d + \delta - 2$ follows that $d_i \leq 5$, hence $d_p \in \{1, 3, 5\}$. If $d_1 \geq 1$, then all the d_i's are positive and we get the claimed four cases.

Since $d_1 + d_{p-2} > 0$, follows that $d_1 \geq -3$, and if $d_1 = -3$, then $d_{p-2}, d_{p-1}, d_p = 5$; hence $p = 6$ and, since the sum yields 6, we must have $(-3, -3, -3, 5, 5, 5)$, which contradicts however $d_2 + d_3 > 0$.

So, the case $d_1 = -3$ is excluded, and similarly the case $d_1 = -1$ is excluded. Since then $d_{p-2}, d_{p-1}, d_p \geq 3$, hence we must have $(-1, -1, -1, 3, 3, 3)$, which again contradicts $d_2 + d_3 > 0$.

The other assertions follow since $L + mH$ is effective, then also $L + mH - \Delta$ is effective, since $(L + mH) \cdot E_i = -1$. Hence $L + mH$ is effective if and only if $L + mH - \Delta = (m+1)H - L$ is effective, equivalently $H^0(\mathcal{F}(m+1)) \neq 0$.

Then we use the exact sequence

$$0 \to \sum_1^p \mathcal{O}_{\mathbb{P}^3}(\frac{-7-d_j}{2}) \to \sum_1^p \mathcal{O}_{\mathbb{P}^3}(\frac{-7+d_i}{2}) \to \mathcal{F} \to 0,$$

which is exact on global sections. □

The same argument shows that, in general, the restriction $\mathcal{F} \otimes \mathcal{O}_H$ of $\mathcal{F}$ to a smooth hyperplane section H has the following resolution:

$$0 \to \sum_1^p \mathcal{O}_{\mathbb{P}^2}(\frac{-7-d_j}{2}) \to \sum_1^p \mathcal{O}_{\mathbb{P}^2}(\frac{-7+d_i}{2}) \to \mathcal{F} \otimes \mathcal{O}_H \to 0.$$

The associated global section sequence is also exact, and we conclude

Remark 4.5 The cohomology groups of the restriction $\mathcal{F}|H := \mathcal{F} \otimes \mathcal{O}_H$ of $\mathcal{F}$ to a smooth hyperplane section H have respective dimensions (according to the degrees $(1,5), (3,3), (1^3,3), (1^6)$):

- $h^0(\mathcal{F}|H) = 0$,
- $h^0(\mathcal{F}(1)|H) = 1, 0, 0, 0$,
- $h^0(\mathcal{F}(2)|H) = 3, 2, 1, 0$,
- $h^1(\mathcal{F}|H) = 12$,
- $h^1(\mathcal{F}(1)|H) = 7, 6, 6, 6$,
- $h^1(\mathcal{F}(2)|H) = 3, 2, 1, 0$.

It follows that $h^0(\mathcal{F}) = 0$, and $h^0(\mathcal{F}(1)) = 0$ except possibly in the first case. This case is however clear, since we have

Lemma 4.6 $h^0(\mathcal{F}(1)) \neq 0$, *equivalently, L is effective, if and only $\mathcal{N}$ is symmetric and we are in the first case of Lemma 4.4, and also if and only if $t = 15$.*

Proof If L is effective, there is a plane H cutting on $\tilde{Y}$ the divisor $2D + \Delta$, hence we have a plane $z = 0$ such that the intersection with Y is twice a curve Γ, and moreover the plane passes through the nodes of $\mathcal{N}$.

Hence the equation $F = 0$ of Y equals the square of a cubic polynomial $g(x_0, x_1, x_2)$, modulo z, hence

$$F(x_0, x_1, x_2, z) = g(x_0, x_1, x_2)^2 + zP(x_0, x_1, x_2, z),$$

for some degree 5 polynomial P and F is the determinant of the symmetrical matrix with entries P, g, g, z.

At the nodes of $\mathcal{N}$, $g = 0$ must be a smooth curve (else the quadratic part of F would not have rank equal to 3), and the singular points of Y are then clearly the 15 points where $z = P = g = 0$. These are the points where the corank of the symmetrical matrix equals 2.

The last equivalence follows from Theorem 1.10, part (ii) of Endraß, [44], showing that $t \geq 15$, always, equality holding if and only if $h^0(\mathcal{F}(1)) \neq 0$. □

Lemma 4.7 *Assume $h^0(\mathcal{F}(1)) = 0$ and $t \leq 31$: then $h^0(\mathcal{F}(2)) \neq 0$, equivalently, $H + L$ is effective. If $h^0(\mathcal{F}(2)) = 0$, then $h^1(\mathcal{O}_S(H + L)) = \frac{1}{4}(t - 35)$.*

In particular, if $h^0(\mathcal{F}(2)) = 0$ and $t = 35$, then the set $\mathcal{N}$ is symmetric and we are in the fourth case of Lemma 4.4.

Proof We apply the Riemann-Roch inequality to the divisor $H + L$:

$$h^0(\mathcal{O}_S(H + L)) = h^0(\mathcal{O}_S(H + L)) + h^0(\mathcal{O}_S(H - L))$$

$$= h^0(\mathcal{O}_S(H + L)) + h^2(\mathcal{O}_S(H + L))$$

$$\geq \chi(\mathcal{O}_S(H + L)) = 11 + \frac{1}{4}(-9 - t).$$

Hence $h^0(\mathcal{O}_S(H + L)) \geq 1$ as soon as $10 + \frac{1}{8}(-18 - 2t) \geq 0 \Leftrightarrow 2t \leq 62$.

The second assertion follows right away from the previous computation, since then $\chi(\mathcal{O}_S(H + L)) = -h^1(\mathcal{O}_S(H + L))$.

For the third assertion, we know that from $h^0(\mathcal{F}(2)) = 0$ and $t = 35$ follows that $h^i(\mathcal{O}_S(H + L)) = 0$ for all i, equivalently, $h^i(\mathcal{F}(1) = 0$ for all i.

The exact cohomology sequence associated to

$$0 \to \mathcal{F}(1) \to \mathcal{F}(2) \to \mathcal{F}(2)|H \to 0$$

implies then that $H^0(\mathcal{F}(1)|H) = 0$, whence we are in the last case for the hyperplane section and we also infer that $h^i(\mathcal{F}(2) = 0$ for all i.

The exact cohomology sequence associated to

$$0 \to \mathcal{F}(2) \to \mathcal{F}(3) \to \mathcal{F}(3)|H \to 0$$

shows then that $h^0(\mathcal{F}(3)) = h^0(\mathcal{F}(3)|H) = 6$, and that these global sections generate $\mathcal{F}(3)$ on the smooth locus, hence everywhere by the reflexivity of $\mathcal{F}$.

We can now just apply theorem 2.23 of [21], or give a direct argument as follows: we can show that there is an exact sequence

$$0 \to \mathcal{O}_{\mathbb{P}^3}(-1)^6 \to \mathcal{O}_{\mathbb{P}^3}^6 \to \mathcal{F}(3) \to 0$$

by the analogous result for the restriction to the hyperplane section. □

Remark 4.8

(i) The analysis of the code of the Barth sextic, together with Remark 3.3 show that there are half-even sets of nodes with $t = 35$ and not symmetric, hence with $h^0(\mathcal{F}(2)) \neq 0$.

It shows moreover that the cases $t = 39, t = 43$ do indeed occur ($t = 39$ had been previously constructed by Barth [7]).

(ii) The same argument as in Lemma 4.7 shows that $h^0(\mathcal{F}(3)) \neq 0$, equivalently, $2H + L$ is effective as soon as $t < 59$.

The cases where $t \geq 47$ can be ruled out for $\nu = 65$ by coding theory, see Chap. 5.

(iii) Endraß [44] proved that $t = 19, 23$ do not occur.

In another vein, we can give an upper bound for t:

Proposition 4.9 *Assume that $H + L$ is effective. Then $t \leq 43$.*

Proof If $H + L$ is effective, there is an effective divisor D such that $2D \equiv 2H + 2L = 3H + \Delta$, and since $D \cdot \Delta = -1$, $D = C + \Delta$, with C effective, hence $2C \equiv 3H - \Delta$.

Consider the polar system, spanned by the partial derivatives of F: it is a subsystem of $5H - \Delta$, and it has no base points, whence $(5H - \Delta) \cdot C \geq 0$.

We get

$$0 \leq (5H - \Delta) \cdot (3H - \Delta) = 90 - 2t \;\Leftrightarrow\; t \leq 45,$$

and we conclude since $t + 1$ is divisible by 4. □

Proposition 4.10 *Assume that $h^0(\mathcal{F}(1)) = 0$ and that $h^0(\mathcal{F}(2)) \geq 2$, observing that*

$$h^0(\mathcal{F}(2)) \geq 2 \Leftrightarrow \; dim|H + L| \geq 1 \Leftrightarrow \; dim|2H - L| \geq 1.$$

Then for a smooth hyperplane section H we have

$$h^0(\mathcal{F}|_H(2)) = 3,$$

unless $t = 27$, $h^0(\mathcal{F}(2)) = 2$ and $|2H - L|$ is a base point free pencil.

Proof By the exact sequence $0 \to \mathcal{F}(1) \to \mathcal{F}(2) \to \mathcal{F}|_H(2) \to 0$ we obtain that $2 \leq h^0(\mathcal{F}|_H(2)) \leq 3$, hence we are just in one of the first two cases of Remark 4.5).

In the second case we have $h^0(\mathcal{F}|_H(2)) = 2$ and the linear system has no base point on H.

Writing therefore

$$|2H - L| = |H + L - \Delta| = F + |M|.$$

where F is the fixed part, we have $FH = 0$.

Therefore the divisor F is supported on the exceptional divisors, and we can write:

$$F = \sum_i a_i E_i.$$

Write $G := F + M$, so that $2G \equiv 3H - \Delta$. We have $GE_i = 1, GH = GM = 9$

$$G^2 = \frac{1}{4}(3H - \Delta)^2 = \frac{1}{4}(54 - 2t).$$

On the other hand first of all

$$1 = GE_i = ME_i + FE_i \Rightarrow ME_i = 1 + 2a_i,$$

secondly

$$G^2 = (M + F)^2 = M^2 + 2MF + F^2 = M^2 + 2MF + \sum_i(-2a_i^2) =$$

$$= M^2 + \sum_i(2a_i(2a_i + 1) - 2a_i^2) = M^2 + 2\sum_i(a_i^2 + a_i).$$

Since $3H \equiv 2G + \Delta \equiv 2M + \sum_i(2a_i + 1)E_i$, and a cubic cannot have a point of multiplicity bigger than 4, it follows that $2a_i + 1 \leq 3 \Leftrightarrow a_i \leq 1$.

By our assumption follows that $t \geq 27$.

Since $M^2 \geq 0$, we obtain a contradiction if $t \geq 31$.

If instead $t = 27$, it follows that $a_i = 0$ for all i, hence F=0, and $M^2 = 0$. □

Proposition 4.11 *Assume that* $t = 31$ *and that* $h^0(\mathcal{F}(2)) = 1$. *Then we are in the symmetric case with degrees* (1, 1, 1, 3).

Proof We have $h^0(\mathcal{F}(1)) = 0$ by Lemma 4.6, so by duality $h^2(\mathcal{F}(2)) = 0$. Since $h^0(\mathcal{F}(2)) = h^2(\mathcal{F}(1)) = 1$ and $\chi(\mathcal{F}(2)) = \chi(\mathcal{F}(2)) = 1$, follows the vanishing $h^1(\mathcal{F}(2)) = h^1(\mathcal{F}(1)) = 0$.

By the exact sequence $0 \to \mathcal{F}(1) \to \mathcal{F}(2) \to \mathcal{F}|_H(2) \to 0$ follows that we are in the third case for $\mathcal{F}|_H$, hence $h^0(\mathcal{F}|_H(1)) = 0$.

By the exact cohomology sequence associated to the sequence $0 \to \mathcal{F} \to \mathcal{F}(1) \to \mathcal{F}|_H(1) \to 0$ follows that $h^1(\mathcal{F}) \leq h^1(\mathcal{F}(1)) = 0$, hence we can apply Corollary 4.2 and obtain that $\mathcal{N}$ is symmetric (the sheaf $\mathcal{F}$ is Cohen-Macaulay), and of the desired form by Lemma 4.4. □

Remark 4.12 It is an interesting question to see whether if $t = 27$ or $t = 31$ then necessarily the half-even set $\mathcal{N}$ is symmetric.

In order to construct symmetric half-even sets of nodes with $t = 31$ we consider three distinct half-even sets of nodes $\mathcal{N}_1, \mathcal{N}_2, \mathcal{N}_3$ with $t = 15$.

We first observe that the sum $\mathcal{N}_1 + \mathcal{N}_2$ of $\mathcal{N}_1$ and $\mathcal{N}_2$, that is, the symmetric difference $(\mathcal{N}_1 \cup \mathcal{N}_2) \setminus (\mathcal{N}_1 \cap \mathcal{N}_2)$, is a strictly even set of weight (= cardinality) $t = 24$.

In fact, its weight is at most 30, and lies in the set $\{24, 32, 40, 56\}$ by Catanese and Tonoli [29].

Hence $|\mathcal{N}_1 \cap \mathcal{N}_2| = 3$.

Since half-even sets of nodes $\mathcal{N}_1, \mathcal{N}_2, \mathcal{N}_3$ with $t = 15$ correspond to nodes lying in a plane section H_i of Y, the above observation shows that $\mathcal{N}_1 \cap \mathcal{N}_2$ consists of three collinear nodes (in the line $H_1 \cap H_2$).

Therefore a priori $\mathcal{N}_1 \cap \mathcal{N}_2 \cap \mathcal{N}_3$ has cardinality 0, 1, 3. Accordingly the half-even set $\mathcal{N}_1 + \mathcal{N}_2 + \mathcal{N}_3$ would have cardinality $t = 27, 31, 39$.

Hence we get $t = 31$ if and only if the three planes H_1, H_2, H_3 are linearly independent and meet in a node of Y. In this case we shall show that $\mathcal{N}_1 + \mathcal{N}_2 + \mathcal{N}_3$ is symmetric.

Proposition 4.13 *Assume that we have a nodal sextic Y with three distinct half-even sets of nodes $\mathcal{N}_1, \mathcal{N}_2, \mathcal{N}_3$ with $t = 15$, such that $\mathcal{N} := \mathcal{N}_1 + \mathcal{N}_2 + \mathcal{N}_3$ has cardinality $t = 31$.*

Denote by L_i the corresponding divisor classes on $S = \tilde{Y}$ such that

$$2L_i \equiv H + \sum_{j \in \mathcal{N}_i} E_j,$$

and L such that

$$2L \equiv H + \sum_{j \in \mathcal{N}} E_j.$$

Then $|2H - L|$ has projective dimension 0, that is, $\mathcal{N}$ is symmetric.

Proof We know that $t = 15$ implies that $H - L_i$ is effective, $|H - L_i| = \{C_i\}$, where C_i maps to a plane cubic in Y, where the plane H_i is everywhere tangent to Y. Since $2C_i + \sum_{j \in \mathcal{N}_i} E_j \equiv H$, it follows that C_i does not contain any exceptional curve and that C_i is reduced.

In fact, let x_0 be a linear form vanishing on the plane H_i: then the equation of Y can be written as

$$F(x) = g_3(x_1, x_2, x_3)^2 + x_0 P_5(x_1, x_2, x_3) + x_0^2 B(x) = 0.$$

The singular points on Y on the plane $x_0 = 0$ are just the points $x_0 = g_3 = P_5 = 0$. These are 15 points if we have a transversal intersection, which is only possible if g_3 has no multiple factors, and then the full transform of $x_0 = 0$ contains each node precisely with multiplicity 1.

Moreover, for $i \neq j$, C_i and C_j have no common component: since C_i spans a plane $H_i \neq H_j$, the only possibility were that C_i and C_j have a common line. But if g_3 has a linear factor x_1, then the line $x_0 = x_1 = 0$ passes through 5 nodes of Y: this is a contradiction to $|\mathcal{N}_i \cap \mathcal{N}_j| = 3$.

We have that

$$C_1+C_2+C_3+\Delta' \in |2H-L|, \ \Delta' = E'+E_{1,2}+E_{2,1}+E_{1,3}+E_{3,1}+E_{2,3}+E_{3,2},$$

where $E' + E_{i,j} + E_{j,i}$ are the exceptional curves corresponding to the three nodes in $\mathcal{N}_i \cap \mathcal{N}_j$.

Clearly then $C_i \cdot E' = 1, C_i \cdot E_{j,k} = 1, 0$ according to $i \in \{j,k\}, \ i \notin \{j,k\}$.

Since $C_i \cdot \Delta' = 5, C_i \cdot H = 3, C_i L = \frac{1}{2}(3+11) = 7$ follows that $C_i \cdot (2H-L) = -1$.

If C_i is irreducible, it follows immediately that C_i is in the fixed part of the linear system $|2H - L|$.

If instead C_i is reducible, write $C_i = R + Q$, where R is a line; then $C_i \cdot (2H - L) = -1$ implies that an irreducible component of C_i is in the fixed part.

If the line R is in the fixed part, let us calculate:

$$(2H - L - R) \cdot Q = (2H - L) \cdot Q - R \cdot Q =$$

$$= (2H - L) \cdot C_i - (2H - L) \cdot R - 2 = (2H - L) \cdot R - 3.$$

We have

$$(2H - L) \cdot R = \frac{1}{2}(3H - \sum_{j \in \mathcal{N}} E_j) \cdot R \leq \frac{3}{2},$$

hence $(2H - L - R) \cdot Q < 0$ and Q is in the fixed part if Q if irreducible, or at least it contains a line in the fixed part.

Hence we may assume without loss of generality (changing the decomposition of C_i) that Q is in the fixed part, and we calculate:

$$(2H - L - Q) \cdot R = (2H - L) \cdot R - 2 \leq \frac{1}{2} \Rightarrow (2H - L - Q) \cdot R < 0,$$

hence we have shown that C_i is in the fixed part of $|2H - L|$.

Therefore

$$|2H - L| = C_1 + C_2 + C_3 + |\Delta'| = \{C_1 + C_2 + C_3 + \Delta'\},$$

and our assertion is proven. □

4.2 Discriminant Nodal Sextic Surfaces

This section is devoted to the study of discriminant sextics, obtained via the projection to $\mathbb{P}^3$, with centre a plane $L \subset X$, of a cubic hypersurface X of $\mathbb{P}^6$.

4.2.1 *Preliminaries on Discriminant Nodal Sextics Surfaces*

We proceed now to some considerations concerning the case of discriminantal nodal sextics, in view of (4) of Theorem 2.16.

The finite set $\mathrm{Sing}(X) \cap L$ is the base locus in $L \cong \mathbb{P}^2$ of the system of quadrics spanned by Q_0, Q_1, Q_2, Q_3, which can only be either a net or a web, and not a pencil: otherwise there would be a line $\{x_1 = x_2 = 0\}$ where A' vanishes, hence this line would be singular for Y.

Theorem 4.14 $\mathrm{Sing}(X) \cap L$ *has cardinality at most* 3. *If the cardinality equals* 3, *then we may choose coordinates such that the matrix*

$$A' := \begin{pmatrix} 0 & x_1 & x_2 \\ x_1 & 0 & x_3 \\ x_2 & x_3 & 0 \end{pmatrix}, \tag{4.1}$$

in particular in this case the cubic $M := \{x \mid \det(A'(x) = 0\}$ *consists of three general planes.*

If the cardinality of $\mathrm{Sing}(X) \cap L$ *equals* 2, *then there are* 3 *possibilities for the cubic* M*: either it consists of a quadric cone (rank* 3 *quadric) and a plane passing through the vertex, or of a rank* 4 *quadric and a tangent plane, or of three general planes.*

Finally, if M *is reducible, then* $\mathrm{Sing}(X) \cap L$ *can have cardinality only* $0, 2, 3$, *while cardinality* 1 *occurs for* M *irreducible.*

There is exactly one case with M *reducible where* $\mathrm{Sing}(X) \cap L$ *is empty:* M *consists of a quadric cone and a plane intersecting it transversally in a smooth conic.*

Proof We have seen that for each such base point (that is, a point of $\mathrm{Sing}(X) \cap L$) we can take coordinates such that the point is e_0, and the two points $x_1 = x_2 = x_0 x_3 = 0$ in $\mathbb{P}^3$ yield nodal singularities of the conic bundle $\mathcal{C}$.

It follows then that the cubic M (everywhere tangent to Y) is singular along the line $x_1 = x_2 = 0$.

There are two cases, the first is when we have a web of conics. Here we can use the list of symmetric cubics which are discriminant of a web of conics, from [22], page 33, and deduce that, since M is singular along a line, we are in cases (i), or (iv), or (vi).

However the enumeration in loc. cit. is incomplete, since the pencil of conics $\lambda x_0^2 + \mu x_1^2 = 0$ was not considered, dually this leads to the web with matrix

$$A' := \begin{pmatrix} 0 & x_0 & x_1 \\ x_0 & 0 & x_2 \\ x_1 & x_2 & x_3 \end{pmatrix}. \tag{4.2}$$

In this case, the base locus consists of 2 points, and M consists of the plane $x_0 = 0$ and a smooth quadric, to which the plane is tangent.

The first of case (i) is the matrix

$$A' := \begin{pmatrix} 0 & 0 & x_1 \\ 0 & x_0 & x_2 \\ x_1 & x_2 & x_3 \end{pmatrix}, \tag{4.3}$$

which must be excluded since then $\det(A) \in (x_1, q_0)^2$, a contradiction.

The second of case (i) is the matrix

$$A' := \begin{pmatrix} x_0 & 0 & x_3 \\ 0 & x_1 & 0 \\ x_3 & 0 & x_2 \end{pmatrix}, \tag{4.4}$$

yielding an empty base locus and $M = \{x_1(x_0x_2 - x_3^2) = 0\}$, a quadric cone and a plane intersecting it in a smooth conic.

While case (iv) yields a point as base locus and cannot be excluded: in this case M is irreducible:

$$A' := \begin{pmatrix} x_0 & 0 & x_2 \\ 0 & x_1 & x_3 \\ x_2 & x_3 & 0 \end{pmatrix}, \tag{4.5}$$

Similarly case (vi),

$$A' := \begin{pmatrix} 0 & x_0 & x_1 \\ x_0 & x_1 & x_2 \\ x_1 & x_2 & x_3 \end{pmatrix}, \tag{4.6}$$

yields a point as base locus and M is irreducible.

Next, we consider the second case, where we have a net (hence M is a cone over a plane curve). We may assume that

$$A' := \begin{pmatrix} 0 & x_1 & x_2 \\ x_1 & b(x) & c(x) \\ x_2 & c(x) & d(x) \end{pmatrix}. \tag{4.7}$$

If the base locus contains three non collinear points, then we may assume $b = d = 0$, and since we have a net $c(x)$ is linearly independent from x_1, x_2, so that we may set $c(x) = x_3$; hence we obtain the desired normal form 4.1, for which the base locus consists of exactly 3 points, and M consists of 3 general planes.

The base locus cannot contain three collinear points, else it would contain a line, and be infinite, a contradiction.

If the base locus contains exactly two points, we may assume $b = 0$, $d \neq 0$, and by rows and columns operations that $c(x) = c(x_2, x_3)$.

Then the equation of M is

$$-d(x)x_1^2 + 2x_1x_2c(x) = x_1(2x_2c(x_2, x_3) - d(x)x_1) = 0.$$

Hence M is the cone over a cubic union of a line and a conic.

If $c \equiv 0$ we may set $d =: x_3$ and we would get that the base locus consists of a simple base point and a double one. But this case is excluded since then $\det(A) \in (x_1, q_1)^2$, and Y does not have isolated singularities.

We may assume then that either $c(x) = x_3$, or $d(x) = x_3$, $c(x) = \lambda x_2$, $\lambda \neq 0$. In the latter case the conic is $2\lambda x_2^2 - x_1x_3 = 0$, it is smooth, and tangent to the line $x_1 = 0$. In the former case we get the conic $2x_2x_3 - x_1d(x) = 0$, intersecting the line $x_1 = 0$ in two distinct points. The conic is smooth unless $d = -\frac{1}{2}\lambda\mu x_1 + \lambda x_2 + \mu x_3$, in which case with rows and columns operations and with new variables we obtain the normal form with $b = 0$, $d = \lambda x_1$, where $\lambda \neq 0$ (since otherwise we would have three base points).

For the assertion concerning the hypothesis that M is reducible, we simply run the list again in the case of a web, and observe that M contains a plane only in the second case of (i), where there are no base points, or in the 'missing case' 4.2 where M is the union of a plane and a smooth quadric and there are two base points.

If the net has two base points, then we saw that M is reducible.

We consider next the case of a net where there are no base points, and show that then M is irreducible.

Otherwise, we can assume that x_3 divides the equation of M.

In this case the net yields a morphism $\phi : \mathbb{P}^2 \to \mathbb{P}^2$ of degree 4, and x_1, x_2, x_3 are coordinates of the dual plane of the target $\mathbb{P}^2$. The line $x_3 = 0$ is dual to a point of the target $\mathbb{P}^2$, and if we take a general other point, the pencil of lines through it pulls back to a general pencil in the net which has 4 distinct base points, hence it has normal form $x_1z_1z_2 + x_2z_3(z_1 + z_2 + z_3) = 0$.

Then the net has the form

$$2x_1z_1z_2 + 2x_2z_3(z_1 + z_2 + z_3) + x_3Q_3(z) = 0,$$

where $Q_3(z)$ does not vanish at the 4 points $e_1, e_2, e_2 - e_3, e_1 - e_3$.

The condition that x_3 divides the discriminant means that setting $x_3 = 0$, we get a discriminant which is identically zero as a polynomial in x_1, x_2. But indeed we get $2x_1x_2(x_2 - x_1)\ mod(x_3)$, a contradiction.

We show now that the case of a net where there is exactly one base point, and for which M is reducible, does not exist.

Case (II): assume that the net contains 2 double lines, say $z_1^2 = 0$ and $z_2^2 = 0$. Then the matrix

$$A' := \begin{pmatrix} 0 & \lambda x_3 & \mu x_3 \\ \lambda x_3 & x_1 + b_3 x_3 & c_3 x_3 \\ \mu x_3 & c_3 x_3 & x_2 + d_3 x_3 \end{pmatrix} \tag{4.8}$$

and we get that $det(A) \in (x_3, q_0)^2$, a contradiction.

Case (I): assume that the net contains exactly 1 double line, say $z_1^2 = 0$, and assume as usual that the base point is e_0.

Then the matrix

$$A' := \begin{pmatrix} 0 & x_1 & x_2 \\ x_1 & x_3 & c(x_1, x_2) \\ x_2 & c(x_1, x_2) & d(x_1, x_2) \end{pmatrix}. \tag{4.9}$$

With row and column operations we obtain first that $c = c(x_2) = \lambda x_2$, and then we reduce to the case where $\lambda = 0$, if we take $x_3 - 2\lambda x_1$ as new third variable.

Hence we may assume $c = 0$, so that

$$\det(A') = -dx_1^2 - x_2^2 x_3,$$

which, being of degree 1 in x_3, is irreducible unless $d = \lambda x_2$.

But in this case the base locus is given by

$$z_0 z_1 = z_1^2 = 2z_0 z_2 + \lambda z_2^2 = 0 \Leftrightarrow z_1 = z_2(2z_0 + \lambda z_2) = 0,$$

and it consists of two points.

Case (0): there is no rank 1 conic in the net. Hence M is the cone over a reducible plane curve C, given together with a nontrivial invertible sheaf $\mathcal{F}$ of 2-torsion and with a basis of $H^0(\mathcal{F}(1))$ ($\mathcal{F}$ is the cokernel of $A' : \mathcal{O}_{\mathbb{P}^2}(-2)^3 \to \mathcal{O}_{\mathbb{P}^2}(-1)^3$). Hence, since $Pic^0(C)$ has nontrivial 2-torsion, C consists either of three general lines, or a line and a smooth conic meeting in two points.

In both cases there is just one normal form since $Pic^0(C) = \mathbb{C}^*$ (see the theorem of [6] about ineffective theta characteristics, and also [21], Proposition 2.28 and Remark 2.29).

In the former case we have the normal form 4.1, since in this case there is no point in the net where the rank of A' drops down to 1; this case however must be excluded since then we have three base points of the net.

In the latter we may take the normal form (again the rank of A' is at each point ≥ 2)

$$A' := \begin{pmatrix} 0 & x_1 & x_2 \\ x_1 & x_3 & x_3 \\ x_2 & x_3 & x_2 \end{pmatrix}. \tag{4.10}$$

The curve C is then

$$x_2\,(-x_1^2 + 2x_1x_3 - x_2x_3) = 0,$$

but we have 2 base points instead of 1, the solutions of

$$z_0z_1 = 2z_0z_2 + z_2^2 = 2z_1z_2 + z_1^2 = 0 \Leftrightarrow z_1 = z_2(2z_0 + z_2) = 0.$$

□

Theorem 4.15 *Let Y be a nodal sextic surface in $\mathbb{P}^3$ with $\nu(Y)$ nodes and a symmetric half-even set $\mathcal{N}$ of cardinality* 31*: then Y is the discriminant of a nodal cubic $X \subset \mathbb{P}^6$ for the projection with centre a plane $L \subset X$.*

X has exactly $\gamma(X) = \nu(Y) - 31 - r$ nodes, where $r \in \{0, 1, 2, 3\}$ is the cardinality of $L \cap \mathrm{Sing}(X)$.

Proof Since the set $\mathcal{N}$ is symmetric, there is a matrix of homogeneous polynomials $A(x)$, where $\deg A_{i,j} = \frac{1}{2}(d_i + d_j)$, and $d_1, d_2, d_3, d_4 = 1, 1, 1, 3$, and such that $\mathcal{N}$ is the set of points where the matrix $A(x)$ has rank $= 2$.

Hence Y is the discriminant of a cubic hypersurface, and the assertion follows directly from Theorem 2.16 and from Theorem 4.14. □

In the next subsection we shall show that, in the case $\nu(Y) = 65$, r can only be 2, 3, that is, the cubic has then 31 or 32 nodes.

4.2.2 65 Nodal Sextics as Cubic Discriminants

By Theorem 5.3, the (extended) code of Barth's sextic is isomorphic to the (extended) code of an arbitrary sextic with 65 nodes.

In particular, any such sextic has 26 half-even sets of cardinality 15. Let $C_1, \ldots, C_{26}$ be the corresponding reduced plane sections.

Let $\pi : \tilde{Y} \to Y$ be the minimal resolution. For a subset $\mathcal{N} \subset \mathrm{Sing}\, Y$ of nodes, we denote by $E_{\mathcal{N}}$ the sum of the corresponding (-2)-curves in $\tilde{Y}$.

We fix an even set $\mathcal{N}$ of cardinality 31 such that there are three half-even sets $\mathcal{N}_\alpha$, $\mathcal{N}_\beta$, $\mathcal{N}_\gamma$ of cardinality 15 whose (code-theoretic) sum is $\mathcal{N}$. The proof of Proposition 4.13 applies in the same way: let $\tilde{H} \in |\pi^*\mathcal{O}_Y(1)|$, and $L \equiv \frac{1}{2}(\tilde{H}+E_{\mathcal{N}})$. Then

$$|2\tilde{H} - L| = \left|\frac{1}{2}(3\tilde{H} - E_{\mathcal{N}})\right| = \{\tilde{C}_\alpha + \tilde{C}_\beta + \tilde{C}_\gamma + E_{\mathcal{N}'}\},$$

where $\tilde{C}_\alpha$ is the proper transform of C_α and $\mathcal{N}'$ is the set of nodes that appear precisely twice in $\mathcal{N}_\alpha + \mathcal{N}_\beta + \mathcal{N}_\gamma$ ($\#\mathcal{N}' = 6$).

We claim that there are $v_0, v_1, v_2 \in H^0(3\tilde{H} - L)$ and $v_3 \in H^0(2\tilde{H} - L)$ such that $\{v_0, v_1, v_2, v_3 \otimes H^0(H)\}$ spans $H^0(3\tilde{H} - L)$, and that there are linear forms $\{\ell_i : i = 0, 1, 2\}$ with the properties

(i) $B_{ii} := v_i^2$ and $B_{i3} := v_i v_3$ are divisible by ℓ_i,
(ii) B_{33} is divisible by $\ell_0\ell_1\ell_2$.

Then the (i, i)-minor of the matrix (B_{ij}) is divisible by $\ell_j\ell_k \times$ (defining equation of $Y)^2$, but has degree $5 + 5 + 3 = 13$, thus $a_{ii} = 0$.

Lemma 4.16

(i) $|3\tilde{H} - L - \tilde{C}_\alpha - \tilde{C}_\beta| = |\tilde{H}| + \tilde{C}_\gamma + E_{\mathcal{N}'}$.
In particular, $H^0(3\tilde{H} - L - \tilde{C}_\alpha - \tilde{C}_\beta)$ *can be identified with the image of* $H^0(\tilde{H}) \otimes H^0(2\tilde{H} - L) \to H^0(3\tilde{H} - L)$.
(ii) There is a decomposition of $\mathcal{N}$ *as the sum of five half-even sets of cardinality 15, among which is* $\mathcal{N}_\alpha$ *(resp. for* $\mathcal{N}_\beta$, $\mathcal{N}_\gamma$*).*

Proof

(i) Let $P_i \in \mathcal{N}'$. Then,

$$(E_i \,.\, \tilde{C}_\gamma) = \begin{cases} 1 \text{ if } P_i \in \mathcal{N}_\gamma \\ 0 \text{ otherwise.} \end{cases}$$

Hence, $(E_i \,.\, \tilde{H} + \tilde{C}_\gamma + E_{\mathcal{N}'}) = (E_i \,.\, \tilde{C}_\gamma) - 2 < 0$. In this way, we prove that $E_{\mathcal{N}'}$ is in the fixed part. Let $R \subset \tilde{C}_\gamma$ be an irreducible component of degree $d = (R \,.\, \tilde{H})$. Then, R contains $5d$ nodes of $\mathcal{N}_\gamma$([44, Prop. 2.4]), hence

$$(R \,.\, \tilde{H} + \tilde{C}_\gamma) = d + \frac{1}{2}(d - 5d) = -d < 0,$$

showing that R is in the fixed part of $|\tilde{H} + \tilde{C}_\gamma|$.
(ii) By theorem 5.3, it suffices to check the statement for Barth's sextic. It can be checked by computer.[1]

□

[1] See the script [33, Mathematica 17].

The previous proposition implies that any decomposition of $\mathcal{N}$ into five 15-half-even sets cannot contain more than one of $\mathcal{N}_\alpha, \mathcal{N}_\beta, \mathcal{N}_\gamma$.

Proposition 4.17 *Let v_0 (resp. v_1, v_2) be the section corresponding to a 5-decomposition of $\mathcal{N}$ containing $\mathcal{N}_\alpha$ (resp. $\mathcal{N}_\beta$, $\mathcal{N}_\gamma$), and let $v_3 \in H^0(2\tilde{H}-L)\setminus\{0\}$. Then, v_0, v_1, v_2 and $W := v_3 \otimes H^0(\tilde{H})$ spans $H^0(3\tilde{H}-L)$.*

Proof The subspace spanned by v_0 and W is 5-dimensional and the corresponding linear system contains $\tilde{C}_\alpha$ in its base locus. Since $(v_1 = 0) \not\supset \tilde{C}_\alpha$, v_0, v_1 and W spans a 6-dimensional space, say K.

To prove $v_2 \notin K$, we first consider the following: let $\mathcal{N}_0 = \mathcal{N}_1 \cap \mathcal{N}_2 \cap \mathcal{N}_3$, then

$$\left|3\tilde{H} - L - E_{\mathcal{N}'\setminus\mathcal{N}_0}\right| = \left|\tilde{H} + \tilde{C}_\alpha + \tilde{C}_\beta + \tilde{C}_\gamma + E_{\mathcal{N}_0}\right|$$

For simplicity, we put $D := \tilde{H} + \tilde{C}_\alpha + \tilde{C}_\beta + \tilde{C}_\gamma + E_{\mathcal{N}_0}$. Suppose $\tilde{C}_\alpha$ is irreducible, then it is in the fixed part since

$$(\tilde{C}_\alpha \,.\, D) = 3 - 6 + 0 + 0 + 1 < 0.$$

If $\tilde{C}_\alpha$ is reducible, let $R \subset \tilde{C}_\alpha$ be an irreducible component that does not contain $\mathcal{N}_0$ (R always exists by [44, Lemma 2.3]). Let $d = (R \,.\, \tilde{H})$, then $(R \,.\, E_{\mathcal{N}_\alpha}) = 5d$ and $(R \,.\, E_{\mathcal{N}_i}) = d$ for $i = \beta, \gamma$. Hence,

$$(R \,.\, D) = d + \left(\frac{d}{2} - \frac{5d}{2}\right) + \left(\frac{d}{2} - \frac{d}{2}\right) + \left(\frac{d}{2} - \frac{d}{2}\right) + 0 = -d < 0$$

thus R is in the fixed part. If $R' \subset \tilde{C}_\alpha$ is the component containing $\mathcal{N}_0$, then $(R' \,.\, D - R) = -d + 1 - (R' \,.\, R) < 0$ (irreducible components of C_α do not intersect at nodes, thus $(R' \,.\, R) > 0$). This shows that $\tilde{C}_\alpha$ is in the fixed part. It follows that

$$|D| = |\tilde{H}| + \tilde{C}_\alpha + \tilde{C}_\beta + \tilde{C}_\gamma + E_{\mathcal{N}_0}.$$

Next we claim that $(v_2 = 0)$ do not contain $E_{\mathcal{N}'\setminus\mathcal{N}_0}$. Suppose not, then $(v_2 = 0)$ contains $\tilde{C}_\gamma + E_{\mathcal{N}'\setminus\mathcal{N}_0}$, thus it comes from $|3\tilde{H} - L - \tilde{C}_\gamma - E_{\mathcal{N}'\setminus\mathcal{N}_0}| = |\tilde{H}| + \tilde{C}_\alpha + \tilde{C}_\beta + E_{\mathcal{N}_0}$. By Lemma 4.16, $v_2 \in W$, a contradiction.

Pick $P_k \in \mathcal{N}' \setminus \mathcal{N}_0$ such that $(v_2 = 0) \not\supset E_k$. Then, $(3\tilde{H} - L - \tilde{C}_\gamma \,.\, E_k) \geq 0$. This implies $(\tilde{C}_\alpha \,.\, E_k) = (\tilde{C}_\beta \,.\, E_k) = 1$ since

$$3\tilde{H} - L - \tilde{C}_\gamma \sim \tilde{H} + \tilde{C}_\alpha + \tilde{C}_\beta + E_{\mathcal{N}'}$$

and $(E_{\mathcal{N}'}.E_k) = -2$. In particular, $(\tilde{C}_\gamma.E_k) = 0$ and $(3\tilde{H} - L.E_k) = 0$. The divisor $(v_0 = 0)$ contains $\tilde{C}_\alpha$, which positively intersects with E_k, but $(3\tilde{H} - L \,.\, E_k) = 0$, thus $(v_0 = 0)$ contains E_k. Similarly, $(v_1 = 0)$ contains E_k. This shows that $|K|$ contains E_k in its fixed part. By choice of k, we have $(v_2 = 0) \not\supset E_k$, thus K and v_2 are linearly independent. □

Let v_0 (and v_1, v_2 similarly) to be the section defined by $\left(\tilde{C}_\alpha + \text{ sum of four } \tilde{C}_i's\right)$, and v_3 by $\tilde{C}_\alpha + \tilde{C}_\beta + \tilde{C}_\gamma + E_{\mathcal{N}'}$. The set $\{v_0, v_1, v_2, v_3\}$ satisfies the described properties, hence $a_{11} = a_{22} = a_{33} = 0$.

Corollary 4.18 *The Barth sextic and every other sextic surface with 65 nodes is the discriminant of the projection p_L of a cubic hypersurface $X \subset \mathbb{P}^6$ with 31 nodes, with centre a plane L intersecting $\mathrm{Sing}(X)$ in three points, and not contained in a three dimensional linear subspace $\Lambda \subset X$.*

In a nodal sextic with 65 nodes, the number of half-even sets of cardinality 31 is 1690. Amongst them, 1300 admit the decomposition as the sum of three half-even sets of cardinality 15. Each of those 1300 half-even sets yields a pair (X, L) as in Corollary 4.18. We observe here that every half-even set of cardinality 31 is either the sum of three half-even sets of cardinality 15, or the sum of a half-even set of cardinality 15 with an even set of cardinality 24.

We consider the remaining 390 cases in which $\mathcal{N}$ is a 31-half-even set that admits a $\mathcal{N}_\alpha + \mathcal{Q}$ decomposition where $\mathcal{N}_\alpha$ is an 15-half-even set, and $\mathcal{Q}$ is a 24-even set cut out by a quadric $V \subset \mathbb{P}^3$. Consider the degree 6 curve Q such that $2Q = Y \cdot V$, and let $\tilde{Q} \subset \tilde{Y}$ be the proper transform of Q. We remark that V is an irreducible quadric; otherwise, V is a union of two planes, thus $\mathcal{Q}$ is a sum of two 15-half-even sets.

To prove that $\mathcal{N}$ is symmetric, we need to study numerical invariants such as $\tilde{Q}^2$. To do this, we first regard Q as a divisor in V, and compare intersection theories between V and Y.

Lemma 4.19 *Let X be a smooth projective variety, and let $Q_1, Q_2 \subset X$ be reduced closed curves. Let $S, S' \subset X$ be smooth projective surfaces containing Q_1, Q_2. Suppose Q_1 and Q_2 share no common irreducible components. Then,*

$$(Q_1 \cdot Q_2)_S = (Q_1 \cdot Q_2)_{S'},$$

where the subscript indicates the ambient surface where the intersection takes place.

Proof Let $Q = Q_1 \cup Q_2$ endowed with the reduced structure.

This follows from the exact sequence

$$0 \to \mathcal{O}_Q \to (\mathcal{O}_{Q_1} \oplus \mathcal{O}_{Q_2}) \to \mathcal{O}_{Q_1 \cap Q_2} \to 0$$

defining the intersection multiplicity, so that

$$(Q_1 \cdot Q_2)_S = h^0(\mathcal{O}_{Q_1 \cap Q_2}) = (Q_1 \cdot Q_2)_{S'}.$$

□

Proposition 4.20 $|2\tilde{H} - L| = \{\tilde{C}_\alpha + \tilde{Q} + E_{\mathcal{N}'}\}$, *where* $\mathcal{N}' = \mathcal{N}_\alpha \cap \mathcal{Q}$.

Proof In the first part of the proof, we claim that $\tilde{C}_\alpha$ is in the fixed part. If $\tilde{C}_\alpha$ is irreducible, it is trivial since $(\tilde{C}_\alpha \,.\, 2\tilde{H} - L) = -1$. Assume $\tilde{C}_\alpha$ is reducible, let $\{R_i\}$ be the set of irreducible components, and let $d_i = (R_i \,.\, \tilde{H})$. Then,

$$\big(R_i \,.\, \tilde{C}_\alpha + \tilde{Q} + E_{\mathcal{N}'}\big) = \Big(\frac{d_i}{2} - \frac{5d_i}{2}\Big) + \Big(d_i - \frac{1}{2}(R_i \,.\, E_Q)\Big) + (R_i \,.\, E_{\mathcal{N}'})$$
$$= -d_i + \frac{1}{2}(R_i \,.\, E_Q).$$

Hence, $(R_i \,.\, E_Q)$ is even and the sum over i equals $(\tilde{C}_\alpha \,.\, E_Q) = 4$.

Case 1. Assume $(R_1 \,.\, E_Q) = 4$. Then, all R_j with $j \neq 1$ are in the fixed part, and

$$(R_1 \,.\, 2\tilde{H} - L - \sum_{j\neq 1} R_j) = -d_1 + 2 - \sum_{j\neq 1}(R_1 \,.\, R_j).$$

By Proposition [44, Lemma 2.3], R_i are proper transform of the planes curves that do not intersect at the nodes of Y. Thus, $(R_1 \,.\, R_j) = d_1 d_j$, so the latter is always negative.

Case 2. Assume $(R_i \,.\, E_Q) = 2$ for $i = 1, 2$. Since $\sum d_i = 3$, there are two subcases: $d_1 = d_2 = 1$ and there is a third component R_3, or (w.l.o.g) $d_1 = 2$. In the first subcase, R_3 is in the fixed part, and $(R_1 \,.\, 2\tilde{H} - L - R_3) = -(R_1 \,.\, R_3) < 0$ thus R_1 is in the fixed part. Similarly, R_2 should be in the fixed part. In the remaining subcase, R_1 is in the fixed part, and then $(R_2 \,.\, 2\tilde{H} - L - R_1) < 0$, so R_2 is in the fixed part. We conclude that $\tilde{C}_\alpha$ is in the fixed part, and so is $E_{\mathcal{N}'}$.

It remains to prove $|\tilde{Q}| = \{\tilde{Q}\}$. Let $Q_1 \subset Q$ be an irreducible component. Again, we split into two cases: whether V is singular or not.

Case 1. Assume V is a quadric cone. In this case $V \simeq \mathbb{P}(1, 1, 2)$. Let $h_V \in |\mathcal{O}_V(1)|$, and let e be an integer determined by $Q_1 \sim e \cdot h_V$. For $H_{\mathbb{P}^3} \in |\mathcal{O}_{\mathbb{P}^3}(1)|$, $H_{\mathbb{P}^3}\big|_V = 2h_V$, and $h_V^2 = 1/2$. Thus,

$$(\tilde{Q}_1 \,.\, \tilde{H}) = (Q_1 \,.\, H_{\mathbb{P}^3})_{\mathbb{P}^3} = (Q_1 \,.\, 2h_V)_V = e.$$

By adjunction formula one may write $\tilde{Q}_1^2$ in terms of e and $p_a(\tilde{Q}_1)$. To estimate the arithmetic genus, we look at the proper transform of Q_1 along the resolution $\mathbb{F}_2 \to V$.

By Endraß [44, Proposition 2.4], Q_1 cuts out at least $4e$ nodes of Y. If we vary Q_1 and do the same procedure for all irreducible components of Q, we find that the number of nodes cut out by Q_1 is exactly $4e$ as the sum of all e equals to $(Q.\tilde{H}) = 6$ and Q cuts out $4 \times 6 = 24$ nodes. This also shows that the vertex v of V cannot be a smooth point of Y; otherwise [44, Proposition 2.4] would imply that Q_1 cuts out $4e + 1$ nodes of Y.

Let $Q_{1,\mathbb{F}_2} \subset \mathbb{F}_2$ be the proper transform of Q_1. We claim that $Q_{1,\mathbb{F}_2}$ and $\tilde{Q}_1$ may be identified via the proper transform along the blowing up $\tilde{\mathbb{P}}^3 \to \mathbb{P}^3$ at $\{P_1, \ldots, P_{65}\}$. By Endraß [44, Lemma 2.3], Q is smooth at the nodes of Y except possibly at v. Hence, blowing up at $\{P_1, \ldots, P_{65}\} \setminus \{v\}$ does not affect both Q and V. So, in the case $v \notin Y$ the claim is clear. If $v \in Y$, then v is a node of Y, so the proper transform of V along $\tilde{\mathbb{P}}^3 \to \mathbb{P}^3$ is nothing but the resolution $\mathbb{F}_2 \to V$.

Let us estimate $p_a(Q_{1,\mathbb{F}_2})$. We have

$$(Q_{1,\mathbb{F}_2})^2_{\mathbb{F}_2} \leq (Q_1^2)_V = \frac{1}{2}e^2$$

and

$$(K_{\mathbb{F}_2} \,.\, Q_{1,\mathbb{F}_2})_{\mathbb{F}_2} = (K_V \,.\, Q_1)_V = -2e,$$

hence $2\,p_a(Q_{1,\mathbb{F}_2}) - 2 \leq \frac{1}{2}e^2 - 2e$. The adjunction formula for $\tilde{Q}_1 \subset \tilde{Y}$ reads

$$\tilde{Q}_1^2 + (2\tilde{H} \,.\, \tilde{Q}_1) = 2\,p_a(\tilde{Q}_1) - 2 \leq \frac{1}{2}e^2 - 2e,$$

thus $\tilde{Q}_1^2 \leq \frac{1}{2}e^2 - 4e$. Let $Q' = Q - Q_1$. In V, we have $Q' \sim (6 - e)h_V$, hence

$$(Q_1 \,.\, Q')_V = \frac{1}{2}e(6 - e).$$

By Lemma 4.19 and the identification $\tilde{Q} = Q_{\mathbb{F}_2}$ in $\tilde{\mathbb{P}}^3$, $(\tilde{Q}_1.\tilde{Q}')_{\tilde{Y}} = (Q_{1,\mathbb{F}_2}.Q'_{\mathbb{F}_2})_{\mathbb{F}_2}$. Thus,

$$(\tilde{Q}_1 \,.\, \tilde{Q}) = (Q_{1,\mathbb{F}_2} \,.\, Q'_{\mathbb{F}_2})_{\mathbb{F}_2} + \tilde{Q}_1^2 \leq \frac{1}{2}e(6 - e) + \frac{1}{2}e^2 - 4e = -e < 0,$$

thus $\tilde{Q}_1$ is in the fixed part of $|\tilde{Q}|$. This proves the case when V is a quadric cone.

Case 2. Now we consider the case if V is nonsingular. Let $Q_1 \in |\mathcal{O}_V(a, b)|$. Since $Q \in |\mathcal{O}_V(3, 3)|$, $0 \leq a, b \leq 3$. The adjunction formula:

$$(Q_1^2)_V + (K_V \,.\, Q_1)_V = 2ab - 2(a + b) = 2\,p_a(Q_1) - 2.$$

Since Q_1 is smooth at the nodes of Y, $p_a(\tilde{Q}_1) = p_a(Q_1)$. So,

$$\tilde{Q}_1^2 = 2ab - 2(a + b) - (2\tilde{H} \,.\, \tilde{Q}_1) = 2ab - 4(a + b).$$

For $Q' = Q - Q_1 \in |\mathcal{O}_V(3 - a, 3 - b)|$, by Lemma 4.19,

$$(\tilde{Q}_1 \,.\, \tilde{Q}') = (Q_1 \,.\, Q')_V = a(3 - b) + b(3 - a),$$

thus

$$(\tilde{Q}_1 \cdot \tilde{Q}) = a(3-b) + b(3-a) + 2ab - 4(a+b) = -(a+b) < 0,$$

thus $\tilde{Q}_1$ is in the fixed part of $|\tilde{Q}|$. This implies $|\tilde{Q}| = \{\tilde{Q}\}$ as desired. □

We want to choose the generators $v_0, v_1, v_2 \in H^0(3\tilde{H} - L)$ and $v_3 \in H^0(2\tilde{H} - L)$, having following property: after suitable reordering of indices in $\{L_i\}$,

$$\begin{aligned}
(v_0 = 0) &= L_0 + L_1 + L_2 + L_3 + L_4 + (\text{some } E_i \text{ terms}) \\
(v_1 = 0) &= L_4 + L_5 + L_6 + L_7 + L_8 + (\text{some } E_i \text{ terms}) \\
(v_2 = 0) &= L_8 + L_9 + L_{10} + L_{11} + L_{12} + (\text{some } E_i \text{ terms}) \\
(v_3 = 0) &= L_0 + Q + E_{\mathcal{N}'}
\end{aligned} \tag{4.11}$$

Then the some entries of the matrix (B_{ij}) have linear factors described as follows:

$\ell_0\ell_4$	ℓ_4		ℓ_0
ℓ_4	$\ell_4\ell_8$	ℓ_8	
	ℓ_8	ℓ_8	
ℓ_0			ℓ_0

The second principal (3,3)-minor of this matrix has determinant of the form

$$\ell_0\ell_4 \begin{vmatrix} \ell_4\ell_8 & \\ & \ell_0 \end{vmatrix} + \ell_4 \begin{vmatrix} \ell_4 & \\ \ell_0 & \ell_0 \end{vmatrix} + \ell_0 \begin{vmatrix} \ell_4 & \ell_4\ell_8 \\ \ell_0 & \end{vmatrix},$$

so it is divisible by $\ell_0\ell_4 \cdot (\text{defining equation of Y})^2$. Due to degree reason, $a_{33} = 0$.

This excludes (7) of Theorem 4.14, hence we conclude by Theorem 4.14 that there are two base points.

Now, we directly check the $\mathfrak{A}_5$-orbits of $31 = 15 + 24$ half-even sets, and found the basis 4.11 directly. The following example is an explicit demonstration; the other orbits can be done by the same procedure.

Example 4.21 We consider the following $\mathcal{N} = \mathcal{N}_7 + \mathcal{Q}$:

$$\mathcal{N} = 1\ 11001111111110011011110110000000010000100000000111111000011000011$$

$$\mathcal{N}_7 = 1\ 11000000000010000100001110000000010000100000011000000000011000011$$

$$\mathcal{Q} = 0\ 00001111111100011111111000000000000000000000011111111000000000000$$

We have $\mathcal{N}' = \mathcal{N}_7 \cap \mathcal{Q} = \{P_{18}, P_{23}, P_{46}, P_{47}\}$. The sections v_i are given by the following decompositions of $\mathcal{N}$:

v_0	$\mathcal{N}_1+\mathcal{N}_7+\mathcal{N}_9+\mathcal{N}_{10}+\mathcal{N}_{25}$	$(2,2,2,2)$
v_1	$\mathcal{N}_1+\mathcal{N}_2+\mathcal{N}_{12}+\mathcal{N}_{13}+\mathcal{N}_{17}$	$(0,0,2,2)$
v_2	$\mathcal{N}_{12}+\mathcal{N}_{14}+\mathcal{N}_{19}+\mathcal{N}_{23}+\mathcal{N}_{24}$	$(0,0,0,0)$
v_3	$\mathcal{N}_7+\mathcal{Q}$	$(2,2,2,2)$

The quadruple in the right hand side is the vanishing order of the corresponding curves at $\mathcal{N}'$; for instance, the quadruple in $(0,0,2,2)$ indicates that the corresponding divisor

$$C_1+C_2+C_{12}+C_{13}+C_{17}$$

has zeros at $(P_{18}, P_{23}, P_{46}, P_{47})$ with orders $(0,0,2,2)$ respectively. From this we may read that v_0, v_1, v_2, v_3 is the desired set of generators. Indeed, all sections in $W := \langle v_0, v_3 \otimes H^0(\tilde{H})\rangle$ have vanishing orders at least $(2,2,2,2)$, so $v_1 \notin W$. Then, all sections in $K := \langle v_1, W\rangle$ have vanishing orders at least $(0,0,2,2)$, hence $v_2 \notin K$.

Corollary 4.22 *Every nodal sextic Y with 65 nodes admits a pair (X, L) where $X \subset \mathbb{P}^6$ is a nodal cubic with 32 nodes and $L \subset X$ is a plane intersecting* Sing X *at two points, such that Y is the discriminant of the projection $p_L\colon X \dashrightarrow \mathbb{P}^3$ with centre L.*

There is no such pair (X, L) with $|X \cap L| \leq 1$ such that Y is the discriminant of the projection $p_L\colon X \dashrightarrow \mathbb{P}^3$ with centre L.

Proposition 4.23 *Let $X \subset \mathbb{P}^6$ be a nodal cubic and let P_1, P_2 be nodes of X. Then the line $\Lambda := P_1 * P_2$ is contained in X and there is a plane L containing P_1, P_2 such that $L \subset X$.*

Proof $\Lambda \cdot X \geq 4$, so $\Lambda \subset X$. Take coordinates

$$(u, x) := (u_0, u_1, x_0, \dots, x_4)$$

such that $\Lambda = \{x = 0\}$ and such that the nodes are $P_1 = e_0$, $P_2 = e_1$.

Then the equation of X writes as

$$F(u,x) = u_0^2\lambda_{0,0}(x) + 2u_0u_1\lambda_{0,1}(x) + u_1^2\lambda_{1,1}(x) + u_0Q_0(x) + u_1Q_1(x) + G(x).$$

Since e_0, e_1 are nodes the linear parts of the Taylor development vanish at them, hence $\lambda_{0,0}(x) \equiv 0$, $\lambda_{1,1}(x) \equiv 0$, hence

$$F(u,x) = 2u_0u_1\lambda_{0,1}(x) + u_0Q_0(x) + u_1Q_1(x) + G(x).$$

Let $L_x := \Lambda * \{(0,x)\}$: then $L_x \subset X$ if and only if

$$\lambda_{0,1}(x) = Q_0(x) = Q_1(x) = G(x) = 0.$$

These are 4 equations in $\mathbb{P}^4$, so they always have a solution (and in general 12 solutions). □

Remark 4.24 Let $X \subset \mathbb{P}^6$ be a nodal cubic and let P_1, P_2, P_3 be (linearly independent) nodes of X. Then if L is the plane containing P_1, P_2, P_3, then we can choose coordinates

$$(v, x) := (v_0, v_1, v_2, x_0, \dots, x_3)$$

such that $L = \{x = 0\}$ and such that the nodes are e_0, e_1, e_2.

Then the equation of X writes as

$$F(v, x) = \lambda v_0 v_1 v_2 + \sum_{i,j} v_i v_j \lambda_{i,j}(x) + \sum_i v_i Q_i(x) + G(x).$$

And $L \subset X$ if and only if $\lambda = 0$.

In particular if we have an equisingular family $\mathcal{X}$ of nodal cubics $X_t \subset \mathbb{P}^6$, and we let L_t be the plane spanned by 3 nodes $P_1(t), P_2(t), P_3(t)$ of X_t, then $L_t \subset X_t$ if and only if $\lambda(t) = 0$, hence we get a subfamily of codimension at most 1.

We can now summarize the previous Corollaries 4.18 and 4.22 in the following

Theorem 4.25 *The Barth's sextic as well as every other sextic surface with 65 nodes*

(i) is the discriminant of the projection p_L of a cubic hypersurface $X \subset \mathbb{P}^6$ with 31 nodes, and with centre a plane L intersecting Sing(X) *in three points and not contained in a three dimensional linear subspace $\Lambda \subset X$,*
(ii) is the discriminant of the projection p_L of a cubic hypersurface $X \subset \mathbb{P}^6$ with 32 nodes, and with centre a plane L intersecting Sing(X) *in two points and not contained in a three dimensional linear subspace $\Lambda \subset X$,*
(iii) cannot be the discriminant of the projection p_L of a cubic hypersurface $X \subset \mathbb{P}^6$ with 33 or more nodes.

$\mathcal{F}(6, 65)$ *is smooth of dimension* 18 *at the point corresponding to the Barth sextic, and at any other point corresponding to an unobstructed surface.*

Taking cubic discriminants of cubic hypersurfaces as in i) or ii) one obtains irreducible components of dimension ≥ 18 *of* $\mathcal{F}(6, 65)$.

Proof

(i): by proposition 4.13 it suffices to show the existence of three distinct half-even sets of nodes $\mathcal{N}_i$, $i = 1, 2, 3$ of cardinality 15 such that $\mathcal{N} := \mathcal{N}_1 + \mathcal{N}_2 + \mathcal{N}_3$ has cardinality $t = 31$.

We use the theorem proven in Appendix of Chap. 5, that all sextic surfaces with 65 nodes have extended code isomorphic to the one of the Barth sextic. Then it suffices to show that the assertion holds for the Barth sextic.

While assertions (ii) and (iii) follow from Theorem 4.14 and Corollary 4.22.

That the Barth sextic is unobstructed is shown in Sect. 2.5.2, while an explicit determinantal equation for the Barth sextic is contained in Sect. 4.4: in this case $L \cap \mathrm{Sing}(X)$ consists of three points since the cubic $M = \{\det(A') = 0\}$ clearly consists of three planes, and we can then invoke Theorem 4.14 and Theorem 2.16.

For the assertion about the dimension, this follows easily from the fact that the dimension of the space of sextics with 65 nodes is at least $83 - 65 = 18$. The smoothness follows from assertion (vii) of Theorem 1.25 that the Barth sextic is unobstructed, which shows indeed that at this point the dimension of $\mathcal{F}(6, 65)$ is exactly 18.

We can use the unobstructedness of the Segre cubic X_0 (see v) of Theorem 1.25) in order to obtain an irreducible component $\mathcal{X}_{32} \subset \mathcal{F}_5(3, 32)$ with number of moduli $m = 3$ and an irreducible component $\mathcal{X}_{31} \subset \mathcal{F}_5(3, 31)$ with number of moduli $m = 4$.

We can now take two nodes $P_1(t)$, $P_2(t)$ of $X_t \in \mathcal{X}_{32}$ and use Proposition 4.23 to infer the existence of a plane $L_t \subset X_t$ passing through $P_1(t)$, $P_2(t)$; in this way we construct a family of dimension at least 18 of sextics which, by Theorem 2.16, are 65-nodal if their singularities are just nodes.

Similarly, choosing three non collinear nodes of $X'_t \in \mathcal{X}_{32}$, we obtain, on a subfamily of $\mathcal{X}_{32}$ of codimension at most 1, by Remark 4.24, that the plane L' spanned by the 3 nodes is contained in the cubic hypersurface X'_t.

In this way we also construct a family of dimension at least 18 of sextics which, by Theorem 2.16, are 65-nodal if their singularities are just nodes. □

Remark 4.26 The previous theorem raises the question about the irreducibility of $\mathcal{F}(6, 65)$, in possible analogy with the irreducibility of $\mathcal{F}(3, 4)$, $\mathcal{F}(4, 16)$, $\mathcal{F}(5, 31)$.

It also raises the question: given a general 32-nodal cubic X and two nodes, is there only a finite number of planes contained in X and passing through the 2 nodes? Then the above construction would yield a family of dimension exactly 18.

Observe that up to now there are four different constructions of a 65-nodal sextic: the Barth equation, the construction by Pettersen [75] via nodal determinantal quartic hypersurfaces in $\mathbb{P}^4$, and the above two constructions via cubic discriminants.

4.2.3 Nodal Sextics as Quartic Discriminants

Let $W \subset \mathbb{P}^4$ be a nodal quartic hypersurface, and choose coordinates $(w, x) := (w, x_0, \ldots, x_3)$ such that $(1, 0)$ is a node P_0 of $W = \{f(w, x) = 0\}$. Then the equation of W writes as

$$f(w, x) = w^2 Q(x) + wB(x) + g(x).$$

Projecting to $\mathbb{P}^3$ with centre the node P_0, we get the discriminant surface

$$Y = \{x | F(x) := 4Q(x)g(x) - B^2(x) = 0\},$$

which is the branch locus of the rational double covering $W \dashrightarrow \mathbb{P}^3$.

Comparing Sing(W) with Sing(Y) we see that

$$\mathrm{Sing}(W) = \{(w, x)|2wQ + B = 0, w^2\nabla Q + w\nabla B + \nabla g = 0\},$$

$$\mathrm{Sing}(Y) = \{x|4g\nabla Q + 4Q\nabla g - 2B\nabla B = 0\}.$$

One observes that if P is a singular point of W, then the line $P_0 * P$ is either contained in W, or it intersects W only in P_0, P, hence P maps to a point of Y. The former situation occurs precisely when $Q(x) = g(x) = B(x) = 0$.

If $Q(x) \neq 0$, then we have at most one singular point P of W on the line $P_0 * (0, x)$, which satisfies $w = -\frac{B}{2Q}$ and then we see that the equation of W at P is the suspension of the one of Y at x, since

$$4Qf=(2wQ)^2+2(2wQ)B+4Qg=(2wQ+B)^2+(4Qg-B^2)=(2wQ+B)^2+F(x).$$

We observe that $w' := (2wQ + B)$ completes the affine x coordinates to a system of local coordinates.

Hence over the open set $Q(x) \neq 0$ we have a bijection between singular points of W and singular points of Y, such that the equation of W is of the form $(w')^2 + F(x) = 0$.

In particular, we have a node $x \in Y$ downstairs if and only if we have a node of W, $P = (-\frac{B}{2Q}, x) \in W$ upstairs.

If instead $Q(x) = 0$, and P is a singular point, then $B(x) = 0$ for P, and also $g(x) = 0$ for P.

Similarly, if $Q(x) = 0$ and $x \in Y$, then $B(x) = 0$. Over the points for which $Q(x) = B(x) = 0$, the equations for the singular points are then simplified to

$$w^2\nabla Q + w\nabla B + \nabla g = 0,$$

respectively

$$g\nabla Q = 0.$$

At this point it is crucial to use the hypothesis that P_0 is a node, ensuring that the quadric Q is a smooth quadric, so that $Q = \nabla Q = 0$ has no solution except P_0.

Hence the singular points of Y with $Q = 0$ are only the points x where $Q = B = g = 0$, and over them we do not have any singular point of W if the three gradients are linearly independent, a condition which is equivalent to the condition that these points x are nodes of Y. Follows:

Proposition 4.27 *If the sextic surface Y is the discriminant of the projection of a quartic hypersurface from a node P_0, then outside the closed set $Q(x) = B(x) = 0$ we have a bijection between the singular points $P = (w, x)$ of W and the singular* *points $x \in \mathrm{Sing}(Y)$, such that the singularity at P is the suspension of the one of Y at x.*

In particular, if Y is nodal, then W is also nodal, and there is a bijection between the singular points of W different from P_0 and the singular points of Y different from the 24 points $\{Q = B = g = 0\}$.

In particular, if Y has ν nodes, then W has $\delta = \nu - 23$ nodes.

Conversely, if W is nodal, again there is a bijection between the nodes of W different from P_0 and the singular points of Y which are not in the set $\{Q = B = g = 0\}$, which are also nodes. While the points in the set $\{Q = B = g = 0\}$ yield nodes of Y if and only if the number of lines Λ passing through P_0 and contained in W is 24.

Proof We have already proven the assertion concerning the points in the open set $Q(x) \neq 0$.

For the others, we use that since P_0 is a node, the quadric Q is smooth. Then we may restrict to consider the points lying in the closed set $\{Q = B = g = 0\}$. If Y is nodal these are 24 nodes, moreover these are nodes if and only if this set consists of 24 distinct points. This is however equivalent to requiring that on W these equations define 24 different lines through P_0. □

Pettersen [75] considered quartics which are determinants of 4×4 matrices with entries linear forms, which have in general 20 singular points (Proposition 1.3.7 ibidem).

With very hard (and also computer supported) work he was able to prove the following result (Theorem 5.5.1 of [75]):

Theorem 4.28 (Pettersen) *There exists a unique irreducible 3-dimensional family of 65-nodal sextic surfaces which are the discriminant of a nodal determinantal quartic hypersurface W with 42 nodes for the projection from one node. This family contains the Barth sextic.*

This theorem (taken cum grano salis replacing the word 'dimension' with 'number of moduli') produces then an irreducible family of dimension 18 of 65-nodal sextics.

Conversely, we have:

Theorem 4.29 *Each 65-nodal sextic Y admits an even set $\mathcal{N}$ of cardinality 24. $\mathcal{N}$ is symmetric hence determines a realization of Y as the discriminant of a quartic hypersurface W for the projection with centre a double point P_0.*

W has 42 nodes if and only if the unique quadric Q cutting on $\tilde{Y}$ the linear system $|L + H|$ is smooth (L is here the divisor such that $2L \equiv \sum_{x \in \mathcal{N}} E_x$).

There exist choices of $\mathcal{N}$ such that either the rank of Q is either 3, or Q is smooth.

In the case where Q has rank 3, P_0 is not a node, and W has as singular points P_0 and 40 nodes, plus possibly another singular point P'.

The open set $\mathfrak{U}$ inside the family of all 65-nodal sextics, consisting of the sextics admitting an even set $\mathcal{N}$ of cardinality 24 with Q smooth, and for which the *corresponding 42-nodal quartic hypersurface W is determinantal, is irreducible of dimension 18 and contains the Barth sextic.*

Proof Since we have seen that all 65-nodal sextics have the same code (Chap. 5), there is an even set $\mathcal{N}$ of cardinality 24.

We have already seen in this section that we may assume that $\mathcal{N}$ is not the sum of two half-even sets (of cardinality 15).

$\mathcal{N}$ is symmetric by Theorem 3.8.2 of [20], hence is associated to a representation

$$Y = \{x|4Q(x)g(x) - B^2(x) = 0\},$$

where the quadric $\{Q(x) = 0\}$ is uniquely defined. We may write $Q = Q_{\mathcal{N}}$ to recall that the quadric depends upon the choice of $\mathcal{N}$.

Observe that $Q = Q_{\mathcal{N}}$ is irreducible, else it is the union of two planes, and $\mathcal{N}$ is the sum of two half-even sets of cardinality 15.

Since $\mathcal{N} = \{Q = B = g = 0\}$ we conclude that, if $\{Q(x) = 0\}$ is smooth, then, by Proposition 4.27, Y is the discriminant of a quartic W with 42 nodes.

The condition that $\{Q(x) = 0\}$ is smooth is $disc(Q) \neq 0$, hence defines an open set $\mathfrak{U}' := \{Y | \exists\, \mathcal{N} \text{ s.t. } disc(Q_{\mathcal{N}}) \neq 0\}$ inside the family of all 65-nodal sextics.

Moreover, the condition that W is determinantal is an open condition by virtue of the following Lemma 4.30.

The rest follows then from Pettersen's Theorem 4.28.

Finally, if $\{Q(x) = 0\}$ is a rank 3 quadric, then, since we assume Y nodal, a point x' where $Q = B = \nabla Q = 0$ would be a double point of Y, hence a node of Y where $g \neq 0$.

Then on the line $P_0 * (0, x')$ the singular points of W different from P_0 are determined by the equation $w\nabla B + \nabla g = 0$. There is then at most one solution P', since $\nabla g = 0$ implies $g = 0$, a contradiction.

If there is no such singular point x', then W shall have 41 nodes plus the singular point P_0.

If such a point x' exists, then W shall have 40 nodes plus the singular point P_0 plus possibly another singular point P' on the line $P_0 * (0, x')$. □

The following result is indeed valid in a higher generality, but we just state in a form suitable to our purposes here.

Lemma 4.30 *Let $W \subset \mathbb{P}^4$ be a linear determinantal quartic $W = \{x|det(a_{ij}(x) = 0\}$, which has only δ nodes as singularities. Then there exists an open set of the nodal Severi variety $\mathcal{F}_3(4, \delta)$ containing W and consisting of nodal linear determinantal quartics.*

Proof Consider the complete intersection variety

$$Z \subset \mathbb{P}^4 \times \mathbb{P}^3,\ Z := \{(x, v)| \sum_j a_{ij}(x)v_j = 0,\ i = 1, 2, 3, 4\}.$$

Among the nodes of W there are the 20 nodes (see for instance [75], Prop. 1.3.7) where the corank of the matrix $A(x) := (a_{ij})_{i,j=1,\dots 4}$ equals 2. Let $\mathcal{N}_r$ be this set of nodes.

Then $\pi : Z \to W$ is the small resolution of the set of nodes $\mathcal{N}_r$, and the direct image

$$\mathcal{F} := \pi_*(\mathcal{O}_Z(0,1)),$$

of the invertible sheaf of bidegrees $(0, 1)$ is a torsion free sheaf admitting the resolution

$$0 \to \mathcal{O}_{\mathbb{P}^4}(-2)^4 \to \mathcal{O}_{\mathbb{P}^4}(-1)^4 \to \mathcal{F} \to 0,$$

where the middle homomorphism is given by the matrix $A(x)$.

Indeed, $H^0(\mathcal{F}(1))$ has dimension 4 , and since $\mathcal{F}$ is Cohen Macaulay, and we have those 4 relations $\sum_j a_{ij}(x)v_j = 0$, then the sequence is exact (alternatively we can use the direct image of the Koszul complex on $P := \mathbb{P}^4 \times \mathbb{P}^3$).

Consider now a neighbourhood of W inside $\mathcal{F}_3(4, \delta)$: then we can take the corresponding small resolutions Z_t of the 20 nodes of W_t which are deformation of the nodes of W lying in $\mathcal{N}_r$.

We use now that $H^i(\mathcal{O}_Z) = 0$ for $i = 1, 2$, because of the Koszul complex on $V := \mathcal{O}_P(-1, -1)^4$ associated to the four sections

$$\sum_j a_{ij}(x)v_j \in H^0(\mathcal{O}_P(1, 1)) :$$

$$0 \to \Lambda^4(V) \to \Lambda^3(V) \to \Lambda^2(V) \to V \to \mathcal{O}_P\mathcal{O}_P \to \mathcal{O}_Z \to 0.$$

By semicontinuity also $H^i(\mathcal{O}_{Z_t}) = 0$ for $i = 1, 2$. Hence $\mathrm{Pic}(Z_t)$ is locally constant, and the line bundle $\mathcal{O}_Z(0, 1)$ extends for Z_t in a neghbourhood. Hence also Z_t admits such a line bundle, which yields an embedding in $P = \mathbb{P}^4 \times \mathbb{P}^3$ as such a complete intersection.

We conclude therefore that also W_t, being the projection of Z_t, is linear determinantal. □

4.2.4 Segre Type Representation of 65-Nodal Sextics

We shall now show that if Y has 65 nodes, then it admits a Segre realization

$$Y = \{\lambda_1(x) \dots \lambda_6(x) - B(x)^2 = 0\}.$$

This follows immediately from Theorem 1.29 once we prove that the Barth sextic (hence, any other 65 nodal sextic) admits a subcode of $\mathcal{K}''$ which is the Segre code.

This in turn follows once we find a Segre realization of the Barth sextic: but this is precisely given by the defining equation

$$(\tau^2x^2 - y^2)(\tau^2y^2 - z^2)(\tau^2z^2 - x^2) - \frac{1}{4}(2\tau + 1)w^2(x^2 + y^2 + z^2 - w^2)^2,$$

where $\tau := \frac{1}{2}(1 + \sqrt{5})$ is the inverse of the golden ratio.

Here the six linear forms $\lambda_i(x)$ are not in general position, since they have a common zero set, but the code is indeed the Segre code since the 15 lines intersection of pairs of planes $\{\lambda_i(x) = \lambda_j(x) = 0\}$ pass through 3 nodes each.

We may ask whether there is a realization like for the Barth sextic:

$$Y = \{x|\lambda_1(x)\dots\lambda_6(x) - \lambda_0(x)^2Q(x)^2 = 0\}.$$

Since Y has the same code $\mathcal{K}'$ as the Barth sextic, we saw that there are codewords $v_1, \dots, v_6$, such that they correspond to half-even sets of cardinality 15 produced by linear forms $\lambda_1(x), \dots, \lambda_6(x)$.

The question boils down to the question whether the divisor of B contains the divisor of a linear form λ_0, equivalently, whether $B(x) = \lambda_i(x) = 0$ defines a plane cubic that splits as a conic and a line Λ_i, and the six lines $\Lambda_1, \dots, \Lambda_6$ are coplanar. However it seems that the above realization might be a special occurrence, by virtue of the following observation.

It can be computationally checked[2] that the extended code of Barth sextic contains precisely 65 Segre subcodes. This leads to further 64 Segre realizations. For instance, one of the realizations consists of six planes

$$(\tau w - \tau^2x + z)(-\tau w - \tau^2x + z)$$
$$(-w + x - y + z)(w + x - y + z)(-w + x + y + z)(w + x + y + z),$$

where $B(x)$ equals[3]

$$\tau^2x^3 - \tau^2w^2x - \bar{\tau}xy^2 - w^2z + \tau^3x^2z - \tau^3y^2z + \bar{\tau}xz^2 + z^3.$$

In this case $B(x)$ is an irreducible cubic, unlike the Segre realization introduced in the beginning. We also remark that in the latter realization the six planes linearly span $\mathbb{P}^4$. This happens for every other Segre realization except for the first one.

It follows from this computational result that the group of projective automorphisms of the Barth surface fixes the point $\{x = y = z = 0\}$, leaves invariant the

[2] See the script [33, Mathematica 20].

[3] See the script [33, Mathematica 21].

plane $\{w = 0\}$ and the sphere $\Sigma := \{x^2 + y^2 + z^2 - w^2 = 0\}$, and permutes the six planes

$$\{\tau x \pm y = 0\}, \{\tau y \pm z = 0\}, \{\tau z \pm x = 0\}.$$

We use this remark in order to give a computational proof of the following result

Theorem 4.31 *The group of automorphisms of the Barth sextic Y has order 120, hence it is the Icosahedral group $\mathbb{Z}/2 \times \mathfrak{A}_5$.*

We first observe that all automorphisms are projective automorphisms since they preserve the canonical divisor, which is equal to $2H$. Since the surface Y has no two torsion in second homology, these automorphisms preserve H, hence they are induced from projectivities.

The following argument is due to Stephen Coughlan.

We start from the defining equation $f(w, x, y, z) = 0$ of the Barth sextic $Y \subset \mathbb{P}^3$, where $F(w, x, y, z) :=$

$$(\tau^2 x^2 - y^2)(\tau^2 y^2 - z^2)(\tau^2 z^2 - x^2) - \frac{1}{4}(2\tau + 1)w^2(x^2 + y^2 + z^2 - w^2)^2$$

and $\tau := \frac{1}{2}(1 + \sqrt{5})$ is the inverse of the golden ratio.

Using this observation, we show that the projective automorphism group of Y has order 120, with the help of a computer.

First change the coordinates to

$$a := \tau x - y,\ b := \tau y - z,\ c := \tau z - x,\ w := w,$$

and define expressions

$$d := \tau(a + b) + c,\ e := \tau(b + c) + a,\ f := \tau(c + a) + b.$$

In these coordinates, the Barth sextic has the nice form

$$abcdef - 1/4\tau w^2(ad + be + cf - \tau w^2)^2,$$

the six permuted planes are

$$\{a = 0\}, \{b = 0\}, \{c = 0\}, \{d = 0\}, \{e = 0\}, \{f = 0\} \tag{4.12}$$

and the invariant sphere is $\{ad + be + cf - \tau w^2\}$.

In order to make the proof of Theorem 4.31 more transparent, we make its statement more explicit.

Proposition 4.32 *The projective automorphism group of the Barth sextic has 120 elements. Moreover, in the coordinates a, b, c, w, a projective automorphism* *preserving the Barth sextic can be written in the form*

$$w \mapsto \pm w,\ a \mapsto p_1\sigma_1,\ b \mapsto p_2\sigma_2,\ c \mapsto p_3\sigma_3$$

where $(\sigma_1, \sigma_2, \sigma_3)$ *are certain triples of three distinct elements from* $\{a, b, c, d, e, f\}$, *and* (p_1, p_2, p_3) *are certain triples in* $\{\pm 1\}^3$.

Proof Let Φ be a projective automorphism of the Barth sextic. There exist scalars $\alpha, \beta, \gamma, \delta, \epsilon, \varphi$ such that

$$\Phi(a) = \alpha\sigma_1, \Phi(b) = \beta\sigma_2, \Phi(c) = \gamma\sigma_3, \Phi(d) = \delta\sigma_4, \Phi(e) = \epsilon\sigma_5, \Phi(f) = \varphi\sigma_6.$$

where $\sigma = (\sigma_1, \dots, \sigma_6)$ is an ordered permutation of $a, \dots, f$.

Since Φ is a homomorphism, we have $\Phi(d) = \Phi(\tau(a + b) + c)$ and similar for $\Phi(e)$, $\Phi(f)$. Substituting the above general form for Φ into these equalities gives three equations

$$\delta\sigma_4 = \tau(\alpha\sigma_1+\beta\sigma_2)+\gamma\sigma_3,\ \epsilon\sigma_5 = \tau(\beta\sigma_2+\gamma\sigma_3)+\alpha\sigma_1,\ \varphi\sigma_6 = \tau(\gamma\sigma_3+\alpha\sigma_1)+\beta\sigma_2.$$

Comparing the coefficients of a, b, c in each of the three equalities gives a homogeneous system of nine linear equations in six unknowns $\alpha, \dots, \varphi$.

Using a computer, we consider each 6-tuple σ, construct its system of linear equations, and determine the space V_σ of solutions. We find that there are 60 permutations σ for which V_σ is spanned by a single vector in $\{\pm 1\}^6$; for the other permutations, V_σ is the zero space and therefore does not correspond to an automorphism.

We examine the cases where V_σ is nonempty and write $p = (p_1, \dots, p_6)$ for the basis vector in $\{\pm 1\}$. We have a 1-parameter family of possible automorphisms corresponding to the space V_σ:

$$\Phi(a) = p_1\alpha\sigma_1, \Phi(b) = p_2\alpha\sigma_2, \Phi(c) = p_3\alpha\sigma_3,$$

$$\Phi(d) = p_4\alpha\sigma_4, \Phi(e) = p_5\alpha\sigma_5, \Phi(f) = p_6\alpha\sigma_6.$$

Next observe that the sphere $ad + be + cf - \tau w^2$ is invariant under Φ, and $\Phi(w) = \lambda w$. Substituting this 1-parameter family and then scaling $\lambda = \pm\alpha$, we get that $\Phi(ad + be + cf - \tau w^2) = \lambda^2(ad + be + cf - \tau w^2)$. Thus there are 60 permutations σ with V_σ non-zero, and for each such V_σ there are two possibilities for λ, this gives 120 automorphisms in total. It is easy to see from the normal form, that in each case, the Barth sextic is also invariant under Φ. □

The proof uses a computer to work out the solutions of the system of linear equations for each of the 720 permutations.

4.3 The Code of the Barth Sextic

We start again from the defining equation $f(w, x, y, z) = 0$ of the Barth sextic $Y \subset \mathbb{P}^3$, where $f(w, x, y, z) :=$

$$(\tau^2 x^2 - y^2)(\tau^2 y^2 - z^2)(\tau^2 z^2 - x^2) - \frac{1}{4}(2\tau + 1)w^2(x^2 + y^2 + z^2 - w^2)^2$$

and $\tau := \frac{1}{2}(1 + \sqrt{5})$ is the inverse of the golden ratio.

In [8], Barth described the icosahedral symmetry of Y. Consider the icosahedron with vertices

$$(1, 0, \pm\tau, \pm 1),\ (1, \pm 1, 0, \pm\tau),\ (1, \pm\tau, \pm 1, 0).$$

Each face of the icosahedron can be labelled by an ordered pair of distinct integers ij, $1 \le i, j \le 5$. Figure 4.1 shows the labelling using the dual dodecahedron.

The action of $\mathfrak{A}_5$ via $\sigma \cdot ij \mapsto \sigma(i)\sigma(j)$ realizes $\mathfrak{A}_5$ as the group of rotational icosahedral symmetries.

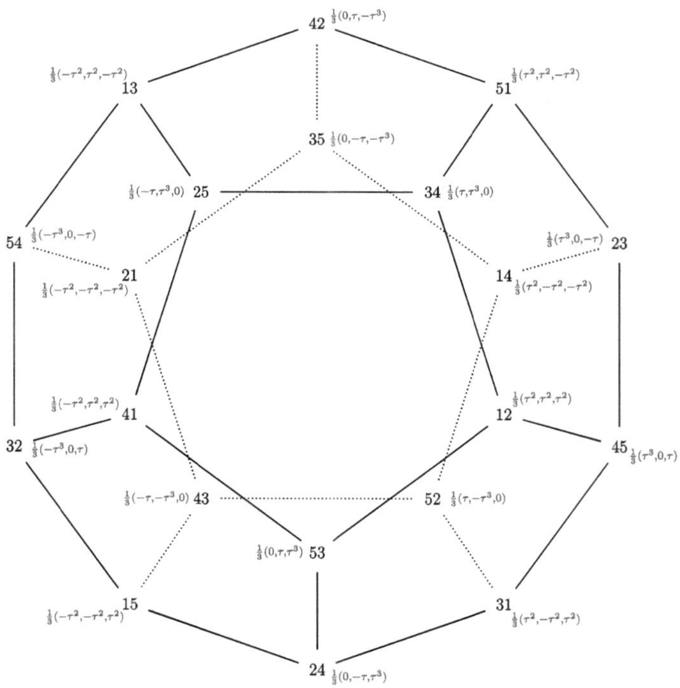

Fig. 4.1 Labelling vertices of the dual dodecahedron

Consider a line which connects two opposite centres of the faces of the icosahedron. Such a line is called a *centre line*. In Fig. 4.1, centre lines are the lines connecting opposite vertices, so they can be labelled by the unordered pair (ij). The parentheses mean that we do not distinguish (ij) and (ji). The *mid lines* are the lines which connect mid points of two opposite edges of the icosahedron. Equivalently, they are the lines connecting mid points of the opposite edges of the dual dodecahedron. They correspond to double transpositions, hence they can be labelled by $(ij)(kl)$, where i, j, k, l are all distinct.

Each mid line contains 3 nodes, and each centre line contains 2 nodes. Since there are 15 mid lines and 10 centre lines, all the $65 = 3 \times 15 + 2 \times 10$ nodes belong to these lines.

(A,B) Each mid line $(ij)(kl)$ intersects the plane $(w = 0)$ at a node, which will be denoted by $\mathrm{A}_{(ij)(kl)}$. In the affine space $(w \neq 0)$, the node closer to ij will be denoted by $B_{ij(kl)}$.

(C) For each centre line (ij), the node close to ij will be denoted by C_{ij}.

The nodes of type A form an orbit of $\mathfrak{A}_5$, they correspond to double pairs (stabilised by a Klein subgroup $(\mathbb{Z}/2)^2$), that is, double transpositions.

Similarly those of type C, form the orbit of ordered pairs, and are stabilised by a 3-cycle. These correspond to cyclic subgroups of order 3 (self normalizing).

Those of type B form the orbit stabilised by a double transposition: one takes an ordered pair and a pair (in this way we get an orbit of 60 elements on which $\mathfrak{A}_5$ acts transitively), but then identifying $ab(cd)$ with $ba(dc)$ the stabiliser becomes $(ab)(cd)$, hence we get the cosets for the subgroup generated by this double transposition.

Clearly, these are three orbits of the $\mathfrak{A}_5$-action on the set of nodes.

A direct computation (which can be performed by hand) leads to the following table, presenting explicit coordinates of the nodes (see Table 4.1).

Using Table 4.1, we may explicitly describe the representation $\mathfrak{A}_5 \to \mathrm{GL}(\mathcal{K})$ via the permutation representations $\rho_A \colon \mathfrak{A}_5 \to \mathfrak{S}_{15}$, $\rho_B \colon \mathfrak{A}_5 \to \mathfrak{S}_{30}$, and $\rho_C \colon \mathfrak{A}_5 \to \mathfrak{S}_{20}$.

Proposition 4.33 *Let* $\sigma = (1\,2\,3)$ *and* $\tau = (1\,4)(2\,5)$ *be generators of* $\mathfrak{A}_5$. *We have*

$$\rho_A(\sigma) = (1\;10\;7)(2\;15\;4)(3\;5\;12)(6\;8\;14)(9\;13\;11)$$
$$\rho_A(\tau) = (1\;11)(3\;14)(4\;7)(5\;13)(6\;10)(12\;15)$$

$$\rho_B(\sigma) = (1\;22\;11)(2\;30\;8)(3\;9\;20)(4\;6\;29)(5\;18\;15)(7\;13\;24)(10\;12\;28)(14\;16\;27)(17\;25\;19)(21\;26\;23)$$
$$\rho_B(\tau) = (1\;19)(2\;6)(3\;27)(4\;11)(5\;23)(7\;28)(8\;15)(9\;25)(10\;22)(12\;16)(13\;26)(14\;18)(20\;29)(24\;30)$$

$$\rho_C(\sigma) = (1\;16\;3)(2\;12\;19)(4\;20\;9)(5\;14\;7)(6\;10\;17)(8\;18\;11)$$
$$\rho_C(\tau) = (1\;13)(2\;6)(3\;17)(4\;10)(5\;15)(7\;19)(8\;12)(9\;16)(11\;14)(18\;20).$$

Due to the relation

Table 4.1 Nodes in Barth sextic

$A_{(12)(34)}$	$A_{(41)(25)}$	$A_{(31)(52)}$	$A_{(51)(34)}$	$A_{(12)(53)}$	$A_{(41)(53)}$	$A_{(31)(24)}$	$A_{(51)(42)}$
$(0,1,\tau,-\bar\tau)$	$(0,-1,\tau,-\bar\tau)$	$(0,1,-\tau,-\bar\tau)$	$(0,1,\tau,\bar\tau)$	$(0,-\bar\tau,1,\tau)$	$(0,\bar\tau,1,\tau)$	$(0,-\bar\tau,-1,\tau)$	$(0,-\bar\tau,1,-\tau)$
$A_{(12)(45)}$	$A_{(41)(32)}$	$A_{(31)(45)}$	$A_{(51)(23)}$	$A_{(23)(45)}$	$A_{(25)(34)}$	$A_{(24)(53)}$	
$(0,\tau,-\bar\tau,1)$	$(0,-\tau,-\bar\tau,1)$	$(0,\tau,\bar\tau,1)$	$(0,\tau,-\bar\tau,-1)$	$(0,1,0,0)$	$(0,0,1,0)$	$(0,0,0,1)$	
$B_{12(34)}$	$B_{41(25)}$	$B_{31(52)}$	$B_{51(34)}$	$B_{21(43)}$	$B_{14(52)}$	$B_{13(25)}$	$B_{15(43)}$
$(2,1,\tau,-\bar\tau)$	$(2,-1,\tau,-\bar\tau)$	$(2,1,-\tau,-\bar\tau)$	$(2,1,\tau,\bar\tau)$	$(2,-1,-\tau,\bar\tau)$	$(2,1,-\tau,\bar\tau)$	$(2,-1,\tau,\bar\tau)$	$(2,-1,-\tau,-\bar\tau)$
$B_{12(53)}$	$B_{41(53)}$	$B_{31(24)}$	$B_{51(42)}$	$B_{21(35)}$	$B_{14(35)}$	$B_{13(42)}$	$B_{15(24)}$
$(2,-\bar\tau,1,\tau)$	$(2,\bar\tau,1,\tau)$	$(2,-\bar\tau,-1,\tau)$	$(2,-\bar\tau,1,-\tau)$	$(2,\bar\tau,-1,-\tau)$	$(2,-\bar\tau,-1,-\tau)$	$(2,\bar\tau,1,-\tau)$	$(2,\bar\tau,-1,\tau)$
$B_{12(45)}$	$B_{41(32)}$	$B_{31(45)}$	$B_{51(23)}$	$B_{21(54)}$	$B_{14(23)}$	$B_{13(54)}$	$B_{15(32)}$
$(2,\tau,-\bar\tau,1)$	$(2,-\tau,-\bar\tau,1)$	$(2,\tau,\bar\tau,1)$	$(2,\tau,-\bar\tau,-1)$	$(2,-\tau,\bar\tau,-1)$	$(2,\tau,\bar\tau,-1)$	$(2,-\tau,-\bar\tau,-1)$	$(2,-\tau,\bar\tau,1)$
$B_{23(45)}$	$B_{32(54)}$	$B_{25(34)}$	$B_{52(43)}$	$B_{24(53)}$	$B_{42(35)}$		
$(1,1,0,0)$	$(1,-1,0,0)$	$(1,0,1,0)$	$(1,0,-1,0)$	$(1,0,0,1)$	$(1,0,0,-1)$		
C_{12}	C_{41}	C_{31}	C_{51}	C_{21}	C_{14}	C_{13}	C_{15}
$(1,1,1,1)$	$(1,-1,1,1)$	$(1,1,-1,1)$	$(1,1,1,-1)$	$(1,-1,-1,-1)$	$(1,1,-1,-1)$	$(1,-1,1,-1)$	$(1,-1,-1,1)$
C_{53}	C_{24}	C_{35}	C_{42}	C_{45}	C_{32}	C_{54}	C_{23}
$(1,0,-\bar\tau,\tau)$	$(1,0,\bar\tau,\tau)$	$(1,0,\bar\tau,-\tau)$	$(1,0,-\bar\tau,-\tau)$	$(1,\tau,0,-\bar\tau)$	$(1,-\tau,0,-\bar\tau)$	$(1,-\tau,0,\bar\tau)$	$(1,\tau,0,\bar\tau)$
C_{34}	C_{25}	C_{43}	C_{52}				
$(1,-\bar\tau,\tau,0)$	$(1,\bar\tau,\tau,0)$	$(1,\bar\tau,-\tau,0)$	$(1,-\bar\tau,-\tau,0)$				

$$f(x+y+z,x,y,z) = -\tau \cdot h(x,y,z)^2,$$

where

$$h(x,y,z) = \tau^2\left(x^2y+y^2z+z^2x\right)+\left(xy^2+yz^2+zx^2\right)+\left(\tau^2+1\right)xyz,$$

the singularities contained in the hyperplane H perpendicular to $(1,-1,-1,-1)$ form a half-even set $Sing(Y)\cap H =$

$$\left\{\begin{array}{c} A_{(31)(52)},\ A_{(51)(42)},\ A_{(41)(32)}, \\ B_{51(34)},\ B_{41(53)},\ B_{31(45)},\ B_{23(45)},\ B_{25(34)},\ B_{42(35)}, \\ C_{41},\ C_{31},\ C_{51},\ C_{24},\ C_{23},\ C_{25} \end{array}\right\}$$

The corresponding codeword of $\mathcal{K}'\backslash\mathcal{K}$ is given by

$$\big(1\,|\,0010\,0001\,0100\,000\,|\,00010000\,01000000\,00100000\,101010\,|\,0111\,0000\,0100\,0001\,0100\big).$$

The $\mathfrak{A}_5$-orbit consists[4] of 20 half-even sets having cardinality 15:

[4] For this and the next computation see the script [33, Mathematica 3].

$$\begin{pmatrix}
1 \mid 0000\;0000\;0110\;001 \mid 01000010\;10010000\;10010000\;000000 \mid 01000010\;1001\;1001\;0000 \\
1 \mid 0000\;0000\;1001\;001 \mid 00001001\;00100100\;00100100\;000000 \mid 00001001\;0110\;1001\;0000 \\
1 \mid 0000\;0000\;1001\;001 \mid 10010000\;01000010\;01000010\;000000 \mid 10010000\;1001\;0110\;0000 \\
1 \mid 0000\;0000\;0110\;001 \mid 00100100\;00001001\;00001001\;000000 \mid 00100100\;0110\;0110\;0000 \\
1 \mid 0000\;1010\;0000\;010 \mid 01000001\;01000001\;00001010\;000000 \mid 00001010\;1100\;0000\;0110 \\
1 \mid 0000\;0101\;0000\;010 \mid 10100000\;10100000\;00010100\;000000 \mid 00010100\;1100\;0000\;1001 \\
1 \mid 0000\;1010\;0000\;010 \mid 00010100\;00010100\;10100000\;000000 \mid 10100000\;0011\;0000\;1001 \\
1 \mid 0000\;0101\;0000\;010 \mid 00001010\;00001010\;01000001\;000000 \mid 01000001\;0011\;0000\;0110 \\
1 \mid 1100\;0000\;0000\;100 \mid 00100001\;11000000\;00100001\;000000 \mid 11000000\;0000\;1100\;0011 \\
1 \mid 0011\;0000\;0000\;100 \mid 11000000\;00100001\;11000000\;000000 \mid 00100001\;0000\;1100\;1100 \\
1 \mid 1100\;0000\;0000\;100 \mid 00010010\;00001100\;00010010\;000000 \mid 00001100\;0000\;0011\;1100 \\
1 \mid 0011\;0000\;0000\;100 \mid 00001100\;00010010\;00001100\;000000 \mid 00010010\;0000\;0011\;0011 \\
1 \mid 0100\;1000\;0010\;000 \mid 10000000\;00000010\;00000100\;101001 \mid 10000110\;0010\;1000\;0100 \\
1 \mid 1000\;0100\;0001\;000 \mid 00000100\;00000001\;10000000\;100110 \mid 10000101\;1000\;0001\;0010 \\
1 \mid 0001\;0010\;1000\;000 \mid 00000010\;10000000\;00000001\;011010 \mid 10000011\;0100\;0010\;1000 \\
1 \mid 0010\;0001\;0100\;000 \mid 00010000\;01000000\;00100000\;101010 \mid 01110000\;0100\;0001\;0100 \\
1 \mid 0100\;1000\;0010\;000 \mid 00001000\;00100000\;01000000\;010110 \mid 01101000\;1000\;0010\;0001 \\
1 \mid 0001\;0010\;1000\;000 \mid 00100000\;00001000\;00010000\;100101 \mid 00111000\;0001\;1000\;0010 \\
1 \mid 0010\;0001\;0100\;000 \mid 00000001\;00000100\;00000010\;010101 \mid 00000111\;0001\;0100\;0001 \\
1 \mid 1000\;0100\;0001\;000 \mid 01000000\;00010000\;00001000\;011001 \mid 01011000\;0010\;0100\;1000
\end{pmatrix}$$

and the $\mathbb{F}_2$-basis is as follows.

$$\begin{pmatrix}
1 \mid 0000\;0000\;0110\;001 \mid 00000110\;10000001\;00011000\;111111 \mid 10011111\;0011\;1100\;1010 \\
0 \mid 1000\;0001\;0001\;010 \mid 00001110\;00001011\;11000001\;100110 \mid 11000100\;1011\;0001\;0100 \\
0 \mid 0100\;0001\;0111\;111 \mid 00001011\;11000010\;11101001\;100110 \mid 00100000\;1101\;1011\;0111 \\
0 \mid 0010\;0001\;0010\;001 \mid 00000111\;10000101\;00011010\;101010 \mid 10011000\;0010\;1000\;1011 \\
0 \mid 0001\;0001\;0100\;100 \mid 00001101\;00010110\;00001110\;010101 \mid 00010101\;0001\;0111\;0010 \\
0 \mid 0000\;1001\;0011\;110 \mid 00000101\;01100011\;10110001\;001111 \mid 11010111\;0110\;0101\;1100 \\
0 \mid 0000\;0101\;0110\;011 \mid 00001100\;10001011\;01011001\;111111 \mid 11011110\;0000\;1100\;1100 \\
0 \mid 0000\;0011\;0101\;101 \mid 00000110\;10110010\;00101011\;001111 \mid 10011111\;0011\;1100\;1010 \\
0 \mid 0000\;0000\;1111\;000 \mid 00001111\;10100101\;00111100\;111111 \mid 10010110\;0101\;0101\;1010 \\
0 \mid 0000\;0000\;0000\;000 \mid 10001000\;00100010\;01000100\;111111 \mid 11101110\;1010\;1010\;0101 \\
0 \mid 0000\;0000\;0000\;000 \mid 01000100\;00010001\;10001000\;111111 \mid 11011101\;1010\;0101\;1010 \\
0 \mid 0000\;0000\;0000\;000 \mid 00100010\;10001000\;00010001\;111111 \mid 10111011\;0101\;1010\;1010 \\
0 \mid 0000\;0000\;0000\;000 \mid 00010001\;01000100\;00100010\;111111 \mid 01110111\;0101\;0101\;0101
\end{pmatrix}$$

Since we know from e.g. [53] that $\dim(\mathcal{K}) \leq 12$, the above matrix is a generator matrix of $\mathcal{K}'$ and we obtain a generator matrix of $\mathcal{K}$ by removing the first row and the first column.[5]

The weight enumerator[6] of $\mathcal{K}$ is given by

[5] Another argument for the fact that the above matrix M is a generator matrix of $\mathcal{K}'$, without using the result of [53], goes as follows. By construction, removing the first row and column generates a subcode of $\mathcal{K}$. It can be easily checked that this subcode is projective, so that we can conclude from Proposition 5.1 that no extension to a 13-dimensional code exists.

[6] See the script [33, Mathematica 4].

$$W(x) = 1x^0 + 390x^{24} + 3055x^{32} + 650x^{40}$$

and the weight enumerator of $\mathcal{K}'\backslash\mathcal{K}$ is given by

$$W(x) = 26x^{16} + 650x^{28} + 1690x^{32} + 1300x^{36} + 300x^{40} + 130x^{44}.$$

Both codes $\mathcal{K}$ and $\mathcal{K}'$ have an automorphism group of order $15\,600 = 2^4 \cdot 3 \cdot 5^2 \cdot 13$ that is isomorphic to $\mathrm{P}\,\Sigma\,\mathrm{L}(2, 25) = \mathrm{PSL}(2, 25) \rtimes \langle F \rangle$, where $F : x \mapsto x^5$ is the Frobenius automorphism of $\mathbb{F}_{25}$, as can be checked by a MAGMA computation[7] and for reasons that we shall now explain.

In fact, we can associate to Y a graph $\mathcal{G}$ with the same automorphism group. To this end let $P_1, \ldots, P_{65}$ be the vertices of $\mathcal{G}$. Two vertices P_i and P_j are joined by an edge if and only if

$$g_{i,j}(t) := f\big(t \cdot P_i + (1 - t) \cdot P_j\big) = ct^2(1 - t)^2(t - x_1)(t - x_2),$$

where $c, x_1, x_2 \in \mathbb{C}$ and $|\{0, 1, x_1, x_2\}| = 4$, i.e., the line $t \cdot P_i + (1-t) \cdot P_j$ intersects the surface Y in the two nodes P_i, P_j and two different additional points.

Every linear automorphism ρ preserving the equation f corresponds to an automorphism $\tilde{\rho}$ of $\mathcal{K}$, i.e., ρ induces a permutations on $Sing(Y)$, which can be written as a permutation $\tilde{\rho}$ of the columns of a generator matrix of $\mathcal{K}$. The converse is not true in general.

In Sect. 4.3.1, we shall prove that the group of linear automorphisms of the Barth surface is indeed $\big\langle \mathfrak{A}_5, (\mathbb{Z}/2)^3 \big\rangle$, where $(\mathbb{Z}/2)^3$ is the group of the coordinate sign changes, and has order 120.

In particular the Barth surface does not admit a linear automorphism of order 13, while the automorphism group of $\mathcal{K}$ contains elements of order 13.

The graph $\mathcal{G}$ is nowadays called *Doro-Hall graph* and is one of the three *locally Petersen graphs*, i.e., graphs where the neighbourhood of every vertex is isomorphic to the Petersen graph (quotients of the dodecahedral graph by the antipodal involution), see [50].

4.3.1 The Automorphism Group of the Barth Sextic

We know that the set N of the 65 nodes is invariant under a group of linear collineations isomorphic to the icosahedral group $\mathfrak{A}_5 \times \mathbb{Z}/2\mathbb{Z}$.

The aim of this subsection is to given a non computational proof of Theorem 4.31, that the icosahedral group $\mathfrak{A}_5 \times \mathbb{Z}/2\mathbb{Z}$ is the full group of automorphisms of the Barth sextic.

---

[7] See the script [33, MAGMA 22].

The action of $\mathfrak{A}_5$ partitions N into three orbits N_A, N_B and N_C, which are the sets of nodes of type A, B and C of size 15, 30 and 20, respectively.

We call a line L a *node-secant* if it contains at least 2 nodes. By abuse of notation we shall say that L is a secant. If L contains exactly k nodes, L is called a k-secant.

4.3.1.1 Secants Through N_A

The nodes of type A span the plane $(w = 0) =: E$. Throughout 4.3.1.1, we shall only consider the nodes and secants inside E. As $w = 0$ for all points in E, the w-coordinate is skipped in computations.

Lemma 4.34 *Let $P \in N_A$.*

(a) The node P is contained in precisely two 2-secants, two 3-secants and two 5-secants.
(b) Let $Q, Q' \in N_A$ be the second nodes on the two 2-secants through P. Then $\overline{QQ'}$ is a 2-secant.

Proof As the nodes of type A form a single orbit under the $\mathfrak{A}_5$-action, it is enough to consider the single node $P = (1, 0, 0)$. For part 4.34, we compute the list of secants through P shown in Table 4.2. For part 4.34, we see that $\{Q, Q'\} = \{(0, 1, 0), (0, 0, 1)\}$, so $\overline{QQ'} = (x = 0)$, which contains no other node of type A. □

In the next steps, we will avoid doing direct coordinate computations. Instead, the conclusions will solely be drawn from the incidence properties of the secants provided by Lemma 4.34. The exception will be the final Lemma 4.43, where the incidence properties are not enough to show the statement, see Remark 4.44.

Lemma 4.35 *The secants of N_A fall into fifteen 2-secants, ten 3-secants and six 5-secants.*

Proof Use Lemma 4.34 and count the flags (P, L) with $P \in N_A$, L a secant and $P \in L$ in two ways. □

Table 4.2 Secants through $P = (1, 0, 0) = A_{(23)(45)}$

secant L through P		further nodes on L	
2-secant	$(z = 0)$	$A_{(25)(34)} = (0, 1, 0)$	
2-secant	$(y = 0)$	$A_{(24)(35)} = (0, 0, 1)$	
3-secant	$(y + \tau^2 z = 0)$	$A_{(31)(52)} = (1, -\tau, -\bar{\tau})$	$A_{(51)(34)} = (1, \tau, \bar{\tau})$
3-secant	$(y - \tau^2 z = 0)$	$A_{(12)(34)} = (1, \tau, -\bar{\tau})$	$A_{(41)(52)} = (-1, \tau, -\bar{\tau})$
5-secant	$(\tau y - z = 0)$	$A_{(12)(53)} = (-\bar{\tau}, 1, \tau)$	$A_{(41)(53)} = (\bar{\tau}, 1, \tau)$
		$A_{(12)(45)} = (\tau, -\bar{\tau}, 1)$	$A_{(41)(32)} = (-\tau, -\bar{\tau}, 1)$
5-secant	$(\tau y + z = 0)$	$A_{(31)(24)} = (-\bar{\tau}, -1, \tau)$	$A_{(51)(42)} = (\bar{\tau}, 1, -\tau)$
		$A_{(31)(45)} = (\tau, \bar{\tau}, 1)$	$A_{(51)(23)} = (\tau, -\bar{\tau}, -1)$

By Lemma 4.34, the 2-secants partition N_A into five parts of size 3, which we will call *triads*. Let $\mathcal{T}$ be the set of the five triads. For $P \in N_A$, let T_P be the triad containing P.

Remark 4.36 In the proof of Lemma 4.34, we have seen that the three nodes $A_{(23)(45)}$, $A_{(25)(34)}$ and $A_{(24)(53)}$ form a triad, which are exactly the nodes of type A where the number 1 does not appear in the label. Since by construction, the $\mathfrak{A}_5$-action is compatible with the labelling, we get that the five triads T_i with $\{1, \dots, 5\}$ are given by the five nodes of type A where the number i does not appear in the label.

Lemma 4.37

(a) *Each* 5*-secant contains exactly one node of every triad.*
(b) *For each* $P \in N_A$*, the two* 3*-secants passing through* P *together contain exactly one node from each triad different from* T_P.

Proof Any pair of nodes in the same triad spans a 2-secant, so no 3- or 5-secant can contain two nodes from the same triad. This implies part 4.37.

For part 4.37, let $P \in N_A$ and let T be one of the four triads different from T_P. For each $Q \in T$, the line $\overline{PQ}$ is a 3-secant or a 5-secant. By part 4.37 and since there are two 5-secants passing through P, for two of the three nodes $Q \in T$ the line $\overline{PQ}$ is a 5-secant. Therefore, for the third node $Q \in T$ the line $\overline{PQ}$ is a 3-secant. □

Lemma 4.38 *Let* T_1, T_2, T_3 *be three different triads. Then there is a unique* 3*-secant containing a node from each triad* T_1, T_2, T_3.

Proof We denote the other two triads by T_4 and T_5. Assume that there are two 3-secants L_1, L_2 containing a node from each triad T_1, T_2, T_3. By Lemma 4.37, $L_1 \cap L_2$ is not a node, so L_1 and L_2 together cover 6 distinct nodes. Denote them by P_{ij} with $P_{ij} \in L_j$, $P_{ij} \in T_i$ where $i \in \{1, 2, 3\}$ and $j \in \{1, 2\}$. Each node P_{ij} is contained in a second 3-secant L_{ij}. By Lemma 4.37, the other two nodes on L_{ij} are contained in T_4 and T_5. In particular, the six 3-secants L_{ij} are pairwise distinct.

Assume that there is a node $Q \in T_4 \cup T_5$ contained in two 3-secants $L_{ij} \neq L_{i'j'}$ with $i, i' \in \{1, 2, 3\}$ and $j, j' \in \{1, 2\}$. Then L_{ij} and $L_{i'j'}$ are two distinct 3-secants through Q both containing a node of T_4 and T_5. This not possible by Lemma 4.37, so the six 3-secants L_{ij} cover 12 distinct nodes in $T_4 \cup T_5$. This is a contradiction to $\#(T_4 \cup T_5) = 6$.

So there is at most one 3-secant containing a node of T_1, T_2 and T_3. Now the claim follows as the number of 3-secants equals the number $\binom{5}{3} = 10$ of 3-sets of triads. □

We will call a 2-set of two disjoint 2-sets of triads a *tetrad* and we will denote the set of tetrads by X. Note that $\#X = \binom{5}{2}\binom{3}{2}/2 = 15$. A node $P \in N_A$ defines the set $\{L_1, L_2\}$ of the two 3-secants passing through P. For $i \in \{1, 2\}$, the 3-secant L_i determines the set $\{T_{i1}, T_{i2}\}$ of the two triads different from T_P which contain a point on L_i. By Lemma 4.37, the set

$$t(P) := \{\{T_{11}, T_{12}\}, \{T_{21}, T_{22}\}\}$$

is a tetrad.

Example 4.39 Let $\mathcal{T} = \{T_1, T_2, T_3, T_4, T_5\}$ with T_i defined as in Remark 4.36 and let $P = A_{(23)(45)} = (1, 0, 0)$. From Table 4.2, the two 3-secants passing through P are $L_1 = (y + \tau^2 z = 0)$ containing the nodes $(1, -\tau, -\bar{\tau}) = A_{(31)(52)} \in T_4$ and $(1, \tau, \bar{\tau}) = A_{(51)(34)} \in T_2$ and $L_2 = (y - \tau^2 z = 0)$ containing the nodes $(1, \tau, -\bar{\tau}) = A_{(12)(34)} \in T_5$ and $(-1, \tau, -\bar{\tau}) = A_{(41)(52)} \in T_3$. Therefore, $t(A_{(23)(45)}) = \{\{T_4, T_2\}, \{T_5, T_3\}\}$.

This example shows that to our surprise, the map t does not match the labelling of the nodes of type A introduced in the construction.

Lemma 4.40 *The map $t : N_P \to X$ is a bijection.*

Proof By $\#X = 15 = \#N_A$, it suffices to show that t is injective. Let $P \in N_A$ and $t(P) = \{\{T_1, T_2\}, \{T_3, T_4\}\} \in X$. We denote the fifth triad by T_5. By Lemma 4.38, there is a unique 3-secant L passing through nodes on T_1, T_2 and T_5, so P is determined as the node $L \cap T_5$. □

Now let G_A be the stabilizer of the point set N_A in $\mathrm{Aut}(E)$. The map t induces a map $t_G : G_A \to \mathfrak{S}(X)$.

Lemma 4.41 *The map $t_G : G_A \to \mathfrak{S}(X)$ is an injective homomorphism of groups.*

Proof As t is defined based on incidence properties of lines spanned by the nodes in N_A, the map t_G is a homomorphism of groups. Let $g \in G_A$ and let $\varphi = t_G(g)$. For $P \in N_A$, $g(P)$ is uniquely determined as $t^{-1}(\varphi(t(P)))$ by Lemma 4.40. Therefore g is uniquely determined by $\varphi = t_G(g)$ and thus t_G is an injection. □

Remark 4.42 We can use Lemma 4.41 to efficiently complete Example 4.39 to a list of all t-values of the nodes of type A. From Example 4.39 we know that

$$t(1, 0, 0) = t(A_{(23)(45)}) = \{\{T_2, T_4\}, \{T_3, T_5\}\}.$$

Since t is a group homomorphism and the $\mathfrak{A}_5$-action is compatible with the labelling of the nodes of type A, for $\sigma = (234) \in \mathfrak{A}_5$ we get that

$$\begin{aligned} t(0, 1, 0) = t(A_{(34)(25)}) &= t(\sigma \cdot A_{(23)(45)}) = \sigma(t(A_{(23)(45)})) \\ &= \sigma(\{\{T_2, T_4\}, \{T_3, T_5\}\}) = \{\{T_2, T_3\}, \{T_4, T_5\}\}. \end{aligned}$$

Using other elements $\sigma \in \mathfrak{A}_5$, we get the list in Table 4.3. For completeness, we also included the list of the triads of the nodes.

Lemma 4.43 $G_A \cong \mathfrak{A}_5$.

Proof Let $\mathcal{T} = \{T_1, T_2, T_3, T_4, T_5\}$ with T_i defined as in Remark 4.36. Assume that there is a $\rho \in G_A$ such that $t_G(\rho)$ is the transposition $T_3 \leftrightarrow T_4$. From Table 4.3 we

Table 4.3 Triads and tetrads of the nodes of type A

$P \in N_A$	T_P	$t(P)$
$A_{(12)(34)} = (1, \tau, -\bar{\tau})$	T_5	$\{\{T_1, T_3\}, \{T_2, T_4\}\}$
$A_{(41)(25)} = (-1, \tau, -\bar{\tau})$	T_3	$\{\{T_1, T_5\}, \{T_2, T_4\}\}$
$A_{(31)(52)} = (1, -\tau, -\bar{\tau})$	T_4	$\{\{T_1, T_2\}, \{T_3, T_5\}\}$
$A_{(51)(34)} = (1, \tau, \bar{\tau})$	T_2	$\{\{T_1, T_4\}, \{T_3, T_5\}\}$
$A_{(12)(53)} = (-\bar{\tau}, 1, \tau)$	T_4	$\{\{T_1, T_5\}, \{T_2, T_3\}\}$
$A_{(41)(53)} = (\bar{\tau}, 1, \tau)$	T_2	$\{\{T_1, T_3\}, \{T_4, T_5\}\}$
$A_{(31)(24)} = (-\bar{\tau}, -1, \tau)$	T_5	$\{\{T_1, T_4\}, \{T_2, T_3\}\}$
$A_{(51)(42)} = (-\bar{\tau}, 1, -\tau)$	T_3	$\{\{T_1, T_2\}, \{T_4, T_5\}\}$
$A_{(12)(45)} = (\tau, -\bar{\tau}, 1)$	T_3	$\{\{T_1, T_4\}, \{T_2, T_5\}\}$
$A_{(41)(32)} = (-\tau, -\bar{\tau}, 1)$	T_5	$\{\{T_1, T_2\}, \{T_3, T_4\}\}$
$A_{(31)(45)} = (\tau, \bar{\tau}, 1)$	T_2	$\{\{T_1, T_5\}, \{T_3, T_4\}\}$
$A_{(51)(23)} = (\tau, -\bar{\tau}, -1)$	T_4	$\{\{T_1, T_3\}, \{T_2, T_5\}\}$
$A_{(23)(45)} = (1, 0, 0)$	T_1	$\{\{T_2, T_4\}, \{T_3, T_5\}\}$
$A_{(25)(34)} = (0, 1, 0)$	T_1	$\{\{T_2, T_3\}, \{T_4, T_5\}\}$
$A_{(24)(35)} = (0, 0, 1)$	T_1	$\{\{T_2, T_5\}, \{T_3, T_4\}\}$

get that

$$\rho(1, 0, 0) = (0, 1, 0), \quad \rho(0, 1, 0) = (1, 0, 0) \quad \text{and} \quad \rho(0, 0, 1) = (0, 0, 1).$$

Therefore, ρ can be represented by a matrix

$$S = \begin{pmatrix} 0 & \alpha & 0 \\ \beta & 0 & 0 \\ 0 & 0 & 1 \end{pmatrix} \quad \text{with} \quad \alpha, \beta \neq 0.$$

Again Table 4.3 shows that the node $(\tau, \bar{\tau}, 1)$ is fixed by ρ, and that the node $(1, \tau, \bar{\tau})$ is mapped to $(\bar{\tau}, 1, \tau)$. The first condition yields $\alpha = -(\tau + 1)$ and $\beta = -(\bar{\tau} + 1)$. Now one checks that S does not map $(1, \tau, \bar{\tau})$ to a scalar multiple of $(\bar{\tau}, 1, \tau)$, which is a contradiction to the second condition.

So using Lemma 4.41, $G_A \cong t(G_A)$ is a proper subgroup of $\mathfrak{S}(\mathcal{T}) \cong \mathfrak{S}_5$. By the known $\mathfrak{A}_5$-invariance of N_A, the only remaining possibility is $G_A \cong \mathfrak{A}_5$. □

Remark 4.44 It can be checked that the set of permutations of N_A preserving the incidences among all secants is indeed isomorphic to the full symmetric group $\mathfrak{S}_5$. For that reason, a proof of Lemma 4.43 must involve some additional property, like in the above proof the concrete coordinate representation of the nodes of type A.

4.3.1.2 The Automorphisms of the Barth Sextic

Let G be the stabilizer of N in $\mathrm{PGL}(\mathbb{C}^4)$.

Lemma 4.45 *The 5-secants in Lemma 4.35 are the only secants that pass through more than four nodes. In particular, G preserves the nodes of type A.*

Proof Since nodes of type B are in the conic $w^2 = x^2 + y^2 + z^2$, no line contains more than two of them. The same is true for nodes of type C since they are contained in $3w^2 = x^2 + y^2 + z^2$. Thus, if a 5-secant L is not contained in E, then it contains precisely one node of type A, two nodes of type B, and two nodes of type C. Using $\mathfrak{A}_5$-symmetry, we may assume that this node is $A_{(23)(45)} = (0, 1, 0, 0)$. Then, L can be parametrized by $t \mapsto (1, t, y_0, z_0)$, but it can be directly checked from Table 4.1 that nodes of type B and C are distinguishable by looking at yz-coordinates. Thus, L cannot join nodes of type B and of type C. Now, it is obvious that a k-secant ($k \geq 5$) contains at least two nodes of type A, hence $k = 5$.

The last statement follows from the fact that the 5-secants in E covers N_A. □

Lemma 4.46 *The point $O := (1, 0, 0, 0)$ is fixed by all elements of G.*

Proof By Lemma 4.45, G preserves the plane E. Let T be the union $\bigcup_{P\in N_A} \overline{OP}$. Then, for any element $\phi \in G$, $\phi T = \bigcup_{P\in N_A} \overline{(\phi O)P}$. We recall that the lines $\{\overline{OP}\}$ are the intersections of two planes among the six planes $(\tau^2 x - y^2)(\tau^2 y^2 - z^2)(\tau^2 z^2 - x^2)$, hence are 3-secants.

We claim that if $Q \notin N$ and $\overline{QP}$ is a 3-secant for every $P \in N_A$, then $Q = O$. To do this, first we consider $P = A_{(23)(45)} = (0, 1, 0, 0)$. Consulting Table 4.1, we find that the 3-secants containing P but not contained in E are determined by the pairs

$$(2, \pm 1, \epsilon_1\tau, \epsilon_2\bar{\tau}), \quad (2, \pm\tau, \epsilon_1\bar{\tau}, \epsilon_2), \quad (2, \pm\bar{\tau}, \epsilon_1, \epsilon_2\tau),$$
$$(1, \pm 1, \epsilon_1, \epsilon_2), \quad (1, \pm 1, 0, 0), \quad (1, \pm\tau, 0, \epsilon_2\bar{\tau}), \quad (1, \pm\bar{\tau}, \epsilon_1\tau, 0),$$

where $\epsilon_1, \epsilon_2 \in \{+1, -1\}$ are fixed signs. The 3-secants through $(0, 0, 1, 0)$ and $(0, 0, 0, 1)$ are obtained by cyclic rotation of xyz-coordinates.

Now, put $Q = (1, x_0, y_0, z_0)$. The line $\overline{QP}$ is parametrized by $t \mapsto (1, t, y_0, z_0)$, and is one of the 3-secants listed above. Hence, up to $\pm$ signs, the pair (y_0, z_0) is one of

$$\tfrac{1}{2}(\tau, \bar{\tau}),\ \tfrac{1}{2}(\bar{\tau}, 1),\ \tfrac{1}{2}(1, \tau),\ (1, 1),\ (0, 0),\ (0, \bar{\tau}),\ (\tau, 0). \tag{4.13}$$

Due to cyclic xyz-symmetry, the pairs (z_0, x_0) and (x_0, y_0) (up to $\pm$ signs) also belong to 4.13. The possibilities for (x_0, y_0, z_0) are reduced to $(0, 0, 0)$, $(\pm 1, \pm 1, \pm 1)$, $\frac{1}{2}(\pm\tau, \pm\bar{\tau}, \pm 1)$ (+cyclic rotations). Among these, only $Q = O$ satisfies $Q \notin N$. □

By Lemma 4.46 we will call the point O the *centre* of the Barth sextic.

Lemma 4.47 *The secants through the centre O consist of fifteen 3-secants of type pattern ABB*[8] *and ten 2-secants of type pattern CC.*

Proof It can be immediately seen that the line joining O and $A_{(ij)(kl)}$ contains $B_{ij(kl)}$ and $B_{ji(lk)}$. A line joining O and C_{ij} also contains C_{ji}. In both cases, the line cannot contain further nodes as there is no 5-secants of type pattern ABBCC (cf. Lemma 4.45 and its proof). □

Remark 4.48 Let $U \subset \mathbb{P}^3$ be the union of non-secant lines which intersect N. In the proof of Lemma 4.47, we observed that $O \notin U$. In fact, a computer-based approach[9] verifies that $U = \mathbb{P}^3 \setminus \{O\}$. To verify, we compute the intersections of all the pairs of secants, and for each intersection point, find the subset of N that are covered by secants passing though this point. The centre is the unique point through which the secants cover the entire N.

This is another interesting property that characterizes the centre.

Lemma 4.49 *The elements of G preserve the type of the nodes.*

Proof By Lemma 4.45, G preserves the nodes of type A. Also, by Lemmas 4.46 and 4.47, G preserves the set $\{\overline{C_{ij}C_{ji}}\}$. This shows that G preserves the nodes of type C, and then also the nodes of type B. □

Proposition 4.50 *The group of linear automorphisms of the set of nodes of the Barth sextic is isomorphic to the icosahedral group $\mathfrak{A}_5 \times \mathbb{Z}/2\mathbb{Z}$.*

Proof By Lemma 4.49, N_A (and hence the plane E spanned by N_A) is invariant under the elements of G. Therefore, the group homomorphism $\Phi : G \to G_E, \phi \mapsto \phi|_E$ is well-defined. By Lemma 4.43 we have $\mathrm{Im}\,\Phi \leq G_E \cong \mathfrak{A}_5$, so $\#\mathrm{Im}\,\Phi \leq 60$.

Let $\phi \in \ker\Phi$. Then ϕ fixes all the points in E. Moreover, the centre $O \notin E$ is fixed by Lemma 4.46. Now as an element of PGL, the collineation ϕ is determined by the image of a single further point $P \notin (E \cup \{O\})$. We choose the node $P = (1, 1, 0, 0)$ of type B. The line $L = \overline{OP}$ is a secant of type pattern ABB (see Lemma 4.47). As the single node Q of type A is fixed by $\phi \in \ker\Phi$, we get that $\phi(L) = \phi(\overline{OQ}) = \overline{OQ} = L$. Therefore, the node P of type B is mapped to one of the two nodes of type B on L, implying that $\#\ker\Phi \leq 2$.

Now by the fundamental theorem on homomorphisms,

$$\#G = \#\ker\Phi \cdot \#\mathrm{Im}\,\Phi \leq 2 \cdot 60 = 120.$$

This finishes the proof as we already know that G contains a subgroup isomorphic to the icosahedral group of order 120. □

---

[8] Type pattern ABB means that such a 3-secant passes through a single node of type A and two nodes of type B.

[9] See the script [33, MAGMA 22].

4.3.2 *The Doro-Hall Graph*

4.3.2.1 Petersen Graph

As we already mentioned, the Petersen graph is constructed by the vertices and edges of the projective dodecahedron, that is, a dodecahedron where opposite points, lines and faces are identified with each other.

As we have also seen, the Petersen graph can also be constructed as the particular Kneser graph with $n = 5$ and $k = 2$.

Recall that the Kneser graph with indices n, k is formed from a set X with n elements, taking as vertex set V the set of subsets of cardinality k of X, and joining two vertices with an edge if and only if the corresponding sets are disjoint. We denote by E the set of edges.

The number of vertices of the Petersen graph is then 10 and the number of edges is 15.

The 45 (unordered) pairs of vertices fall into 15 at distance 1 (edges) and 30 at distance 2, in particular the Petersen graph has diameter 2.

The Petersen graph is the unique strongly regular graph with the parameters $v = 10$, $k = 3$, $\lambda = 0$ and $\mu = 1$. This means that the Petersen graph has $v = 10$ vertices, each vertex has $k = 3$ neighbours. For every neighbouring pair of vertices, there are $\lambda = 0$ common neighbours, and for any non-neighbouring pair or vertices, there is $\mu = 1$ common neighbour.

Moreover, the Petersen graph has diameter 2, and since its automorphism group is $\mathfrak{S}_5$, one sees easily that it is distance-transitive (that is, it acts transitively on the ordered pairs of vertices having a fixed distance).

The girth (the minimal length of a cycle) of the Petersen graph is 5. Each edge is contained in four 5-cycles, and the total number of 5-cycles is 12.

The independence number (maximum number of a set of vertices not connected by an edge) of the Petersen graph is 4. It has 5 maximal independent sets, which in the Kneser representation are given by the sets of 4 unordered pairs containing a common element $x \in X$. Let us denote by I_x the independent set corresponding to $x \in X$; hence the set of all 5 independent sets $\mathcal{I}$ is in bijection with X. The stabilizer of any maximal independent set is therefore $\mathfrak{S}_4$ (the stabilizer of x in $\mathfrak{S}_5$).

Each vertex $a = \{x, y\}$ is contained in two maximal independent sets, namely I_x and I_y. In a maximal independent set, any pair of vertices is at distance 2. Each pair of vertices a, b of the Petersen graph with $d(a, b) = 2$ is contained in a unique maximal independent set.[10]

[10] $d(a, b) = 2$ implies $\#(a \cap b) = 1$, so $a \cap b = \{x\}$ with $x \in X$ and the unique maximal independent set containing $\{a, b\}$ is I_x.

4.3.2.2 Doro-Hall Graph

In [50], the connected local Petersen graphs (that is, the graphs such that, for any vertex, the subgraph having as vertex set the set of its neighbours is a Petersen graph) have been classified.

There are only three such graphs, having respectively 21, 63 and 65 vertices. We shall refer to the largest of these graphs as the *Doro-Hall graph* (it is often simply called the Hall-graph).

For an overview on the Doro-Hall graph, see [16, Prop. 12.2.2] and the subsequent paragraph.

In [50, Thm. 1.2], it has been shown that there is a unique distance-regular graph on 65 vertices with intersection array $\{10, 6, 4; 1, 2, 5\}$. In [50], the original discovery of this graph is attributed to a manuscript "Two new distance-transitive graphs" by Stephen Doro, which apparently has never been published.

We call it the *Doro-Hall graph* – to acknowledge credit to the original discoverer. The graph has been rediscovered independently in [48].

By Gordon [48, Prop. 6], combined with the uniqueness result of [50, Thm. 1.2], the automorphism group of Γ is isomorphic to $\mathrm{P\Sigma L}(2, 25)$. The graph Γ is distance-transitive, where already the normal index-2 subgroup $\mathrm{PSL}(2, 25)$ is acting transitively on all ordered pairs of vertices of fixed distance.

Moreover, Γ is a commuting involutions graph, meaning that, for any $x \in V$, there is a unique nontrivial automorphism τ_x (which necessarily is an involution) of Γ pointwise stabilizing x and all neighbours of x, and moreover any two distinct vertices of $x, y \in V$ are neighbours if and only if τ_x and τ_y commute.

The following constructions of Γ are known.

- Let $F : x \mapsto x^5$ be the Frobenius automorphism of $\mathbb{F}_{25}$.

  Then $G = \mathrm{P\Sigma L}(2, 25) = \mathrm{PSL}(2, 25) \rtimes \langle F \rangle$ is a group of order $15600 = 2^4 \cdot 3 \cdot 5^2 \cdot 13$.

  The set V of vertices of the Hall graph is the conjugacy class F^G of F in G. It is of size 65. The set E of edges is given by the 325 unordered pairs of distinct commuting elements in F^G [50].
- Let $q \in \mathbb{F}_5[x_1, x_2, x_3, x_4]$ be a non-degenerate elliptic quadratic form[11] and $Q = \{\langle x \rangle \in \mathbb{P}^3_{\mathbb{F}_5} \mid q(x) = 0\}$ the corresponding quadric.

  The vertex set V is defined as the set of all points $\langle x \rangle \in \mathbb{P}^3_{\mathbb{F}_5}$ with $x \in \mathbb{F}_5^4$ such that $q(x) \in \{\pm 1\}$.

  The edge set E is defined as the set of pairs $\{x, y\} \subset V$ such that the line $L = x * y$ is a secant of Q, i.e. L intersects Q in 2 distinct points (that is, these points are in $Q \subset \mathbb{P}^3_{\mathbb{F}_5}$) [16, Sec. 12.2].
- An inversive plane of order 5 is a combinatorial 3-(26, 6, 1) design. The set of 130 blocks is invariant under the action of $\mathbb{P}SL(2, 25)$, partitioning the design in two orbits of size 65. Now let V be one of the orbits and define two blocks in V to be neighbours if they are disjoint [48].

[11] We may take $q(x) = x_1 x_2 + x_3^2 + 2x_4^2$.

For each vertex, the distance distribution of the other 64 vertices is $(1^{10}2^{30}3^{24})$.

4.3.2.3 From the Doro-Hall Graph to the Code $\mathcal{K}$ of the Barth Sextic

The code $\mathcal{K}$ of the Barth sextic is indeed isomorphic to the code associated to the Doro-Hall graph as in the following Theorem 4.51: for each vertex, the characteristic function of the set of 24 vertices at distance 3 yields a codeword in $\mathbb{F}_2^{65}$ of Hamming weight 24. The linear hull of these 65 codewords is the code $\mathcal{K}$ consisting of the (strictly) even set of nodes of the Barth sextic. By construction, this code is invariant under the action of the $\mathrm{P\Sigma L}(2, 25)$ in the Doro construction of the Doro-Hall graph.

Theorem 4.51 shows also how $\mathcal{K}'$ is obtained from the Doro-Hall graph; moreover in Chap. 5 we give another description of how $\mathcal{K}'$ can be derived from $\mathcal{K}$ by coding theoretic means. In particular, $\mathcal{K}'$ is invariant under the action of $\mathrm{P\Sigma L}(2, 25)$ too, where the extra coordinate is always left constant. In fact, $\mathrm{P\Sigma L}(2, 25)$ is the full automorphism group of $\mathcal{K}$ as well as of $\mathcal{K}'$.

4.3.2.4 From the Barth Sextic to the Doro-Hall Graph

Indeed, the Doro-Hall graph structure can also be seen from the geometry of the Barth sextic.

The following has been found computationally.[12]

By the $\mathfrak{A}_5$-action, the 65 nodes of the Barth surface Y_B are partitioned into 3 orbits of respective lengths 15 (Type A), 20 (Type B) and 30 (Type C). For the line L connecting an ordered pair of nodes, x and y, of the Barth sextic, one of the following possibilities occurs:

- Type X: L intersects Y_B in two further points.
- Type Y: L intersects Y_B in a further point with multiplicity 2.
- Type Z: x has multiplicity 4 and y has multiplicity 2 on L (Type Z^+) or the other way round (Type Z^-).
- Type 0: L is contained in Y_B.

The frequencies of the types of the lines passing through a node x depend on the type of x. We get the following table:

Type of x	Type X	Type Y	Type Z^+	Type Z^-	Type 0
A	10	46	0	0	8
B	10	42	6	6	0
C	10	46	4	4	0

[12] See the script [33, MAGMA 23].

Let us call two nodes neighbours if their connecting line is of type X: then we obtain a graph, which is indeed isomorphic to the Doro-Hall graph.[13]

We remark that there are exactly 6 lines connecting a pair of nodes which are completely contained in Y_B. Every such line contains 5 nodes, all of type A. Each node of type A is contained in two such lines.

We have the following synthetic description of the codes of the Barth sextic in terms of the Doro-Hall graph:

Theorem 4.51 *The projection $\mathcal{K}''$ of the extended code $\mathcal{K}'$ of the Barth sextic is isomorphic to the code generated by the characteristic functions of independent sets of maximal cardinality of the Doro-Hall graph (a set $\mathcal{N}$ is independent if no edge joins two vertices in $\mathcal{N}$).*

The code $\mathcal{K}$ is generated by the sums of pairs of such characteristic functions (which have weight 15*): these sums yield* 325 *vectors of weight* 24.

Another generating set for $\mathcal{K}$ is given by the other vectors of weight 24*: these are the* 65 *characteristic functions of the spheres of radius* 3 *in the Doro-Hall graph $\mathcal{G}$ (set of vertices at distance* 3 *from a given vertex v).*

Proof We know that the extended code $\mathcal{K}'$ is generated by the codewords corresponding to the half-even sets of cardinality 15.

Observe that in a given half-even set of cardinality 15, the line connecting any two nodes is not an edge of $\mathcal{G}$.

This is because, setting $x = 0$ the equation of the corresponding plane, the equation of f writes as $f = x\phi + g_3^2$, so that the plane section with $\{x = 0\}$ is the plane cubic curve $\{x = g = 0\}$ counted with multiplicity 2, and the 15 nodes are the intersections of the cubic curve with the quintic surface ϕ; hence each such line intersects in three points.

So the corresponding vertices of $\mathcal{G}$ form an independent set (a subset of vertices whose induced subgraph is edgeless) of cardinality 15.

It can be checked that these sets are independent sets of maximal cardinality, and indeed every independent set of 15 vertices corresponds to a half-even set of cardinality 15, as shown by the Mathematica built-in library about the (Doro-) Hall graph, including the independence polynomial:[14]

$$1 + 65x + 1755x^2 + 25805x^3 + 227890x^4 + 1259583x^5 + 4414930x^6 +$$

$$+ 9762935x^7 + 13342485x^8 + 10860980x^9 + 4966663x^{10} +$$

$$+ 1164540x^{11} + 120380x^{12} + 4980x^{13} + 390x^{14} + 26x^{15},$$

[13] See the script [33, Mathematica 10].

[14] See the script [33, Mathematica 10].

where the coefficient of x^k is the number of independent sets consisting of k vertices. In particular, 15 is the maximum cardinality of an independent set, and there are only 26 of them, namely, the 26 half-even sets of cardinality 15.[15]

We have seen that if $\mathcal{N}_1, \mathcal{N}_2$ are half-even sets of cardinality 15, then their intersection has cardinality 3 (the line intersection of the two planes meets the surface in 3 points), hence the symmetric difference $(\mathcal{N}_1 \cup \mathcal{N}_2) \setminus (\mathcal{N}_1 \cap \mathcal{N}_2)$ is a strictly even set of cardinality 24. These are $\frac{1}{2}26 \cdot 25 = 325$ vectors, because such sets of cardinality 24 determine uniquely a couple of planes containing 15 nodes (in fact, a line cannot contain 6 nodes of the sextic, otherwise each plane containing the line would cut the line with mutiplicity 2, hence the line would be contained in the singular set of Y); these 325 vectors generate the code $\mathcal{K}$.

Our knowledge of the weight enumerator for our code (see Chap. 5) shows that there are exactly 390 vectors of weight 24, and it can be verified via computer calculations[16] that the remaining 65 are the characteristic functions of the spheres of radius 3 and that they generate the code $\mathcal{K}$. Observe in fact that, given a vertex v of $\mathcal{G}$, there are exactly 10 vertices at distance 1 from v, 30 at distance 2, and 24 at distance 3 (so 3 is the diameter of $\mathcal{G}$). □

Remark 4.52

(I) The code $\mathcal{K}$ of the Barth sextic is an irreducible representation of its symmetry group $\mathrm{P}\,\Sigma\,\mathrm{L}(2, 25)$.

In fact, let $\mathcal{C}$ be a subcode of $\mathcal{K}$: if $\mathcal{C}$ contains a vector of weight 24, then either it contains one of the 65 characteristic functions of a sphere of radius 3, or it contains one of the 325 sums of pairs of characteristic functions of independent sets of maximal cardinality. In both cases $\mathcal{C}$ would contain the whole orbit if $\mathcal{C}$ is $\mathrm{P}\,\Sigma\,\mathrm{L}(2, 25)$-invariant, hence we would have $\mathcal{C} = \mathcal{K}$.

There remains to exclude that $\mathcal{C}$ can only have the weights 32, 40.

$\mathrm{P}\,\Sigma\,\mathrm{L}(2, 25)$-invariance implies that the effective length of $\mathcal{C}$ is 65, hence by Bonisoli's theorem it cannot be a one weight code (because none of 65, 13, 5 is a power of 2 diminished by 1).

Therefore $\mathcal{C}$ would have to admit both weights 32, 40. If we use that the vectors in $\mathcal{K}$ of weight 40 form a single orbit of cardinality 650 (this is again verified by computer calculations,[17]) then the dimension of $\mathcal{C}$ is at least 10 and we can use the Griesmer bound to compute $n \geq 67$, a contradiction.

Another possibility, using Part II) of this remark, would be to exclude that $\mathcal{C}$ cannot be one of the $\mathfrak{A}_5$-submodules U_{43}, U_{44}.

(II) The inclusion $\mathcal{K} \subset V_A \oplus V_B \oplus V_C$ produces six projections, and the kernels are $\mathfrak{A}_5$-submodules.

---

[15] The coefficient of x^{14} coincides with the number of even sets of cardinality 24: pure coincidence?

[16] See the script [33, Mathematica 10].

[17] See the script [33, MAGMA 22].

MAGMA calculations[18] show the following: denoting by

$$\mathcal{K}_A = ker(\mathcal{K} \rightarrow V_A) =: U_{42}, \quad \mathcal{K}_C = ker(\mathcal{K} \rightarrow V_C) =: U_{41},$$

these are irreducible $\mathfrak{A}_5$-submodules of dimension 4, while $\mathcal{K}_B = 0$.

These calculations also show that the only nontrivial $\mathfrak{A}_5$-submodules are four 4-dimensional irreducible $\mathfrak{A}_5$-submodules, $U_{41}, U_{42}, U_{43}, U_{44}$, and four 8-dimensional submodules,

$$U_{81} = \langle U_{42}, U_{43}, U_{44} \rangle, U_{8j} = \langle U_{41}, U_{4j} \rangle, \quad j = 2, 3, 4.$$

The only inclusions of a 4-dimensional submodule U_{4i} into an 8-dimensional submodule U_{8j} are those which are derived from the definitions of the U_{8j}'s.

$\mathcal{K}$ is the direct sum $U_{41} \oplus U_{42} \oplus U_{43}$ of the first three irreducible $\mathfrak{A}_5$-submodules.

Moreover the only submodules which do not admit the weight 24 are exactly U_{43}, U_{44}. More precisely the weight enumerators of these submodules are

$$\begin{aligned}
W_{U_{41}}(x) &= 1 + 15x^{24} \\
W_{U_{42}}(x) &= 1 + 10x^{24} + 5x^{32} \\
W_{U_{43}}(x) &= 1 + 10x^{32} + 5x^{40} \\
W_{U_{44}}(x) &= 1 + 10x^{32} + 5x^{40} \\
W_{U_{81}}(x) &= 1 + 30x^{24} + 175x^{32} + 50x^{40} \\
W_{U_{82}}(x) &= 1 + 40x^{24} + 155x^{32} + 60x^{40} \\
W_{U_{83}}(x) &= 1 + 45x^{24} + 145x^{32} + 65x^{40} \\
W_{U_{84}}(x) &= 1 + 45x^{24} + 145x^{32} + 65x^{40}
\end{aligned}$$

4.4 Determinantal Equations of the Barth Sextic

In this section we present an explicit description of Barth's sextic $Y \subset \mathbb{P}^3$ as a determinant of the matrix A whose diagonal degrees are $(1, 1, 1, 3)$. In view of Corollary 4.2 and Lemma 4.4, the first step is to find a symmetric half-even set of cardinality 31. To do this, we consider the $\mathfrak{A}_5$-orbit of the linear form $\ell_1 := w - x - y - z$:

[18] See the script [33, MAGMA 22].

$$\begin{array}{ll}
\ell_1 = w - x - y - z & \ell_2 = w + x - y + z \\
\ell_3 = w - x + y + z & \ell_4 = w + x + y - z \\
\ell_5 = w + x + y + z & \ell_6 = \tau w - \tau^2 x + z \\
\ell_7 = \bar{\tau} w - \bar{\tau}^2 y + z & \ell_8 = \bar{\tau} w - \bar{\tau}^2 x + y \\
\ell_9 = w - x + y - z & \ell_{10} = \tau w + \tau^2 x - z \\
\ell_{11} = \bar{\tau} + \bar{\tau}^2 x + y & \ell_{12} = \bar{\tau} w - \bar{\tau}^2 y - z \\
\ell_{13} = w + x - y - z & \ell_{14} = \bar{\tau} w + \bar{\tau}^2 y - z \\
\ell_{15} = \bar{\tau} w - \bar{\tau}^2 x - y & \ell_{16} = \tau w - \tau^2 x - z \\
\ell_{17} = w - x - y + z & \ell_{18} = \bar{\tau} w + \bar{\tau}^2 x - y \\
\ell_{19} = \bar{\tau} w + \bar{\tau}^2 y + z & \ell_{20} = \tau w + \tau^2 x + z.
\end{array}$$

In Sect. 4.3, we have seen that the plane section $H_1 := (\ell_1 = 0) \cap Y$ is the double curve, say $2C_1$, passing through a half-even set of nodes $\mathcal{N}_1$ of cardinality 15. The ideal I_{C_1} of C_1, regarded as an ideal of $\mathbb{C}[w, x, y, z]$, is generated by ℓ_1 and

$$h_1 = \tau^2(x^2y + y^2z + z^2x) + (xy^2 + yz^2 + zx^2) + (\tau^2 + 1)xyz$$

We may also consider $H_i, C_i, \mathcal{N}_i, h_i$ for $i = 2, \ldots, 20$. It can be directly checked that $\mathcal{N} := \mathcal{N}_1 + \mathcal{N}_9 + \mathcal{N}_{17}$ has cardinality 31. By Proposition 4.13, $\mathcal{N}$ is symmetric.

Let $S \to Y$ be the minimal resolution, let $L_i \subset S$ be the proper transform of C_i, and let E_* be the sum of (-2)-curves over a given set of nodes $* \subset Y$. We have

$$2H - L \equiv L_1 + L_9 + L_{17} + E_{\mathcal{M}}$$

where $\mathcal{M}$ is the set of nodes that appear more than once in $\{\mathcal{N}_1, \mathcal{N}_9, \mathcal{N}_{17}\}$. On the other hand, we have three additional expression of $\mathcal{N}$ as the sum of half even sets

$$\begin{array}{l}
\{\mathcal{N}_1, \mathcal{N}_2, \mathcal{N}_4, \mathcal{N}_{12}, \mathcal{N}_{19}\}, \\
\{\mathcal{N}_9, \mathcal{N}_4, \mathcal{N}_5, \mathcal{N}_8, \mathcal{N}_{18}\}, \\
\{\mathcal{N}_{17}, \mathcal{N}_2, \mathcal{N}_5, \mathcal{N}_6, \mathcal{N}_{10}\}.
\end{array} \tag{4.14}$$

For each expression, we associate a section in $H^0(3H - L)$. For instance,

$$3H - L \equiv L_1 + L_2 + L_4 + L_{12} + L_{19} + E_{\mathcal{M}_1} + E_{\mathcal{M}'_1}$$

where $\mathcal{M}_1$ is the set of nodes that appear 2–3 times and $\mathcal{M}'_1$ is the set of nodes that appear 2–5 times in $\{\mathcal{N}_1, \mathcal{N}_2, \mathcal{N}_4, \mathcal{N}_{12}, \mathcal{N}_{19}\}$. Since the right hand side is effective, it corresponds to a section $v_1 \in H^0(3H - L)$. Similarly, we may take v_2 (resp. v_3) corresponding to $\{\mathcal{N}_9, \mathcal{N}_4, \mathcal{N}_5, \mathcal{N}_8, \mathcal{N}_{18}\}$ (resp. $\{\mathcal{N}_{17}, \mathcal{N}_2, \mathcal{N}_5, \mathcal{N}_6, \mathcal{N}_{10}\}$). Also, let $v_4 \in H^0(2H - L)$ be a nonzero section.

We consider the double cover $f : Z \to S$ branched over $H + E_{\mathcal{N}} \equiv 2L$ and the graded algebra

$$R := \bigoplus_{m \geq 0} H^0(S, f_* \mathcal{O}_Z(m)),$$

which admits eigenspace decomposition into $R^+ := \bigoplus_{m \geq 0} H^0(S, \mathcal{O}_S(mH))$ and $R^- := \bigoplus_{m \geq 0} H^0(S, \mathcal{O}_S(mH - L))$. Then, $\{v_1, \ldots, v_4\}$ is a generating set of R^-, and by Catanese [21, Section 2], $B_{ij} := v_i v_j$ form an adjoint matrix of $A(x)$. The divisor form of B_{ij} can be explicitly carried out; for instance,

$$\begin{aligned} \operatorname{Div} B_{14} &= \operatorname{Div} v_1 + \operatorname{Div} v_4 \\ &= (L_1 + L_9 + L_{17}) + (L_1 + L_2 + L_4 + L_{12} + L_{19}) + \sum_{\mathcal{M}} E_{\mathcal{M}} \end{aligned}$$

where the last term indicates a sum of (-2)-curves. It can be decomposed into $2L_1$ and $(L_9 + L_{17} + L_2 + L_4 + L_{12} + L_{19})$, which are respectively a hyperplane section and a cubic section. We may check[19] that the ideal $I_{C_9} \cap I_{C_{17}} \cap I_{C_2} \cap I_{C_4} \cap I_{C_{12}} \cap I_{C_{19}}$ contains a unique cubic equation

$$w^2 y + \tau^3 x^2 y - y^3 - \tau^2 w^2 z - \bar{\tau} x^2 z + \bar{\tau} y^2 z - \tau^3 y z^2 + \tau^2 z^3,$$

showing that B_{14} is the product of ℓ_1 and this cubic (up to constant multiple). In similar way, we may determine other B_{ij} as follows.

(i) B_{i4}, $i < 4$:

$$\begin{aligned} B_{14} &= \lambda_{14}(w - x - y - z)(w^2 y + \tau^3 x^2 y - y^3 - \tau^2 w^2 z \\ &\qquad - \bar{\tau} x^2 z + \bar{\tau} y^2 z - \tau^3 y z^2 + \tau^2 z^3) \\ B_{24} &= \lambda_{24}(w - x + y - z)(w^2 x - x^3 + \tau^2 w^2 y - \bar{\tau} x^2 y \\ &\qquad - \tau^3 x y^2 - \tau^2 y^3 + \tau^3 x z^2 + \bar{\tau} y z^2) \\ B_{34} &= \lambda_{34}(w - x - y + z)(w^2 x - x^3 + \bar{\tau}^3 x y^2 + \bar{\tau}^2 w^2 z \\ &\qquad - \tau x^2 z + \tau y^2 z - \bar{\tau}^3 x z^2 - \bar{\tau}^2 z^3). \end{aligned}$$

(ii) B_{ij}, $i < j < 4$:

$$\begin{aligned} B_{12} = \lambda_{12}(w + x + y - z)(w^4 - w^2 x^2 + \tau w^2 x y - \tau x^3 y \\ - 2w^2 y^2 - \sqrt{5} x^2 y^2 - \bar{\tau}^2 x y^3 + y^4 \\ - w^2 x z + \bar{\tau}^3 x^3 z - \bar{\tau} w^2 y z - \tau^2 x^2 y z \end{aligned}$$

[19] See the script [33, Macaulay2 15].

$$- 3xy^2z + \tau^2y^3z - w^2z^2 + 2x^2z^2$$
$$+ \bar{\tau}^2xyz^2 + \sqrt{5}y^2z^2 + \tau^3xz^3 + \bar{\tau}yz^3)$$

$$\begin{aligned}
B_{13} = \lambda_{13}(w + x - y + z)(w^4 &- w^2x^2 - w^2xy + \tau^3x^3y \\
&- w^2y^2 + 2x^2y^2 + \bar{\tau}^3xy^3 + \bar{\tau}w^2xz \\
&- \bar{\tau}x^3z - \tau w^2yz - \bar{\tau}^2x^2yz \\
&+ \tau^2xy^2z + \tau y^3z - 2w^2z^2 + \sqrt{5}x^2z^2 \\
&- 3xyz^2 - \sqrt{5}y^2z^2 - \tau^2xz^3 + \bar{\tau}^2yz^3 + z^4)
\end{aligned}$$

$$\begin{aligned}
B_{23} = \lambda_{23}(w + x + y + z)(w^4 &- 2w^2x^2 + x^4 + \bar{\tau}w^2xy \\
&- \tau^2x^3y - w^2y^2 + \sqrt{5}x^2y^2 - \bar{\tau}xy^3 \\
&+ \tau w^2xz - \bar{\tau}^2x^3z + w^2yz + 3x^2yz \\
&+ \bar{\tau}^2xy^2z - \tau^3y^3z - w^2z^2 - \sqrt{5}x^2z^2 \\
&+ \tau^2xyz^2 + 2y^2z^2 - \tau xz^3 - \bar{\tau}'^3yz^3)
\end{aligned}$$

(iii) B_{ii}:

$$\begin{aligned}
B_{11} &= \lambda_{11}\ell_1\ell_2\ell_4\ell_{12}\ell_{19} \\
B_{22} &= \lambda_{22}\ell_9\ell_4\ell_5\ell_8\ell_{18} \\
B_{33} &= \lambda_{33}\ell_{17}\ell_2\ell_5\ell_6\ell_{10} \\
B_{44} &= \lambda_{44}\ell_1\ell_9\ell_{17}.
\end{aligned}$$

To determine λ_{ij}, we impose the condition $B_{ij}(P) = 1$ for a fixed $P \in Y$. For instance, we may pick $P = [2 : 1 : 1 : 1] \in Y$.

Now, it can be shown that $\det(B_{ij}) = (\text{constant}) \cdot f^3$, where f is the Barth's sextic equation, and $\operatorname{adj}(B_{ij}) = (\text{constant}) \cdot f^2A$, where A is the matrix

$$A = \begin{pmatrix} 0 & a_{12} & a_{13} & q_1 \\ a_{12} & 0 & a_{23} & q_2 \\ a_{13} & a_{23} & 0 & q_3 \\ q_1 & q_2 & q_3 & G \end{pmatrix},$$

with

$$a_{12}(w, x, y, z) = 3\sqrt{5}(w - x - y + z)$$
$$a_{13}(w, x, y, z) = -3\sqrt{5}(w - x + y - z)$$

$$\begin{aligned}
a_{23}(w,x,y,z) &= 15(w-x-y-z)\\
q_1(w,x,y,z) &= (-x+\tau y+\bar{\tau} z)(w+x+y+z)\\
q_2(w,x,y,z) &= \sqrt{5}(\tau x-\bar{\tau} y-z)(w+x-y+z)\\
q_3(w,x,y,z) &= \sqrt{5}(-\bar{\tau} x+y+\tau z)(w+x+y-z)\\
G(w,x,y,z) &= -\frac{2}{3}(w+x+y-z)(w+x-y+z)(w+x+y+z).
\end{aligned}$$

A final observation is that one could try to construct surfaces with many nodes also in higher degree by taking discriminants of cubic hypersurfaces with sufficiently many nodes.

However, while as we saw for degree $d = 4, 5$ we get exactly nodal maximizing surfaces by choosing a cubic hypersurface with the maximal number of nodes ($16 = 10 + 6$, $31 = 15 + 16$), we have shown here that we cannot take for degree $d = 6$ such a cubic, since $35 + 31 = 66 > 65$.

For $d = 7$ we would get $56 + 52 = 108 > 104$, which is an upper bound for $\mu(7)$ (see for instance [67]), so we should take cubics with at most 48 nodes.

The case $d = 8$, since, as shown in Lemma 2.1, for a cubic hypersurface in $\mathbb{P}^8$ the 4-dimensional linear subspaces $L \subset X$ pass through some singular points of X, shows the necessity of taking (as it happens already for sextics with 65 nodes) subspaces passing through several nodes of X.

Chapter 5
Codes of Nodal Sextics with Many Nodes

Sascha Kurz

Abstract Using a computer search the unique possible code for a sextic with 65 nodes is classified. More precisely, the known coding theoretic properties of the strict code leave three numerical candidates. Incorporating known facts about half-even sets of nodes, i.e. properties of the extended code, yields the Doro-Hall code described in Sect. 4.3 as the unique possibility. For a sextic surface with 64 nodes there remain five candidates.

The aim of this chapter is to determine the possible strict codes $\mathcal{K}$ for nodal sextics in $\mathbb{P}^3$ with ν nodes, when ν is *large*, i.e., $\nu \in \{64, 65\}$. Clearly for the effective length n we have $n \leq \nu$ and the B-inequality (Remark 1.20) gives $k \geq \nu - 53$, i.e., $k \geq 12$ for $\nu = 65$ and $k \geq 11$ for $\nu = 64$. The set of possible non-zero weights for the strict code $\mathcal{K}$ of a nodal sextic in $\mathbb{P}^3$ is $\{24, 32, 40, 56\}$, see [29].

Proposition 5.1 *If $\mathcal{C}$ is a binary linear code with effective length at most* 65, *dimension $k \geq$* 12, *and non-zero weights in* $\{24, 32, 40, 56\}$, *then $k =$* 12 *and $\mathcal{K}$ is isomorphic to one of the following three cases:*

(1)

$$\begin{pmatrix}
001100001110000000111110100001111111001001010010000110000000000000 \\
101001111111000000011011101000010011010001101100000001000000000000 \\
000100111011100011110111001000010000110000110110100000100000000000 \\
010001111111111001100110000100110010001000110100000100010000000000 \\
110001110000001011100111101100001110010001100001010000001000000000 \\
000000011000110111100011010011101110010001011110000000000100000000 \\
010011110001111101010000110100100011101110111111111000000010000000 \\
001000110111101100001111110000000001100110000111100000000001000000 \\
000111110001100011000000000110001111110000110000111100000000100000 \\
000000000111110000001111111111000000000111111000000011000000010000 \\
000000000000011111111111111100000000000000111111110000000000001000 \\
000000000000000000000000000011111111111111111111111000000000000100
\end{pmatrix}$$

$$w_{\mathcal{C}}(x, y) = x^{65}y^0 + 630x^{41}y^{24} + 3087x^{33}y^{32} + 378x^{25}y^{40}$$

S. Kurz (✉)
Mathematisches Institut der Universität Bayreuth, Bayreuth, Germany
e-mail: sascha.kurz@uni-bayreuth.de

F. Catanese (ed.), *Varieties of Nodal Surfaces, Coding Theory and Discriminants of Cubic Hypersurfaces*, Lecture Notes of the Unione Matematica Italiana 31,
https://doi.org/10.1007/978-3-032-05899-7_5

$$
(2)\quad \begin{pmatrix}
10000110110100110000010101100110001010011000001000010000011001100\\
00011001111101100000000001000000011000000011000111010011001110010\\
00001010111110100001000010100100001000111100100001100000001100110\\
00001111101000010000101000110010011100011000010000001001010101100\\
00010011111110110101011001100100001010111011100000001100101100 10\\
00010010110110110000100101000000000000110011000010000000011111110\\
01010101010110100100010100100011101001011101011010000000000000000\\
00001100110011001000111101110101011010011110010000000000000000000\\
00001100001111000111000001110100100101111110001110000000000000000\\
00001111111111110110110010010011000110001000001000000000000000000\\
00000011110011000110001100001110100110001111110110000000000000000\\
00111111111111110110111101100000011001100111100110000000000000000
\end{pmatrix}
$$

$$w_{\mathcal{C}}(x, y) = x^{65}y^{0} + 502x^{41}y^{24} + 3087x^{33}y^{32} + 506x^{25}y^{40}$$

$$
(3)\quad \begin{pmatrix}
10000100000000110110010001110100111101010001011110010100000000000\\
10100100011000001001000110100110111111001000001100011010000000000\\
01000010011100011000000100110100110000011111011110001001000000000\\
11110100001110110100000011010110100001011100000100001000100000000\\
01101011000001100011010001000011001010001111000010111000010000000\\
00101001110111101011000001011000000110111001001000100000001000000\\
00011000111111100000111110001000100010001010101001100000000100000\\
00000111001011100111110001010100000000111100110000001100000000010000\\
000111110001111000000001111001101111110000111100111111000000001000\\
00000000111111100000000000001111000111000000000111111111000000000100\\
00000000000000011111111111111110000000111111111111111111000000000010\\
00000000000000000000000000000001111111111111111111111111000000000001
\end{pmatrix}
$$

$$w_{\mathcal{C}}(x, y) = x^{65}y^{0} + 390x^{41}y^{24} + 3055x^{33}y^{32} + 650x^{25}y^{40}$$

Proof We have applied the software package `LinCode`, see [14, 64], to exhaustively enumerate all codes with the specified properties. The computations were performed on a linux cluster of the University of Bayreuth and took less than a CPU year in total. □

We remark that more computational details on the enumeration in the proof of Proposition 5.1 can be found in [65] and that the problem of the determination of the possible codes of the Barth sextic actually initiated the development of the software package `LinCode` [14, 64]. Using coding theoretic methods and recent algorithmic improvements the subsequent uniqueness result for the code of nodal sextics with 65 nodes in Theorem 5.3 can nowadays be determined in less than two hours on a single core, see [66].

The codes of Proposition 5.1 are the only possible candidates for the strict code $\mathcal{K}$ of a nodal sextic in $\mathbb{P}^3$ with $\mathbb{P}^3$ with $\nu = 65$ nodes, where the code in case (3) is isomorphic to the code $\mathcal{K}_B$ of the Barth sextic, see Sect. 4.3. To further thin out this list we use known properties of the extended code $\mathcal{K}'$, i.e., the minimum distance is at least 16 and the weights are divisible by four, see e.g. [44, Theorem 1.10]. Moreover, $\mathcal{K}'\backslash\mathcal{K}$ does not contain codewords of weight 20 or 24, see [44, Corollary 1.11].

Lemma 5.2 *Let $\mathcal{C}$ be a linear code of effective length n and $\mathcal{C}'$ be spanned by the codewords of $\mathcal{C}$ and an additional codeword c' such that all non-zero weights in $\mathcal{C}'\backslash\mathcal{C}$ are either* 16 *or at least* 28*. If $\mathcal{C}'$ has effective length $n+\delta$, all codewords have a weight of at most Λ, and $w(c') = \gamma$, then there exist $y_c \in \{0, 1\}$ for all $c \in \mathcal{C}$ with $w(c) \neq 0$ and $x_i \in \{0, 1\}$ for all $1 \leq i \leq n$ such that*

$$\sum_{i=1}^{n} x_i = \gamma - \delta \tag{5.1}$$

and

$$\frac{\gamma + w(c)}{2} - 8 - \left(\tfrac{\Lambda}{2} - 8\right) y_c \le \sum_{1 \le i \le n\,:\,c_i=1} x_i \le \frac{\gamma + w(c)}{2} - 8 - 6y_c, \tag{5.2}$$

for all $c \in C$ *with* $w(c) \neq 0$.

Proof W.l.o.g. we assume that $\mathcal{C}'$ is spanning and that $c_i = 0$ for all $n + 1 \le i \le n + \delta$ and all codewords $c \in \mathcal{C}$. Since $\mathcal{C}'$ has effective length $n + \delta$, we have $c_i' = 1$ for all $n + 1 \le i \le n + \delta$. Now let us choose $x_i = c_i'$ for $1 \le i \le n$. Since c' has weight γ we have $\sum_{i=1}^{n} x_i = \gamma - \delta$. For each $c \in \mathcal{C}$ with $w(c) \neq 0$ we set $y_c = 0$ if $w(c' + c) = 16$ and $y_c = 1$ otherwise. Note that

$$w(c' + c) = \gamma + w(c) - 2 \sum_{1 \le i \le n\,:\,c_i=1} x_i,$$

which is equivalent to

$$\sum_{1 \le i \le n\,:\,c_i=1} x_i = \frac{\gamma + w(c)}{2} - \frac{w(c' + c)}{2}.$$

If $y_c = 0$ then inequality (5.2) is given by

$$\frac{\gamma + w(c)}{2} - 8 \le \sum_{1 \le i \le n\,:\,c_i=1} x_i \le \frac{\gamma + w(c)}{2} - 8,$$

which is equivalent to $w(c' + c) = 16$. For $y_c = 1$ inequality 5.2 is given by

$$\frac{\gamma + w(c)}{2} - \Lambda/2 \le \sum_{1 \le i \le n\,:\,c_i=1} x_i \le \frac{\gamma + w(c)}{2} - 14,$$

which is equivalent to $28 \le w(c' + c) \le \Lambda$. □

Given a code $\mathcal{C}$ and the parameters γ, δ, and Λ we can computationally check whether the system of Lemma 5.2 has a solution using an integer linear programming solver like e.g. CPLEX. As an example we spell out all conditions for $\gamma = \Lambda = 44$, $\delta = 1$ and $\mathcal{C} = \mathcal{K}_B$ (case (3) of Proposition 5.1):

$$\sum_{i=1}^{65} x_i = 43$$

$$
\begin{aligned}
6y_c + \sum_{1\le i\le 65\,:\,c_i=1} x_i &\le 34 \ \ \forall c\in\mathcal{C}: w(c)=40,\\
14y_c + \sum_{1\le i\le 65\,:\,c_i=1} x_i &\ge 34 \ \ \forall c\in\mathcal{C}: w(c)=40,\\
6y_c + \sum_{1\le i\le 65\,:\,c_i=1} x_i &\le 30 \ \ \forall c\in\mathcal{C}: w(c)=32,\\
14y_c + \sum_{1\le i\le 65\,:\,c_i=1} x_i &\ge 30 \ \ \forall c\in\mathcal{C}: w(c)=32,\\
6y_c + \sum_{1\le i\le 65\,:\,c_i=1} x_i &\le 26 \ \ \forall c\in\mathcal{C}: w(c)=24,\\
14y_c + \sum_{1\le i\le 65\,:\,c_i=1} x_i &\ge 26 \ \ \forall c\in\mathcal{C}: w(c)=24,\\
x_i &\in \{0,1\} \ \ \forall 1\le i\le 65,\\
y_c &\in \{0,1\} \ \ \forall c\in\mathcal{C}: w(c)\ne 0.
\end{aligned}
$$

Theorem 5.3 *For nodal sextics in $\mathbb{P}^3$ with $\nu = 65$ nodes, the codes $\mathcal{K}$ and $\mathcal{K}'$ are isomorphic to the codes $\mathcal{K}_B$ and $\mathcal{K}'_B$ of the Barth sextic, see Sect. 4.3.*

Proof Due to Remark 1.20 the dimension of $\mathcal{K}$ is at least 12 and from [29] we conclude that the non-zero weights in $\mathcal{K}$ are contained in $\{24, 32, 40, 56\}$. Thus we can assume that $\mathcal{K}$ is isomorphic to one of the three cases listed in Proposition 5.1. Note that the effective length is $62 + i$ in case (i), where $1 \le i \le 3$, and that the dimension is 12 in all cases.

From Proposition 1.19 we conclude that the dimension of $\mathcal{K}'$ is 13. The effective length of $\mathcal{K}'$ lies between $63 + i$ and 66. As shown in [44] the weights of the codewords in $\mathcal{K}'\backslash\mathcal{K}$ are contained in

$$\{16, 28, 32, 36, 40, 44, 48, 52, 56, 60, 64\} =: W.$$

So, if $\mathcal{K}$ is isomorphic to the code in case (i) in Proposition 5.1, then the existence of the extended code $\mathcal{K}'$ implies that the system of Lemma 5.2 has a feasible solution for at least one of the choices $1 \le \delta \le 4 - i$ and $\Lambda = \gamma \in W$. Using CPLEX we computationally check that there are no feasible solutions except when $i = 3$, $\delta = 1$, and $\Lambda = \gamma = 44$. Thus, $\mathcal{K}$ is isomorphic to $\mathcal{K}_B$. For $i = 3$, $\delta = 1$, and $\Lambda = \gamma = 44$ we let CPLEX enumerate all solutions. There are exactly 130 of them, which all yield the same code $\mathcal{C}'$ with

$$
\begin{aligned}
w_{\mathcal{C}'}(x,y) - w_{\mathcal{K}_B}(x,y) = {} & 26x^{50}y^{16} + 650x^{38}y^{28} + 1690x^{34}y^{32}\\
& + 1300x^{30}y^{36} + 300x^{26}y^{40} + 130x^{22}y^{44},
\end{aligned}
$$

so that $\mathcal{K}'$ is isomorphic to $\mathcal{K}'_B$. □

Also for nodal sextics in $\mathbb{P}^3$ with $\nu = 64$ nodes we can apply the same techniques to obtain a list of candidates for the strict codes.

Theorem 5.4 *If $\mathcal{K}$ is the strict code of a nodal sextic in $\mathbb{P}^3$ with $\nu = 64$ nodes, then $\mathcal{K}$ is isomorphic to one of the codes of the following five cases:*

(a)

$$\begin{pmatrix}
1000011001100000111111001011100000011011001100110000100000000000\\
1111000011100010111010111000000000010111000011010010010000000000\\
0111001000100010100010010001101001010010001110111101001000000000\\
0101001001100000000110001110100011001101000111010111000100000000\\
0011011010000000100110011010000101100011100001111110000010000000\\
0010111110010000001110111001011100100000100001001011000001000000\\
0010000011000110100001110000110011011110011111100000000000100000\\
0001111110000100111111101111110000111110001110100011100000010000\\
0000000001111100000000000111111000000000111111001111110000001000\\
0000000000000001111111111111110000000000000000111111110000000100\\
0000000000000000000000000000000111111111111111111111110000000010
\end{pmatrix}$$

$$w_{\mathcal{K}}(x, y) = x^{64}y^0 + 310x^{40}y^{24} + 1551x^{32}y^{32} + 186x^{24}y^{40}$$

(b)

$$\begin{pmatrix}
0001010110111111110000101000110000010100110100011010010000000000\\
0100100000011101100100010001100010011100110111101001101000000000\\
1001100100010000101001010110000100010111011010111001000100000000\\
0101000000011100011001011100001000001110101000111011100010000000\\
1100011000100110011000000110000011011100001010101111000001000000\\
0001001001100010000010010110101000001101010011010011111000100000\\
0011000100011111000000111011101111100010001000110000000100010000\\
0011000011111111000000000111011111110111011100000111111100001000\\
0000111111111111000000000001110001111000011111111111100000000100\\
0000000000000001111111111111111000000001111111111111110000000010\\
0000000000000000000000000000000111111111111111111111110000000001
\end{pmatrix}$$

$$w_{\mathcal{K}}(x, y) = x^{64}y^0 + 246x^{40}y^{24} + 1551x^{32}y^{32} + 250x^{24}y^{40}$$

(c)

$$\begin{pmatrix}
0000011010000000010001101111100011101001011110001100110000000000\\
0111001100100010000001110100000010111001001011011010101000000000\\
0000101110101010110110100111100101110100100101000000000100000000\\
0111101110001010001100101011001010000010101001101000000010000000\\
1001110100000011010000001101000001010110110001001110011001000000\\
1101010010011110010001000110111110000001100110000010001000100000\\
0101010001111111000101100000111010000100000001100011110000010000\\
0011001111111111000011100000001001011100011111011111111000001000\\
0000111111111111100000001111111100011110000000011111111100000100\\
0000000000000001111111111111111000000001111111111111110000000010\\
0000000000000000000000000000000111111111111111111111110000000001
\end{pmatrix}$$

$$w_{\mathcal{K}}(x, y) = x^{64}y^0 + 246x^{40}y^{24} + 1551x^{32}y^{32} + 250x^{24}y^{40}$$

(d)

$$\begin{pmatrix}
0001010101011101111100001101010100010000011010001010110011111110\\
0100010001010001011000010010001110000110010010010001010011001011\\
0000000100010000111010110110101011101110110000011011011111100110\\
0101000000101010101101011110001100011111110101010100011111010000\\
0101010100000010111111100100000110111001111110100101011011110000\\
1100000000110011001100000000000000000110000110011111111111000011\\
0011000000011001111001111000000000000001100110000111100000110011\\
0000110000110000000110011000000000110000111111000000110011001111\\
0000001100110000000011110000000000110011000000111100110000111111\\
0000000011110000111100000000000000111100001111111100111100000000\\
0000000000001111111111111000000000000111111111110000000000000000
\end{pmatrix}$$

$$w_{\mathcal{K}}(x, y) = x^{64}y^0 + 246x^{40}y^{24} + 1551x^{32}y^{32} + 250x^{24}y^{40}$$

$$
(e) \quad \begin{pmatrix}
0001100001001110011000111010111101111111000100111001001001001 1001 \\
0000000001010110000000000001101001011010011001100111000011001 10101 \\
0000111101010110010101010100001111001111000110011000110011001 01011 \\
0000111101010110001111001001011000000000000000000001010010101111000 \\
1001100101011001000011111010101001011010010101010000000000000000 \\
0101101000000000010101010101010100111100011010011001100100000000 \\
0011110001011001010110100000000000110011011000011100110010000000 \\
0000000011000011000011110000111100111100001111000011110000000000 \\
0000000000111100000011111111000000111100001111001100001100000000 \\
0000000000000001111111111111111000000001111111111111111100000000 \\
0000000000000000000000000000000111111111111111111111111100000000
\end{pmatrix}
$$

$$w_{sK}(x, y) = x^{64}y^0 + 243x^{40}y^{24} + 1559x^{32}y^{32} + 244x^{24}y^{40} + x^8y^{56}$$

Proof Due to Remark 1.20 the dimension k of $\mathcal{K}$ is at least 11 and from [29] we conclude that the non-zero weights in $\mathcal{K}$ are contained in $\{24, 32, 40, 56\}$. If $k \geq 12$, then we can read off the possibilities for $\mathcal{K}$ from Proposition 5.1, i.e. cases (1) and (2). However, we have already seen that those two cases do not admit an extended code $\mathcal{K}'$, see the proof of Theorem 5.3.

For the remaining cases we can assume that $k = 11$. Using the software package `LinCode` we enumerate all 357 codes with dimension 11, length 64, and non-zero weights in $\{24, 32, 40, 56\}$. To sift out this list we consider the extended code $\mathcal{K}'$. From Proposition 1.19 we conclude that the dimension of $\mathcal{K}'$ is 12. As shown in [44] the weights of the codewords in $\mathcal{K}'\backslash\mathcal{K}$ are contained in

$$\{16, 28, 32, 36, 40, 44, 48, 52, 56, 60, 64\} =: W.$$

So, if a candidate for $\mathcal{K}$ has effective length n, then the existence of the extended code $\mathcal{K}'$ implies that the system of Lemma 5.2 has a feasible solution for at least one of the choices $1 \leq \delta \leq 65 - n$ and $\Lambda = \gamma \in W$. Using `CPLEX` we computationally check that there are no feasible solutions except for the five codes listed in cases (a)-(e). □

We can check by computer that the Barth code $\mathcal{K}_B$ has a unique shortening to length 64 which is isomorphic to case (d). So, by Theorem 1.23 this code is realized by a nodal sextic surface. For the other four cases it remains an open problem whether they are realized by a nodal sextic surface.

We remark that the approach of the proofs of Theorem 5.3 and Theorem 5.4 is, in principle, also feasible for nodal sextics in $\mathbb{P}^3$ with $\nu \leq 63$ nodes. The candidates for $\mathcal{K}$ can be easily enumerated using `LinCode`. However, for $\nu = 63$ there are more than 70 000 candidates, so that further coding theoretic criteria obtained from algebraic geometry methods are desirable. The latter wish is also valid when trying to address the open problem of septics with the maximum number of nodes since the current upper bound can be attained on the code side of the problem, see [66].

Bibliography

1. Altman, A.B., Kleiman, S.L.: Foundations of the theory of Fano schemes. Compos. Math. **34**(1), 3–47 (1977)
2. Aktaş, Ç.G.: Classification of simple quartics up to equisingular deformation. Hiroshima Math. J. **47**(1), 87–112 (2017)
3. Aktaş, Ç.G.: Deformation classification of quartic surfaces with simple singularities. Rev. Mat. Iberoam. **40**(2), 503–536 (2024)
4. Akyol, A., Degtyarev, A.: Geography of irreducible plane sextics. Proc. Lond. Math. Soc. **111**(6), 1307–1337 (2015)
5. Artin, M.: On isolated rational singularities of surfaces. Am. J. Math. **88**, 129–136 (1966)
6. Barth, W.: Moduli of vector bundles on the projective plane. Invent. Math. **42**, 63–91 (1977)
7. Barth, W.: Counting singularities of quadratic forms on vector bundles. In: Vector Bundles and Differential Equations (Proceedings of Conference, Nice, 1979). Progress in Mathematics, vol. 7, pp. 1–19. Birkhäuser, Boston (1980)
8. Barth, W.: Two projective surfaces with many nodes, admitting the symmetries of the icosahedron. J. Algebraic Geom. **5**(1), 173–186 (1996)
9. Beauville, A.: Sur le nombre maximum de points doubles d'une surface dans $\mathcal{P}^3(\mu(5) = 31)$. In: Journées de Géometrie Algébrique d'Angers, Juillet 1979/Algebraic Geometry, Angers, 1979, pp. 207–215, Sijthoff - Noordhoff, Alphen aan den Rijn (1980)
10. Beauville, A.: Le théorème de Torelli pour les surfaces K3: fin de la démonstration. In: Geometry of K3 Surfaces: Moduli and Periods (Palaiseau, 1981/1982). Astérisque No. 126, pp. 111–12 (1985)
11. Bolker, E.D.: The finite Radon transform. In: Integral geometry, proceedings of AMS-IMS-SIAM joint summer research conferences, Brunswick/Maine 1984. Contemp. Math. **63**, 27–50 (1987)
12. Bonisoli, A.: Every equidistant linear code is a sequence of dual Hamming codes. Ars Combin. **18**, 181–186 (1983)
13. Bosma, W., Cannon, J., Playoust, C.: The Magma algebra system. I. The user language. J. Symbol. Comput. **24**(3–4), 235–265 (1997). Computational algebra and number theory (London, 1993)
14. Bouyukliev, I., Bouyuklieva, S., Kurz, S.: Computer classification of linear codes. IEEE Trans. Inf. Theory **67**, 7807–7814 (2021)

F. Catanese (ed.), *Varieties of Nodal Surfaces, Coding Theory and Discriminants of Cubic Hypersurfaces*, Lecture Notes of the Unione Matematica Italiana 31,
https://doi.org/10.1007/978-3-032-05899-7

15. Brieskorn, E.: Die Auflösung der rationalen Singularitäten holomorpher Abbildungen. Math. Ann. **178**, 255–270 (1968)
16. Brouwer, A.E., Cohen, A.M., Neumaier, A.: Distance-regular graphs. In: Ergebnisse der Mathematik und ihrer Grenzgebiete, vol. 18. Springer, Berlin (1989)
17. Bruce, J.W.: An upper bound for the number of singularities on a projective hypersurface. Bull. London Math. Soc. **13**(1), 47–50 (1981)
18. Burns, D., Wahl, J.M.: Local contributions to global deformations of surfaces. Invent. Math. **26**(1), 67–88 (1974)
19. Campana, F.: Orbifoldes à première classe de Chern nulle. In: The Fano Conference, pp. 339–351. University of Torino, Turin (2004)
20. Casnati, G., Catanese, F.: Even sets of nodes are bundle symmetric. J. Differ. Geom. **47**(2), 237–256 (1997). Erratum. J. Differential Geom. 50, no. 2, 415, (1998)
21. Catanese, F.: Babbage's conjecture, contact of surfaces, symmetric determinantal varieties and applications. Invent. Math. **63**(3), 433–465 (1981)
22. Catanese, F.: On the rationality of certain moduli spaces related to curves of genus 4. In: 'Algebraic Geometry', Proceedings, Ann Arbor 1981, Springer LNM 1008, pp. 30–50. Springer, Berlin (1983)
23. Catanese, F.: On the moduli spaces of surfaces of general type. J. Differ. Geom. **19**(2), 483–515 (1984)
24. Catanese, F.: Moduli of algebraic surfaces. In: Theory of Moduli (Montecatini Terme, 1985). Lecture Notes in Mathematics, vol. 1337, pp. 1–83. Springer, Berlin (1988)
25. Catanese, F.: Generalized Kummer surfaces and differentiable invariants of Noether-Horikawa surfaces. I. In: Manifolds and Geometry (Pisa, 1993). Symposium in Mathematics, vol. 36, pp. 132–177. Cambridge University Press, Cambridge (1996)
26. Catanese, F.: Canonical symplectic structures and deformations of algebraic surfaces. Commun. Contemp. Math. **11**(3), 481–493 (2009)
27. Catanese, F.: Kummer quartic surfaces, strict self-duality, and more. In: Dedieu, T., et al. (eds.) The Art of Doing Algebraic Geometry, a Volume Dedicated to Ciro Ciliberto on the Occasion of His 70th Birthday. Trends in Mathematics, pp. 55–92. Birkhäuser. Cham (2023)
28. Catanese, F., Ceresa, G.: Constructing sextic surfaces with a given number d of nodes. J. Pure Appl. Algebra **23**(1), 1–12 (1982)
29. Catanese, F., Tonoli, F.: Even sets of nodes on sextic surfaces. J. Eur. Math. Soc. **9**(4), 705–737 (2007)
30. Catanese, F., Keum, J.-H., Oguiso, K.: Some remarks on the universal cover of an open K3 surface. Math. Ann. **325**(2), 279–286 (2003)
31. Cayley, A.: A memoir on cubic surfaces. Philos. Trans. R. Soc. London **159**, 231–326 (1869)
32. Chiantini, L., Ciliberto, C.: On the Severi varieties on surfaces in $\mathcal{P}^3$. J. Algebraic Geom. **8**(1), 67–83 (1999)
33. Cho, Y., Coughlan, S., Frapporti, D., Kiermaier, M.: Scripts for "Varieties of Nodal surfaces, coding theory and Discriminants of cubic hypersurfaces" (2022). arXiv: 2206.05492
34. Chmutov, S.V.: Examples of projective surfaces with many singularities. J. Algebraic Geom. **1**(2), 191–196 (1992)
35. Ciliberto, C., Dedieu, T.: On universal Severi varieties of low genus K3 surfaces. Math. Z. **271**(3–4), 953–960 (2012)
36. Collino, A.: The Abel-Jacobi isomorphism for the cubic fivefold. Pac. J. Math. **122**(1), 43–55 (1986)
37. Conway, J.H., Sloane, N.J.A.: Sphere Packings, Lattices and Groups. Grundlehren der Mathematischen Wissenschaften, vol. 290, 2nd edn. Springer, New York (1993)
38. Cremona, L.: Mémoire de géométrie pure sur les surfaces du troisième ordre. J. Reine Angew. Math. **68**, 1–133 (1868)
39. Debarre, O., Manivel, L.: Sur la variété des espaces linéaires contenus dans une intersection complète. Math. Ann. **312**(3), 549–574 (1998)
40. Degtyarev, A.: On deformations of singular plane sextics. J. Algebraic Geom. **17**, 101–135 (2008)

41. Dettweiler, M., Friedman, M., Teicher, M., Topol-Amram, M.: The fundamental group for the complement of Cayley's singularities. Beiträge Algebra Geom. **50**(2), 469–482 (2009)
42. Dimca, A.: On the syzygies and Hodge theory of nodal hypersurfaces. Ann. Univ. Ferrara, Sez. VII, Sci. Mat. **63**(1), 87–101 (2017)
43. Dolgachev, I.: Integral quadratic forms: applications to algebraic geometry (after V. Nikulin). In: Bourbaki Seminar, Vol. 1982/83, , Astérisque, 105–106, pp. 251–278. Society of Mathematics France, France (1983)
44. Endraß, S.: Minimal even sets of nodes. J. die reine und Angewandte Mathematik **503**, 87–108 (1998)
45. Endraß, S.: On the divisor class group of double solids. Manuscript. Math. **99**(3), 341–358 (1999)
46. Fulton, W., Harris, J.: Representation Theory, A First Course, Readings in Mathematics. Graduate Texts in Mathematics, vol. 129. Springer, New York (1991)
47. Gallarati, D.: Ricerche sul contatto di superficie algebriche lungo curve. Acad. Roy. Belgique, Cl. Sci. Mém. Coll. **32**(3), 78pp. (1960)
48. Gordon, L.M.: Richard Levingston. The construction of some automorphic graphs. Geom. Ded. **10**(1–4), 261–267 (1981)
49. Goryunov, V.V.: Symmetric quartics with many nodes. In: Singularities and Bifurcations, Advances in Soviet Mathematics, vol. 21, pp. 147–161. American Mathematical Society, Providence (1994)
50. Hall, J.I.: Locally Petersen graphs. J. Graph Theory **4**(2), 173–187 (1980)
51. Harris, J.: On the Severi problem. Invent. Math. **84**(3), 445–461 (1986)
52. Igusa, J.-I.: On the graded ring of Theta-constants. Am. J. Math. **86**(1), 219–246 (1964)
53. Jaffe, D.B., Ruberman, D.: A sextic surface cannot have 66 nodes. J. Algebraic Geom. **6**(1), 151–168 (1997)
54. James, G.D.: On Witt's theorem for unimodular quadratic forms. Pac. J. Math. **26**, 303–316 (1968)
55. James, G.D.: The Representation Theory of the Symmetric Groups. Lecture Notes in Mathematics, vol. 682. Springer, Berlin (1978)
56. James, G., Liebeck, M.: Representations and Characters of Groups. Cambridge Mathematical Textbooks. Cambridge University Press, Cambridge, x+419pp. (1993)
57. Jessop, C.M.: Quartic Surfaces with Singular Points. Cambridge University Press, Cambridge (1916)
58. Jungnickel, D., Tonchev, V.D.: On Bonisoli's theorem and the block codes of Steiner triple systems. Designs Codes Cryptograph. **86**(3), 449–462 (2018)
59. Kalker, A.A.C.M.: Cubic fourfolds with fifteen ordinary double points. Thesis, Leiden University (1986)
60. Kiermaier, M., Kurz, S.: On the lengths of divisible codes. IEEE Trans. Inf. Theory **66**(7), 4051–4060 (2020)
61. Kloosterman, R.: Nodal surfaces with obstructed deformations. Geom. Dedicata **190**, 143–150 (2017)
62. Kummer, E.E.: On surfaces of degree four containing sixteen singular points. (Über die Flächen vierten Grades mit sechzehn singulären Punkten.) Berl. Monatsber. **1864**, 246–260 (1864)
63. Kung, J.P.S.: Radon transforms in combinatorics and lattice theory. In: Combinatorics and Ordered Sets, Proceedings of AMS-IMS-SIAM Joint Summer Research Conference, Arcata, CA, 1985. Contemporary Mathematics, vol. 57, pp. 33–74 (1986)
64. Kurz, S.: LinCode – computer classification of linear codes (2019). arXiv:1912.09357
65. Kurz, S.: Classification of 8-divisible binary linear codes with minimum distance 24 (2020). arXiv:2012.06163
66. Kurz, S.: Nodal surfaces in $\mathbb{P}^3$ and coding theory (2025, preprint). arXiv 2505.17531
67. Labs, O.: Hypersurfaces with many singularities: history-constructions - algorithms - visualization. Ph.D. Thesis, Johannes Gutenberg-Universität Mainz (2005)
68. Macaulay, F.S.: The Algebraic Theory of Modular Systems. XIV u. 112 S. (8°). University Press, Cambridge (1916)

69. MacWilliams, F.J., Sloane, N.J.A.: The Theory of Error-Correcting Codes. Elsevier, Amsterdam (1977)
70. Milnor, J.: Singular Points of Complex Hypersurfaces. Annals of Mathematics Studies, vol. 61, iii+122 pp. Princeton University Press, Princeton/University of Tokyo Press, Tokyo (1968)
71. Miyaoka, Y.: The maximal number of quotient singularities on surfaces with given numerical invariants. Math. Ann. **268**(2), 159–171 (1984)
72. Mumford, D.: Prym Varieties. I. Contributions to Analysis (a Collection of Papers Dedicated to Lipman Bers), pp. 325–350. Academic Press, New York (1974)
73. Nikulin, V.V.: Kummer surfaces. Izv. Akad. Nauk SSSR Ser. Mat. **39**(2), 278–293, 471 (1975)
74. Nikulin, V.V.: Integral symmetric bilinear forms and some of their geometric applications. Izv. Akad. Nauk SSSR Ser. Mat. **43**(1), 111–177, 238 (1979)
75. Pettersen, K.F.: On nodal determinantal quartic hypersurfaces in $\mathcal{P}^4$. Ph.D. Thesis, University of Oslo (1998)
76. Pjateckii-Shapiro, I.I., Shafarevic, I.R.: Der Satz von Torelli für algebraische Flächen vom Typ K3. Izv. Akad. Nauk SSSR Ser. Mat. **35**, 530–572 (1971)
77. Pignatelli, R., Tonoli, F.: On Wahl's proof of $\mu(6) = 65$. Asian J. Math. **13**(3), 307–310 (2009)
78. Pignatelli, R.: The Code of Barth's Sextic Surface. Preliminary Notes, Bayreuth (2009)
79. Rohn, K.: Die Flächen vierter Ordnung hinsichtlich ihrer Knotenpunkte und ihrer Gestaltung. Gekrönte Preisschrift der Jablonowski'schen Gesellschaft, vol. 58. S. Hirzel, Leipzig (1886)
80. Rohn, K.: Die Flächen vierter Ordnung hinsichtlich ihrer Knotenpunkte und ihrer Gestaltung. Math. Ann. **29**(1), 81–96 (1887)
81. Saint-Donat, B.: Projective models of K-3 surfaces. Am. J. Math. **96**, 602–639 (1974)
82. Schläfli, L.: On the distributionn of surfaces of the third order into species, in reference to the absence or presense of singular points, and the reality of their lines. Philos. Trans. R. Soc. London **6**, 201–241 (1863)
83. Segre, C.: Sulle varietá cubiche dello spazio a quattro dimensioni e su certi sistemi di rette e certe superficie dello spazio ordinario. Torino Mem. **39**, 48pp. (1888)
84. Segre, B.: Sul massimo numero di nodi delle superficie algebriche. Atti Acc. Ligure **10**(1), 15–22 (1952)
85. Sernesi, E.: Deformations of Algebraic Schemes. Grundlehren der Mathematischen Wissenschaften, vol. 334. Springer, Berlin, xii+339pp. (2006)
86. Severi, F.: Vorlesungen über algebraische Geometrie (Übersetzung von E. Löffler.) XVI u. 408 S. B. G. Teubner, Leipzig u. Berlin (1921)
87. Shimada, I.: On elliptic K3 surfaces. Michigan Math. J. **47**(3), 423–446 (2000)
88. Tjurina, G.N.: Resolution of singularities of flat deformations of double rational points. Funkcional. Anal. i Prilozen. **4**(1), 77–83 (1970)
89. Togliatti, E.G.: Sulle forme cubiche dello spazio a cinque dimensioni aventi il massimo numero finito di punti doppi. Mathematical Papers dedicated to Luigi Berzolari 577–593 (1936)
90. Togliatti, E.G.: Ancora sulle forme cubiche dello spazio a 5 dimensioni aventi il massimo numero finito di punti doppi. Atti primo Congr. Un. Mat. Ital., Firenze **1937**, 254–258 (1937)
91. Togliatti, E.G.: Una notevole superficie del quinto ordine con soli punti doppi isolati. Vierteljschr. Naturforsch. Ges. Zürich, vol. 85, pp. 127–132. Beibl, Beiblatt (Festschrift Rudolf Fueter) (1940)
92. Tonchev, V.D.: The uniformly packed binary [27,21,3] and [35,29,3] codes (English). Discrete Math. **149**(1–3), 283–288 (1996)
93. van der Geer, G.: On the geometry of a Siegel modular threefold. Math. Ann. **260**(3), 317–350 (1982)
94. van Geemen, B., Zhao, Y.: Genus three curves and 56 nodal sextic surfaces. J. Algebraic Geom. **27**(4), 583–592 (2018)
95. van Straten, D.: A quintic hypersurface in $\mathcal{P}^4$ with 130 nodes. Topology **32**(4), 857–864 (1993)
96. Verra, A.: The degree of the Gauss map for a general Prym theta-divisor. J. Algebraic Geom. **10**(2), 219–246 (2001)

97. Vinberg, È.B.: The two most algebraic K3 surfaces. Math. Ann. **265**, 1–21 (1983)
98. Wahl, J.M.: Equisingular deformations of normal surface singularities. I. Ann. Math. **104**(2), 325–356 (1976)
99. Wahl, J.M.: Nodes on sextic hypersurfaces in $\mathcal{P}^3$. J. Differ. Geom. **48**(3), 439–444 (1998)

MIX
Papier aus verantwortungsvollen Quellen
Paper from responsible sources
FSC® C105338

If you have any concerns about our products,
you can contact us on
ProductSafety@springernature.com

In case Publisher is established outside the EU,
the EU authorized representative is:
Springer Nature Customer Service Center GmbH
Europaplatz 3, 69115 Heidelberg, Germany

Printed by Libri Plureos GmbH
in Hamburg, Germany